SCHAUM'S OUTLINE OF

THEORY AND PROBLEMS

OF

BEGINNING STATISTICS

•

LARRY J. STEPHENS, Ph.D.

Professor of Mathematics
University of Nebraska at Omaha

SCHAUM'S OUTLINE SERIES
McGRAW-HILL

New York   San Francisco   Washington, D.C.   Auckland   Bogotá   Caracas   Lisbon
London   Madrid   Mexico City   Milan   Montreal   New Dehli
San Juan   Singapore   Sydney   Tokyo   Toronto

To My Mother and Father, Rosie, and Johnie Stephens

LARRY J. STEPHENS is Professor of Mathematics at the University of Nebraska at Omaha. He received his bachelor's degree from Memphis State University in Mathematics, his master's degree from the University of Arizona in Mathematics, and his Ph.D. degree from Oklahoma State University in Statistics. Professor Stephens has over 40 publications in professional journals. He has over 25 years of experience teaching Statistics. He has taught at the University of Arizona, Christian Brothers College, Gonzaga University, Oklahoma State University, the University of Nebraska at Kearney, and the University of Nebraska at Omaha. He has published numerous computerized test banks to accompany elementary Statistics texts. He has worked for NASA, Livermore Radiation Laboratory, and Los Alamos Laboratory. Since 1989, Dr. Stephens has consulted with and conducted Statistics seminars for the engineering group at 3M, Valley, Nebraska plant.

McGraw-Hill

A Division of The McGraw·Hill Companies

Schaum's Outline of Theory and Problems of
BEGINNING STATISTICS

4 5 6 7. 8 9 10 11 12 13 14 15 16 17 18 19 20 PRS PRS 9 0 2 1 0

ISBN 0-07-061259-5

Sponsoring Editor: Barbara Gilson
Production Supervisor: Clara Stanley
Editing Supervisor: Maureen B. Walker

Minitab is a registered trademark of Minitab Inc.

Library of Congress Cataloging-in-Publication Data

Stephens, Larry J.
 Schaum's outline of theory and problems of beginning statistics /
Larry J. Stephens.
   p. cm. – (Schaum's outline series)
  Includes index.
  ISBN 0-07-061259-5 (pbk.)
 1. Mathematical statistics—Outlines, syllabi, etc.
 2. Mathematical statistics—Problems, exercises, etc. I. Title.
 II. Series.
 QA276.19.S74 1998
 519.5'076—dc21   97-45979
                CIP
                AC

Preface

Statistics is a required course for undergraduate college students in a number of majors. Students in the following disciplines are often required to take a course in *beginning statistics:* allied health careers, biology, business, computer science, criminal justice, decision science, engineering, education, geography, geology, information science, nursing, nutrition, medicine, pharmacy, psychology, and public administration. This outline is intended to assist these students in the understanding of Statistics. The outline may be used as a supplement to textbooks used in these courses or a text for the course itself.

The author has taught such courses for over 25 years and understands the difficulty students encounter with statistics. I have included examples from a wide variety of current areas of application in order to motivate an interest in learning statistics. As we leave the twentieth century and enter the twenty-first century, an understanding of statistics is essential in understanding new technology, world affairs, and the ever-expanding volume of knowledge. Statistical concepts are encountered in television and radio broadcasting, as well as in magazines and newspapers. Modern newspapers, such as *USA Today,* are full of statistical information. The sports section is filled with descriptive statistics concerning players and teams performance. The money section of *USA Today* contains descriptive statistics concerning stocks and mutual funds. The life section of *USA Today* often contains summaries of research studies in medicine. An understanding of statistics is helpful in evaluating these research summaries.

The nature of the *beginning statistics* course has changed drastically in the past 30 or so years. This change is due to the technical advances in computing. Prior to the 1960s statistical computing was usually performed on mechanical calculators. These were large cumbersome computing devices (compared to today's hand-held calculators) that performed arithmetic by moving mechanical parts. Computers and computer software were no comparison to today's computers and software. The number of statistical packages available today numbers in the hundreds. The burden of statistical computing has been reduced to simply entering your data into a data file and then giving the correct command to perform the statistical method of interest.

One of the most widely used statistical packages in academia as well as industrial settings is the package called Minitab (Minitab Inc., 3081 Enterprise Drive, State College, PA 16801-3008). I wish to thank Minitab Inc. for granting me permission to include Minitab output, including graphics, throughout the text. Most modern Statistics textbooks include computer software as part of the text. I have chosen to include Minitab because it is widely used and is very friendly. Once a student learns the various data file structures needed to use Minitab, and the structure of the commands and subcommands, this knowledge is readily transferable to other statistical software.

The outline contains all the topics, and more, covered in a *beginning statistics* course. The only mathematical prerequisite needed for the material found in the outline is arithmetic and some basic algebra. I wish to thank my wife, Lana, for her understanding during the preparation of the book. I wish to thank my friend Stanley Wileman for all the computer help he has given me during the preparation of the book. I wish to thank Dr. Edwin C. Hackleman of Delta Software, Inc. for his timely assistance as compositor of the final camera-ready manuscript. Finally, I wish to thank the staff at McGraw-Hill for their cooperation and helpfulness.

LARRY J. STEPHENS

Contents

CONTENTS

Chapter 1

Introduction

STATISTICS

Statistics is a discipline of study dealing with the collection, analysis, interpretation, and presentation of data. Statistical methodology is utilized by pollsters who sample our opinions concerning topics ranging from art to zoology. Statistical methodology is also utilized by business and industry to help control the quality of goods and services that they produce. Social scientists and psychologists use statistical methodology to study our behaviors. Because of its broad range of applicability, a course in statistics is required of majors in disciplines such as sociology, psychology, criminal justice, nursing, exercise science, pharmacy, education, and many others. To accommodate this diverse group of users, examples and problems in this outline are chosen from many different sources.

DESCRIPTIVE STATISTICS

The use of graphs, charts, and tables and the calculation of various statistical measures to organize and summarize information is called *descriptive statistics*. Descriptive statistics help to reduce our information to a manageable size and put it into focus.

EXAMPLE 1.1 The compilation of batting average, runs batted in, runs scored, and number of home runs for each player, as well as earned run average, won/lost percentage, number of saves, etc., for each pitcher from the official score sheets for major league baseball players is an example of descriptive statistics. These statistical measures allow us to compare players, determine whether a player is having an "off year" or "good year," etc.

EXAMPLE 1.2 The publication entitled *Crime in the United States* published by the Federal Bureau of Investigation gives summary information concerning various crimes for the United States. The statistical measures given in this publication are also examples of descriptive statistics and they are useful to individuals in law enforcement.

INFERENTIAL STATISTICS: POPULATION AND SAMPLE

The complete collection of individuals, items, or data under consideration in a statistical study is referred to as the *population*. The portion of the population selected for analysis is called the *sample*. *Inferential statistics* consists of techniques for reaching conclusions about a population based upon information contained in a sample.

EXAMPLE 1.3 The results of polls are widely reported by both the written and the electronic media. The techniques of inferential statistics are widely utilized by pollsters. Table 1.1 gives several examples of populations and samples encountered in polls reported by the media. The methods of inferential statistics are used to make inferences about the populations based upon the results found in the samples and to give an indication about the reliability of these inferences. The results of a poll of 600 registered voters might be reported as follows: Forty percent of the voters approve of the president's economic policies. The margin of error for the survey is 4%. The survey indicates that an estimated 40% of all registered voters approve of the economic policies, but it might be as low as 36% or as high as 44%.

Table 1.1

Population	Sample
All registered voters	A telephone survey of 600 registered voters
All owners of handguns	A telephone survey of 1000 handgun owners
Households headed by a single parent	The results from questionnaires sent to 2500 households headed by a single parent
The CEOs of all private companies	The results from surveys sent to 150 CEO's of private companies

EXAMPLE 1.4 The techniques of inferential statistics are applied in many industrial processes to control the quality of the products produced. In industrial settings, the population may consist of the daily production of toothbrushes, computer chips, bolts, and so forth. The sample will consist of a random and representative selection of items from the process producing the toothbrushes, computer chips, bolts, etc. The information contained in the daily samples is used to construct control charts. The control charts are then used to monitor the quality of the products.

EXAMPLE 1.5 The statistical methods of inferential statistics are used to analyze the data collected in research studies. Table 1.2 gives the samples and populations for several such studies. The information contained in the samples is utilized to make inferences concerning the populations. If it is found that 245 of 350 or 70% of prison inmates in a criminal justice study were abused as children, what conclusions may be inferred concerning the percent of all prison inmates who were abused as children? The answers to this question are found in Chapters 8 and 9.

Table 1.2

Population	Sample
All prison inmates	A criminal justice study of 350 prison inmates
Legal aliens living in the United States	A sociological study conducted by a university researcher of 200 legal aliens
Alzheimer patients in the United States	A medical study of 75 such patients conducted by a university hospital
Adult children of alcoholics	A psychological study of 200 such individuals

VARIABLE, OBSERVATION, AND DATA SET

A characteristic of interest concerning the individual elements of a population or a sample is called a *variable*. A variable is often represented by a letter such as x, y, or z. The value of a variable for one particular element from the sample or population is called an *observation*. A *data set* consists of the observations of a variable for the elements of a sample.

EXAMPLE 1.6 Six hundred registered voters are polled and each one is asked if they approve or disapprove of the president's economic policies. The variable is the registered voter's opinion of the president's economic policies. The data set consists of 600 observations. Each observation will be the response "approve" or the response "do not approve." If the response "approve" is coded as the number 1 and the response "do not approve" is coded as 0, then the data set will consist of 600 observations, each one of which is either 0 or 1. If x is used to represent the variable, then x can assume two values, 0 or 1.

EXAMPLE 1.7 A survey of 2500 households headed by a single parent is conducted and one characteristic of interest is the yearly household income. The data set consists of the 2500 yearly household incomes for the individuals in the survey. If y is used to represent the variable, then the values for y will be between the smallest and the largest yearly household incomes for the 2500 households.

EXAMPLE 1.8 The number of speeding tickets issued by 75 Nebraska state troopers for the month of June is recorded. The data set consists of 75 observations.

QUANTITATIVE VARIABLE: DISCRETE AND CONTINUOUS VARIABLE

A *quantitative variable* is determined when the description of the characteristic of interest results in a numerical value. When a measurement is required to describe the characteristic of interest or it is necessary to perform a count to describe the characteristic, a quantitative variable is defined. A *discrete variable* is a quantitative variable whose values are countable. Discrete variables usually result from counting. A *continuous variable* is a quantitative variable that can assume any numerical value over an interval or over several intervals. A continuous variable usually results from making a measurement of some type.

EXAMPLE 1.9 Table 1.3 gives several discrete variables and the set of possible values for each one. In each case the value of the variable is determined by counting. For a given box of 100 diabetic syringes, the number of defective needles is determined by counting how many of the 100 are defective. The number of defectives found must equal one of the 101 values listed. The number of possible outcomes is finite for each of the first four variables; that is, the number of possible outcomes are 101, 31, 501, and 51 respectively. The number of possible outcomes for the last variable is infinite. Since the number of possible outcomes is infinite and countable for this variable, we say that the number of outcomes is *countably infinite.*

Sometimes it is not clear whether a variable is discrete or continuous. Test scores expressed as a percent, for example, are usually given as whole numbers between 0 and 100. It is possible to give a score such as 75.57565. However, this is not done in practice because teachers are unable to evaluate to this degree of accuracy. This variable is usually regarded as continuous, although for all practical purposes, it is discrete. To summarize, due to measurement limitations, many continuous variables actually assume only a countable number of values.

Table 1.3

Discrete variable	Possible values for the variable
The number of defective needles in boxes of 100 diabetic syringes	0, 1, 2, . . . , 100
The number of individuals in groups of 30 with a type A personality	0, 1, 2, . . . , 30
The number of surveys returned out of 500 mailed in sociological studies	0, 1, 2, . . . , 500
The number of prison inmates in 50 having finished high school or obtained a GED who are selected for criminal justice studies	0, 1, 2, . . . , 50
The number of times you need to flip a coin before a head appears for the first time	1, 2, 3, . . . (there is no upper limit since conceivably one might need to flip forever to obtain the first head)

EXAMPLE 1.10 Table 1.4 gives several continuous variables and the set of possible values for each one. All three continuous variables given in Table 1.4 involve measurement, whereas the variables in Example 1.9 all involve counting.

Table 1.4

Continuous variable	Possible values for the variable
The length of prison time served for individuals convicted of first degree murder	All the real numbers between a and b, where a is the smallest amount of time served and b is the largest amount
The household income for households with incomes less than or equal to $20,000	All the real numbers between a and $20,000, where a is the smallest household income in the population
The cholesterol reading for those individuals having cholesterol readings equal to or greater than 200 mg/dl	All real numbers between 200 and b, where b is the largest cholesterol reading of all such individuals

QUALITATIVE VARIABLE

A *qualitative variable* is determined when the description of the characteristic of interest results in a nonnumerical value. A qualitative variable may be classified into two or more categories.

EXAMPLE 1.11 Table 1.5 gives several examples of qualitative variables along with a set of categories into which they may be classified.

Table 1.5

Qualitative variable	Possible categories for the variable
Marital status	Single, married, divorced, separated
Gender	Male, female
Crime classification	Misdemeanor, felony
Pain level	None, low, moderate, severe
Personality type	Type A, type B

The possible categories for qualitative variables are often coded for the purpose of performing computerized statistical analysis. Marital status might be coded as 1, 2, 3, or 4, where 1 represents single, 2 represents married, 3 represents divorced, and 4 represents separated. The variable gender might be coded as 0 for female and 1 for male. The categories for any qualitative variable may be coded in a similar fashion. Even though numerical values are associated with the characteristic of interest after being coded, the variable is considered a qualitative variable.

NOMINAL, ORDINAL, INTERVAL, AND RATIO LEVELS OF MEASUREMENT

There are four *levels of measurement* or *scales of measurements* into which data can be classified. The *nominal scale* applies to data that are used for category identification. The *nominal level of measurement* is characterized by data that consist of names, labels, or categories only. *Nominal scale data* cannot be arranged in an ordering scheme. The arithmetic operations of addition, subtraction, multiplication, and division are not performed for nominal data.

EXAMPLE 1.12 Table 1.6 gives several qualitative variables and a set of possible nominal level data values. The data values are often encoded for recording in a computer data file. Blood type might be recorded as 1, 2, 3, or 4; state of residence might be recorded as 1, 2, . . . , or 50; and type of crime might be recorded as 0 or 1, or 1 or 2, etc. Similarly, color of road sign could be recorded as 1, 2, 3, 4, or 5 and religion could be recorded as 1, 2, or 3. There is no order associated with these data and arithmetic operations are not performed. For example, adding Christian and Moslem (1 + 2) does not give other (3).

Table 1.6

Qualitative variable	Possible nominal level data values associated with the variable
Blood type	A, B, AB, O
State of residence	Alabama, . . . , Wyoming
Type of crime	Misdemeanor, felony
Color of road signs in the state of Nebraska	Red , white, blue, brown, green
Religion	Christian, Moslem, other

The *ordinal scale* applies to data that can be arranged in some order, but differences between data values either cannot be determined or are meaningless. The *ordinal level of measurement* is characterized by data that applies to categories that can be ranked. *Ordinal scale data* can be arranged in an ordering scheme.

EXAMPLE 1.13 Table 1.7 gives several qualitative variables and a set of possible ordinal level data values. The data values for ordinal level data are often encoded for inclusion in computer data files. Arithmetic operations are not performed on ordinal level data, but an ordering scheme exists. A full-size automobile is larger than a subcompact, a tire rated excellent is better than one rated poor, no pain is preferable to any level of pain, the level of play in major league baseball is better than the level of play in class AA, and so forth.

Table 1.7

Qualitative variable	Possible ordinal level data values associated with the variable
Automobile size description	Subcompact, compact, intermediate, full-size
Product rating	Poor, good, excellent
Socioeconomic class	Lower, middle, upper
Pain level	None, low, moderate, severe
Baseball team classification	Class A, class AA, class AAA , major league

The *interval scale* applies to data that can be arranged in some order and for which differences in data values are meaningful. The *interval level of measurement* results from counting or measuring. *Interval scale data* can be arranged in an ordering scheme and differences can be calculated and interpreted. The value zero is arbitrarily chosen for interval data and does not imply an absence of the characteristic being measured. Ratios are not meaningful for interval data.

EXAMPLE 1.14 Stanford-Binet IQ scores represent interval level data. Joe's IQ score equals 100 and John's IQ score equals 150. John has a higher IQ than Joe; that is, IQ scores can be arranged in order. John's IQ score is 50 points higher than Joe's IQ score; that is, differences can be calculated and interpreted. However, we cannot conclude that John is 1.5 times (150/100 = 1.5) more intelligent than Joe. An IQ score of zero does not indicate a complete lack of intelligence.

EXAMPLE 1.15 Temperatures represent interval level data. The high temperature on February 1 equaled 25°F and the high temperature on March 1 equaled 50°F. It was warmer on March 1 than it was on February 1. That is, temperatures can be arranged in order. It was 25° warmer on March 1 than on February 1. That is, differences may be calculated and interpreted. We cannot conclude that it was twice as warm on March 1 than it was on February 1. That is, ratios are not readily interpretable. A temperature of 0°F does not indicate an absence of warmth.

EXAMPLE 1.16 Test scores represent interval level data. Lana scored 80 on a test and Christine scored 40 on a test. Lana scored higher than Christine did on the test; that is, the test scores can be arranged in order. Lana scored 40 points higher than Christine did on the test; that is, differences can be calculated and interpreted. We cannot conclude that Lana knows twice as much as Christine about the subject matter. A test score of 0 does not indicate an absence of knowledge concerning the subject matter.

The *ratio scale* applies to data that can be ranked and for which all arithmetic operations including division can be performed. Division by zero is, of course, excluded. The *ratio level of measurement* results from counting or measuring. *Ratio scale data* can be arranged in an ordering scheme and differences and ratios can be calculated and interpreted. Ratio level data has an absolute zero and a value of zero indicates a complete absence of the characteristic of interest.

EXAMPLE 1.17 The grams of fat consumed per day for adults in the United States is ratio scale data. Joe consumes 50 grams of fat per day and John consumes 25 grams per day. Joe consumes twice as much fat as John per day, since 50/25 = 2. For an individual who consumes 0 grams of fat on a given day, there is a complete absence of fat consumed on that day. Notice that a ratio is interpretable and an absolute zero exists.

EXAMPLE 1.18 The number of 911 emergency calls in a sample of 50 such calls selected from a 24-hour period involving a domestic disturbance is ratio scale data. The number found on May 1 equals 5 and the number found on June 1 equals 10. Since 10/5 = 2, we say that twice as many were found on June 1 than were found on May 1. For a 24-hour period in which no domestic disturbance calls were found, there is a complete absence of such calls. Notice that a ratio is interpretable and an absolute zero exists.

SUMMATION NOTATION

Many of the statistical measures discussed in the following chapters involve sums of various types. Suppose the number of 911 emergency calls received on four days were 411, 375, 400, and 478. If we let x represent the number of calls received per day, then the values of the variable for the four days are represented as follows: $x_1 = 411$, $x_2 = 375$, $x_3 = 400$, and $x_4 = 478$. The sum of calls for the four days is represented as $x_1 + x_2 + x_3 + x_4$ which equals $411 + 375 + 400 + 478$ or 1664. The symbol Σx, read as *"the summation of x,"* is used to represent $x_1 + x_2 + x_3 + x_4$. The uppercase Greek letter Σ (pronounced sigma) corresponds to the English letter S and stands for the phrase "the sum of." Using the *summation notation*, the total number of 911 calls for the four days would be written as $\Sigma x = 1664$.

EXAMPLE 1.19 The following five values were observed for the variable x: $x_1 = 4$, $x_2 = 5$, $x_3 = 0$, $x_4 = 6$, and $x_5 = 10$. The following computations illustrate the usage of the summation notation.

$$\Sigma x = x_1 + x_2 + x_3 + x_4 + x_5 = 4 + 5 + 0 + 6 + 10 = 25$$

$$(\Sigma x)^2 = (x_1 + x_2 + x_3 + x_4 + x_5)^2 = (25)^2 = 625$$

$$\Sigma x^2 = x_1^2 + x_2^2 + x_3^2 + x_4^2 + x_5^2 = 4^2 + 5^2 + 0^2 + 6^2 + 10^2 = 177$$

$$\Sigma(x - 5) = (x_1 - 5) + (x_2 - 5) + (x_3 - 5) + (x_4 - 5) + (x_5 - 5)$$

$$\Sigma(x - 5) = (4 - 5) + (5 - 5) + (0 - 5) + (6 - 5) + (10 - 5) = -1 + 0 - 5 + 1 + 5 = 0$$

EXAMPLE 1.20 The following values were observed for the variables x and y: $x_1 = 1$, $x_2 = 2$, $x_3 = 0$, $x_4 = 4$, $y_1 = 2$, $y_2 = 1$, $y_3 = 4$, and $y_4 = 5$. The following computations show how the summation notation is used for two variables.

$$\Sigma xy = x_1y_1 + x_2y_2 + x_3y_3 + x_4y_4 = 1 \times 2 + 2 \times 1 + 0 \times 4 + 4 \times 5 = 24$$

$$(\Sigma x)(\Sigma y) = (x_1 + x_2 + x_3 + x_4)(y_1 + y_2 + y_3 + y_4) = (1 + 2 + 0 + 4)(2 + 1 + 4 + 5) = 7 \times 12 = 84$$

$$(\Sigma x^2 - (\Sigma x)^2/4) \times (\Sigma y^2 - (\Sigma y)^2/4) = (1 + 4 + 0 + 16 - 7^2/4) \times (4 + 1 + 16 + 25 - 12^2/4) = (8.75) \times (10) = 87.5$$

COMPUTERS AND STATISTICS

The techniques of descriptive and inferential statistics involve lengthy repetitive computations as well as the construction of various graphical constructs. These computations and graphical constructions have been simplified by the development of computer software. These computer software programs are referred to as *statistical software packages*, or simply *statistical packages*. These statistical packages are large computer programs which perform the various computations and graphical constructions discussed in this outline plus many other ones beyond the scope of the outline. Statistical packages are currently available for use on mainframes, minicomputers, and microcomputers.

There are currently available numerous statistical packages. Four widely used statistical packages are: MINITAB, BMDP, SPSS, and SAS. Many of the figures found in the following chapters are MINITAB generated. MINITAB is a registered trademark of Minitab, Inc., 3081 Enterprise Drive, State College, PA 16801. Phone: 814-238-3280; fax: 814-238-4383; telex: 881612. The author would like to thank Minitab Inc. for their permission to use output from MINITAB throughout the outline.

Solved Problems

DESCRIPTIVE STATISTICS AND INFERENTIAL STATISTICS: POPULATION AND SAMPLE

1.1 Classify each of the following as descriptive statistics or inferential statistics.

(a) The average points per game, percent of free throws made, average number of rebounds per game, and average number of fouls per game as well as several other measures for players in the NBA are computed.

(b) Ten percent of the boxes of cereal sampled by a quality technician are found to be under the labeled weight. Based on this finding, the filling machine is adjusted to increase the amount of fill.

(c) *USA Today* gives several pages of numerical quantities concerning stocks listed in AMEX, NASDAQ, and NYSE as well as mutual funds listed in MUTUALS.

(d) Based on a study of 500 single parent households by a social researcher, a magazine reports that 25% of all single parent households are headed by a high school dropout.

Ans. (a) The measurements given organize and summarize information concerning the players and is therefore considered descriptive statistics.

(b) Because of the high percent of boxes of cereal which are under the labeled weight in the sample, a decision is made to increase the weight per box for each box in the population. This is an example of inferential statistics.

(c) The tables of measurements such as stock prices and change in stock prices are descriptive in nature and therefore represent descriptive statistics.

(d) The magazine is stating a conclusion about the population based upon a sample and therefore this is an example of inferential statistics.

1.2   Identify the sample and the population in each of the following scenarios.

(a) In order to study the response times for emergency 911 calls in Chicago, fifty "robbery in progress" calls are selected randomly over a six-month period and the response times are recorded.

(b) In order to study a new medical charting system at Saint Anthony's Hospital, a representative group of nurses is asked to use the charting system. Recording times and error rates are recorded for the group.

(c) Fifteen hundred individuals who listen to talk radio programs of various types are selected and information concerning their education level, income level, and so forth is recorded.

Ans.   (a) The 50 "robbery in progress" calls is the sample, and all "robbery in progress" calls in Chicago during the six-month period is the population.

(b) The representative group of nurses who use the medical charting system is the sample and all nurses who use the medical charting system at Saint Anthony's is the population.

(c) The 1500 selected individuals who listen to talk radio programs is the sample and the millions who listen nationally is the population.

VARIABLE, OBSERVATION, AND DATA SET

1.3   In a sociological study involving 35 low-income households, the number of children per household was recorded for each household. What is the variable? How many observations are in the data set?

Ans.   The variable is the number of children per household. The data set contains 35 observations.

1.4   A national survey was mailed to 5000 households and one question asked for the number of handguns per household. Three thousand of the surveys were completed and returned. What is the variable and how large is the data set?

Ans.   The variable is the number of handguns per household and there are 3000 observations in the data set.

1.5   The number of hours spent per week on paper work was determined for 200 middle level managers. The minimum was 0 hours and the maximum was 27 hours. What is the variable? How many observations are in the data set?

Ans.   The variable is the number of hours spent per week on paper work and the number of observations equals 200.

QUANTITATIVE VARIABLE: DISCRETE AND CONTINUOUS VARIABLE

1.6   Classify the variables in problems 1.3, 1.4, and 1.5 as continuous or discrete.

Ans. The number of children per household is a discrete variable since the number of values this variable may assume is countable. The values range from 0 to some maximum value such as 10 or 15 depending upon the population.

The number of handguns per household is countable, ranging from 0 to some maximum value and therefore this variable is discrete.

The time spent per week on paper work by middle level managers may be any real number between 0 and some upper limit. The number of values possible is not countable and therefore this variable is continuous.

1.7 A program to locate drunk drivers is initiated and roadblocks are used to check for individuals driving under the influence of alcohol. Let n represent the number of drivers stopped before the first drunk driver is found. What are the possible values for n? Classify n as discrete or continuous.

Ans. The number of drivers stopped before finding the first drunk driver may equal 1, 2, 3, . . . , up to an infinitely large number. Although not likely, it is theoretically possible that an extremely large number of drivers would need to be checked before finding the first drunk driver. The possible values for n are all the positive integers. N is a discrete variable.

1.8 The KSW computer science aptitude test consists of 25 questions. The score reported is reflective of the computer science aptitude of the test taker. How would the score likely be reported for the test? What are the possible values for the scores? Is the variable discrete or continuous?

Ans. The score reported would likely be the number or percent of correct answers. The number correct would be a whole number from 0 to 25 and the percent correct would range from 0 to 100 in steps of size 4. However if the test evaluator considered the reasoning process used to arrive at the answers and assigned partial credit for each problem, the scores could range from 0 to 25 or 0 to 100 percent continuously. That is, the score could be any real number between 0 and 25 or any real number between 0 and 100 percent. We might say that for all practical purposes, the variable is discrete. However, theoretically the variable is continuous.

QUALITATIVE VARIABLE

1.9 Which of the following are qualitative variables?

(a) The color of automobiles involved in several severe accidents
(b) The length of time required for rats to move through a maze
(c) The classification of police administrations as city, county, or state
(d) The rating given to a pizza in a taste test as poor, good, or excellent
(e) The number of times subjects in a sociological research study have been married

Ans. The variables given in (a), (c), and (d) are qualitative variables since they result in nonnumerical values. They are classified into categories. The variables given in (b) and (e) result in numerical values as a result of measuring and counting, respectively.

1.10 The pain level following surgery for an intestinal blockage was classified as none, low, moderate, or severe for several patients. Give three different numerical coding schemes that might be used for the purpose of inclusion of the responses in a computer data file. Does this coding change the variable to a quantitative variable?

Ans.   The responses none, low, moderate, or severe might be coded as 0, 1, 2, or 3 or 1, 2, 3, or 4 or as 10, 20, 30, or 40. There is no limit to the number of coding schemes that could be used. Coding the variable does not change it into a quantitative variable. Many times coding a qualitative variable simplifies the computer analysis performed on the variable.

NOMINAL, ORDINAL, INTERVAL, AND RATIO LEVELS OF MEASUREMENT

1.11 Indicate the scale of measurement for each of the following variables: racial origin, monthly phone bills, Fahrenheit and centigrade temperature scales, military ranks, time, ranking of a personality trait, clinical diagnoses, and calendar numbering of the years.

Ans.   racial origin: nominal   time: ratio
monthly phone bills: ratio   ranking of personality trait: ordinal
temperature scales: interval   clinical diagnoses: nominal
military ranks: ordinal   calendar numbering of the years: interval

1.12 Which scales of measurement would usually apply to qualitative data?

Ans.   nominal or ordinal

SUMMATION NOTATION

1.13 The following values are recorded for the variable x: $x_1 = 1.3$, $x_2 = 2.5$, $x_3 = 0.7$, $x_4 = 3.5$. Evaluate the following summations: Σx, Σx^2, $(\Sigma x)^2$, and $\Sigma(x - .5)$.

Ans. $\Sigma x = x_1 + x_2 + x_3 + x_4 = 1.3 + 2.5 + 0.7 + 3.5 = 8.0$

$\Sigma x^2 = x_1^2 + x_2^2 + x_3^2 + x_4^2 = 1.3^2 + 2.5^2 + 0.7^2 + 3.5^2 = 20.68$

$(\Sigma x)^2 = (8.0)^2 = 64.0$

$\Sigma(x - .5) = (x_1 - .5) + (x_2 - .5) + (x_3 - .5) + (x_4 - .5) = 0.8 + 2.0 + 0.2 + 3.0 = 6.0$

1.14 The following values are recorded for the variables x and y: $x_1 = 25.67$, $x_2 = 10.95$, $x_3 = 5.65$, $y_1 = 3.45$, $y_2 = 1.55$, and $y_3 = 3.50$. Evaluate the following summations: Σxy, $\Sigma x^2 y^2$, and $\Sigma xy - \Sigma x \Sigma y$.

Ans. $\Sigma xy = x_1 y_1 + x_2 y_2 + x_3 y_3 = 25.67 \times 3.45 + 10.95 \times 1.55 + 5.65 \times 3.50 = 125.31$

$\Sigma x^2 y^2 = x_1^2 y_1^2 + x_2^2 y_2^2 + x_3^2 y_3^2 = 25.67^2 \times 3.45^2 + 10.95^2 \times 1.55^2 + 5.65^2 \times 3.50^2 = 8522.26$

$\Sigma xy - \Sigma x \Sigma y = 125.31 - 42.27 \times 8.50 = -233.99$

1.15 The sum of four values for the variable y equals 25, that is, $\Sigma y = 25$. If it is known that $y_1 = 2$, $y_2 = 7$, and $y_3 = 6$, find y_4.

Ans. $\Sigma y = 25 = 2 + 7 + 6 + y_4$, or $25 = 15 + y_4$. From this, we see that y_4 must equal 10.

Supplementary Problems

DESCRIPTIVE STATISTICS AND INFERENTIAL STATISTICS: POPULATION AND SAMPLE

1.16 Classify each of the following as descriptive statistics or inferential statistics.

(a) The *Nielsen Report on Television* utilizes data from a sample of viewers to give estimates of average viewing time per week per viewer for all television viewers.

(b) The U.S. National Center for Health Statistics publication entitled *Vital Statistics of the United States* lists the leading causes of death in a given year. The estimates are based upon a sampling of death certificates.

(c) The *Omaha World Herald* lists the low and high temperatures for several American cities.

(d) The number of votes a presidential candidate receives are given for each state following the presidential election.

(e) The *National Household Survey on Drug Abuse* gives the current percentage of young adults using different types of drugs. The percentages are based upon national samples.

Ans.   (a) inferential statistics   (b) inferential statistics   (c) descriptive statistics
           (d) descriptive statistics   (e) inferential statistics

1.17 Classify each of the following as a sample or a population.

(a) all diabetics in the United States

(b) a group of 374 individuals selected for a *New York Times/CBS* news poll

(c) all owners of Ford trucks

(d) all registered voters in the state of Arkansas

(e) a group of 22,000 physicians who participate in a study to determine the role of aspirin in preventing heart attacks

Ans.   (a) population   (b) sample   (c) population   (d) population   (e) sample

VARIABLE, OBSERVATION, AND DATA SET

1.18 Changes in systolic blood pressure readings were recorded for 325 hypertensive patients who were participating in a study involving a new medication to control hypertension. Larry Doe is a patient in the study and he experienced a drop of 15 units in his systolic blood pressure. What statistical term is used to describe the change in systolic blood pressure readings? What does the number 325 represent? What term is used for the 15-unit drop is systolic blood pressure?

Ans.   The change in blood pressure is the variable, 325 is the number of observations in the data set, and 15-unit drop in blood pressure is an observation.

1.19 Table 1.8 gives the fasting blood sugar reading for five patients at a small medical clinic. What is the variable? Give the observations that comprise this data set.

Table 1.8

Patient name	Fasting blood sugar reading
Sam Alcorn	135
Susan Collins	157
Larry Halsey	168
Bill Samuels	120
Lana Williams	160

Ans. The variable is the fasting blood sugar reading for a patient. The observations are 135, 157, 168, 120, and 160.

1.20 A sociological study involving a minority group recorded the educational level of the participants. The educational level was coded as follows: less than high school was coded as 1, high school was coded as 2, college graduate was coded as 3, and postgraduate was coded as 4. The results were:

```
1  1  2  3  4  3  2  2  2  2  1  1  1  2  2  1  2  3  3  2  2  1  1
2  2  2  2  2  2  1  1  3  3  2  2  2  2  2  1  2  2  2  2  2  1  3
```

What is the variable? How many observations are in the data set?

Ans. The variable is the educational level of a participant. There are 46 observations in the data set.

QUANTITATIVE VARIABLE: DISCRETE AND CONTINUOUS VARIABLE

1.21 Classify the variables in problems 1.18, 1.19, and 1.20 as discrete or continuous.

Ans. The variables in problems 1.18 and 1.19 are continuous. The variable in problem 1.20 is not a quantitative variable.

1.22 A die is tossed until the face 6 turns up on a toss. The variable x equals the toss upon which the face 6 first appears. What are the possible values that x may assume? Is x discrete or continuous?

Ans. x may equal any positive integer, and it is therefore a discrete variable.

1.23 Is it possible for a variable to be both discrete and continuous?

Ans. no

QUALITATIVE VARIABLE

1.24 Give five examples of a qualitative variable.

Ans.   1. Classification of government employees   4. Medical specialty of doctors
         2. Motion picture ratings   5. ZIP code
         3. College student classification

1.25 Which of the following is not a qualitative variable? hair color, eye color, make of computer, personality type, and percent of income spent on food

Ans. percent of income spent on food

NOMINAL, ORDINAL, INTERVAL, AND RATIO LEVELS OF MEASUREMENT

1.26 Indicate the scale of measurement for each of the following variables: religion classification; movie ratings of 1, 2, 3, or 4 stars; body temperature; weights of runners; and consumer product ratings given as poor, average, or excellent.

Ans.   religion: nominal   weights of runners: ratio
         movie ratings: ordinal   consumer product ratings: ordinal
         body temperature: interval

1.27 Which scales of measurement would usually apply to quantitative data?

Ans.   interval or ratio

SUMMATION NOTATION

1.28   The following values are recorded for the variable x: $x_1 = 15$, $x_2 = 25$, $x_3 = 10$, and $x_4 = 5$. Evaluate the following summations: Σx, Σx^2, $(\Sigma x)^2$, and $\Sigma(x - 5)$.

Ans. $\Sigma x = 55$, $\Sigma x^2 = 975$, $(\Sigma x)^2 = 3025$, $\Sigma(x - 5) = 35$

1.29   The following values are recorded for the variables x and y: $x_1 = 17$, $x_2 = 28$, $x_3 = 35$, $y_1 = 20$, $y_2 = 30$, and $y_3 = 40$. Evaluate the following summations: Σxy, $\Sigma x^2 y^2$, and $\Sigma xy - \Sigma x \Sigma y$.

Ans. $\Sigma xy = 2580$, $\Sigma x^2 y^2 = 2{,}781{,}200$, $\Sigma xy - \Sigma x \Sigma y = -4{,}620$

1.30   Given that $x_1 = 5$, $x_2 = 10$, $y_1 = 20$, and $\Sigma xy = 200$, find y_2.

Ans. $y_2 = 10$

Chapter 2

Organizing Data

RAW DATA

Information obtained by observing values of a variable is called *raw data*. Data obtained by observing values of a qualitative variable are referred to as *qualitative data*. Data obtained by observing values of a quantitative variable are referred to as *quantitative data*. Quantitative data obtained from a discrete variable are also referred to as *discrete data* and quantitative data obtained from a continuous variable are called *continuous data*.

EXAMPLE 2.1 A study is conducted in which individuals are classified into one of sixteen personality types using the Myers-Briggs type indicator. The resulting raw data would be classified as qualitative data.

EXAMPLE 2.2 The cardiac output in liters per minute is measured for the participants in a medical study. The resulting data would be classified as quantitative data and continuous data.

EXAMPLE 2.3 The number of murders per 100,000 inhabitants is recorded for each of several large cities for the year 1994. The resulting data would be classified as quantitative data and discrete data.

FREQUENCY DISTRIBUTION FOR QUALITATIVE DATA

A *frequency distribution* for qualitative data lists all categories and the number of elements that belong to each of the categories.

EXAMPLE 2.4 A sample of rural county arrests gave the following set of offenses with which individuals were charged:

rape	robbery	burglary	arson	murder	robbery	rape	manslaughter
arson	theft	arson	burglary	theft	robbery	theft	theft
theft	burglary	murder	murder	theft	theft	theft	manslaughter
manslaughter							

The variable, type of offense, is classified into the categories: rape, robbery, burglary, arson, murder, theft, and manslaughter. As shown in Table 2.1, the seven categories are listed under the column entitled Offense, and each occurrence of a category is recorded by using the symbol / in order to tally the number of times each offense occurs. The number of tallies for each offense is counted and listed under the column entitled Frequency. Occasionally the term absolute frequency is used rather than frequency.

Table 2.1

Offense	Tally	Frequency
Rape	//	2
Robbery	///	3
Burglary	///	3
Arson	///	3
Murder	///	3
Theft	⫻⫻ ///	8
Manslaughter	///	3

14

RELATIVE FREQUENCY OF A CATEGORY

The *relative frequency* of a category is obtained by dividing the frequency for a category by the sum of all the frequencies. The relative frequencies for the seven categories in Table 2.1 are shown in Table 2.2. The sum of the relative frequencies will always equal one.

PERCENTAGE

The *percentage* for a category is obtained by multiplying the relative frequency for that category by 100. The percentages for the seven categories in Table 2.1 are shown in Table 2.2. The sum of the percentages for all the categories will always equal 100 percent.

Table 2.2

Offense	Relative frequency	Percentage
Rape	2/25 = .08	.08 × 100 = 8%
Robbery	3/25 = .12	.12 × 100 = 12%
Burglary	3/25 = .12	.12 × 100 = 12%
Arson	3/25 = .12	.12 × 100 = 12%
Murder	3/25 = .12	.12 × 100 = 12%
Theft	8/25 = .32	.32 × 100 = 32%
Manslaughter	3/25 = .12	.12 × 100 = 12%

BAR GRAPH

A *bar graph* is a graph composed of bars whose heights are the frequencies of the different categories. A bar graph displays graphically the same information concerning qualitative data that a frequency distribution shows in tabular form.

EXAMPLE 2.5 The distribution of the primary sites for cancer is given in Table 2.3 for the residents of Dalton County.

Table 2.3

Primary site	Frequency
Digestive system	20
Respiratory	30
Breast	10
Genitals	5
Urinary tract	5
Other	5

To construct a bar graph, the categories are placed along the horizontal axis and frequencies are marked along the vertical axis. A bar is drawn for each category such that the height of the bar is equal to the frequency for that category. A small gap is left between the bars. The bar graph for Table 2.3 is shown in Fig. 2-1. Bar graphs can also be constructed by placing the categories along the vertical axis and the frequencies along the horizontal axis. See problem 2.5 for a bar graph of this type.

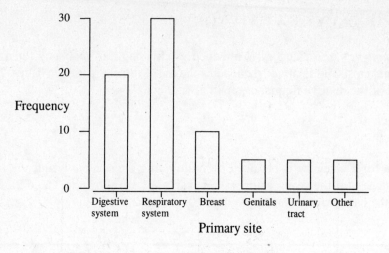

Fig. 2-1

PIE CHART

A *pie chart* is also used to graphically display qualitative data. To construct a pie chart, a circle is divided into portions that represent the relative frequencies or percentages belonging to different categories.

EXAMPLE 2.6 To construct a pie chart for the frequency distribution in Table 2.3, construct a table that gives angle sizes for each category. Table 2.4 shows the determination of the angle sizes for each of the categories in Table 2.3. The 360° in a circle are divided into portions that are proportional to the category sizes. The pie chart for the frequency distribution in Table 2.3 is shown in Fig. 2-2.

Table 2.4

Primary site	Relative frequency	Angle size
Digestive system	.26	$360 \times .26 = 93.6°$
Respiratory	.40	$360 \times .40 = 144°$
Breast	.13	$360 \times .13 = 46.8°$
Genitals	.07	$360 \times .07 = 25.2°$
Urinary tract	.07	$360 \times .07 = 25.2°$
Other	.07	$360 \times .07 = 25.2°$

Primary cancer sites

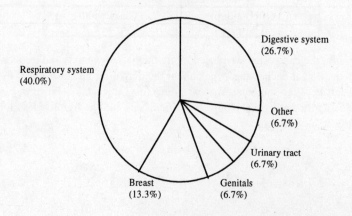

Fig. 2-2

FREQUENCY DISTRIBUTION FOR QUANTITATIVE DATA

There are many similarities between frequency distributions for qualitative data and frequency distributions for quantitative data. Terminology for frequency distributions of quantitative data is discussed first, and then examples illustrating the construction of frequency distributions for quantitative data are given. Table 2.5 gives a frequency distribution of the Stanford–Binet intelligence test scores for 75 adults.

Table 2.5

IQ score	Frequency
80–94	8
95–109	14
110–124	24
125–139	16
140–154	13

IQ score is a quantitative variable and according to Table 2.5, eight of the individuals have an IQ score between 80 and 94, fourteen have scores between 95 and 109, twenty-four have scores between 110 and 124, sixteen have scores between 125 and 139, and thirteen have scores between 140 and 154.

CLASS LIMITS, CLASS BOUNDARIES, CLASS MARKS, AND CLASS WIDTH

The frequency distribution given in Table 2.5 is composed of five *classes*. The classes are: 80–94, 95–109, 110–124, 125–139, and 140–154. Each class has a *lower class limit* and an *upper class limit*. The lower class limits for this distribution are 80, 95, 110, 125, and 140. The upper class limits are 94, 109, 124, 139, and 154.

If the lower class limit for the second class, 95, is added to the upper class limit for the first class, 94, and the sum divided by 2, the *upper boundary* for the first class and the *lower boundary* for the second class is determined. Table 2.6 gives all the boundaries for Table 2.5.

If the lower class limit is added to the upper class limit for any class and the sum divided by 2, the *class mark* for that class is obtained. The class mark for a class is the midpoint of the class and is sometimes called the *class midpoint* rather than the class mark. The class marks for Table 2.5 are shown in Table 2.6.

The difference between the boundaries for any class gives the *class width* for a distribution. The class width for the distribution in Table 2.5 is 15.

Table 2.6

Class limits	Class boundaries	Class width	Class marks
80–94	79.5–94.5	15	87.0
95–109	94.5–109.5	15	102.0
110–124	109.5–124.5	15	117.0
125–139	124.5–139.5	15	132.0
140–154	139.5–154.5	15	147.0

When forming a frequency distribution, the following general guidelines should be followed:
1. *The number of classes should be between 5 and 15*
2. *Each data value must belong to one, and only one, class.*
3. *When possible, all classes should be of equal width.*

EXAMPLE 2.7 Group the following weights into the classes 100 to under 125, 125 to under 150, and so forth:

111	120	127	129	130	145	145	150	153	155	160
161	165	167	170	171	174	175	177	179	180	180
185	185	190	195	195	201	210	220	224	225	230
245	248

The weights 111 and 120 are tallied into the class 100 to under 125. The weights 127, 129, 130, 145 and 145 are tallied into the class 125 to under 150 and so forth until the frequencies for all classes are found. The frequency distribution for these weights is given in Table 2.7

Table 2.7

Weight	Frequency
100 to under 125	2
125 to under 150	5
150 to under 175	10
175 to under 200	10
200 to under 225	4
225 to under 250	4

When a frequency distribution is given in this form, the class limits and class boundaries may be considered to be the same. The class marks are 112.5, 137.5, 162.5, 187.5, 212.5, and 237.5. The class width is 25.

EXAMPLE 2.8 The price for 500 aspirin tablets is determined for each of twenty randomly selected stores as part of a larger consumer study. The prices are as follows:

| 2.50 | 2.95 | 2.65 | 3.10 | 3.15 | 3.05 | 3.05 | 2.60 | 2.70 | 2.75 |
| 2.80 | 2.80 | 2.85 | 2.80 | 3.00 | 3.00 | 2.90 | 2.90 | 2.85 | 2.85 |

Suppose we wish to group these data into seven classes. Since the maximum price is 3.15 and the minimum price is 2.50, the spread in prices is 0.65. Each class should then have a width equal to approximately 1/7 of 0.65 or .093. There is a lot of flexibility in choosing the classes while following the guidelines given above. Table 2.8 shows the results if a class width equal to 0.10 is selected and the first class begins at the minimum price.

Table 2.8

Price	Tally	Frequency
2.50 to 2.59	/	1
2.60 to 2.69	//	2
2.70 to 2.79	//	2
2.80 to 2.89	//// /	6
2.90 to 2.99	///	3
3.00 to 3.09	////	4
3.10 to 3.19	//	2

The frequency distribution might also be given in a form such as that shown in Table 2.9. The two different ways of expressing the classes shown in Tables 2.8 and 2.9 will result in the same frequencies.

Table 2.9

Price	Frequency
2.50 to less than 2.60	1
2.60 to less than 2.70	2
2.70 to less than 2.80	2
2.80 to less than 2.90	6
2.90 to less than 3.00	3
3.00 to less than 3.10	4
3.10 to less than 3.20	2

SINGLE-VALUED CLASSES

If only a few unique values occur in a set of data, the classes are expressed as a single value rather than an interval of values. This typically occurs with discrete data but may also occur with continuous data because of measurement constraints.

EXAMPLE 2.9 A quality technician selects 25 bars of soap from the daily production. The weights in ounces of the 25 bars are as follows:

4.75	4.74	4.74	4.77	4.73	4.75	4.76	4.77
4.72	4.75	4.77	4.74	4.75	4.77	4.72	4.74
4.75	4.75	4.74	4.76	4.75	4.75	4.74	4.75
4.77							

Since only six unique values occur, we will use single-valued classes. The weight 4.72 occurs twice, 4.73 occurs once, 4.74 occurs six times, 4.75 occurs nine times, 4.76 occurs twice, and 4.77 occurs five times. The frequency distribution is shown in Table 2.10.

Table 2.10

Weight	Frequency
4.72	2
4.73	1
4.74	6
4.75	9
4.76	2
4.77	5

HISTOGRAMS

A *histogram* is a graph that displays the classes on the horizontal axis and the frequencies of the classes on the vertical axis. The frequency of each class is represented by a vertical bar whose height is equal to the frequency of the class. A histogram is similar to a bar graph. However, a histogram utilizes classes or intervals and frequencies while a bar graph utilizes categories and frequencies.

EXAMPLE 2.10 A histogram for the aspirin prices in Table 2.9 is shown in Fig. 2-3.

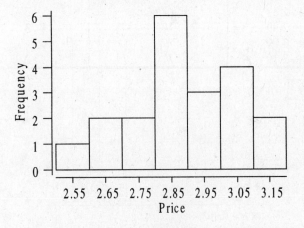

Fig. 2-3

A *symmetric histogram* is one that can be divided into two pieces such that each is the mirror image of the other. One of the most commonly occurring symmetric histograms is shown in Fig. 2-4. This type of histogram is often referred to as a *mound-shaped histogram* or a *bell-shaped histogram*. A symmetric histogram in which each class has the same frequency is called a *uniform* or r*ectangular histogram*. A *skewed to the right histogram* has a longer tail on the right side. The histogram shown in Fig. 2-5 is skewed to the right. A *skewed to the left histogram* has a longer tail on the left side. The histogram shown in Fig. 2-6 is skewed to the left.

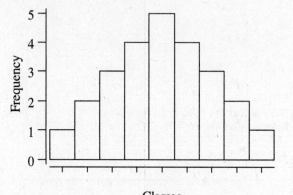

Fig. 2-4

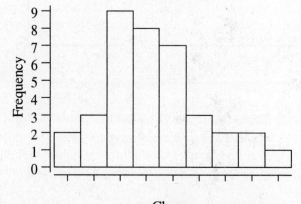

Fig. 2-5

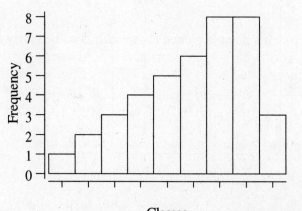

Fig. 2-6

CUMULATIVE FREQUENCY DISTRIBUTIONS

A *cumulative frequency distribution* gives the total number of values that fall below various class boundaries of a frequency distribution.

EXAMPLE 2.11 Table 2.11 shows the frequency distribution of the contents in milliliters of a sample of 25 one-liter bottles of soda. Table 2.12 shows how to construct the cumulative frequency distribution that corresponds to the distribution in Table 2.11.

Table 2.11

Content	Frequency
970 to less than 990	5
990 to less than 1010	10
1010 to less than 1030	5
1030 to less than 1050	3
1050 to less than 1070	2

CUMULATIVE RELATIVE FREQUENCY DISTRIBUTIONS

A *cumulative relative frequency* is obtained by dividing a cumulative frequency by the total number of observations in the data set. The cumulative relative frequencies for the frequency distribution given in Table 2.11 are shown in Table 2.12. *Cumulative percentages* are obtained by multiplying cumulative relative frequencies by 100. The cumulative percentages for the distribution given in Table 2.11 are shown in Table 2.12.

Table 2.12

Contents less than	Cumulative frequency	Cumulative relative frequency	Cumulative percentage
970	0	0/25 = 0	0%
990	5	5/25 = .20	20%
1010	5 + 10 = 15	15/25 = .60	60%
1030	15 + 5 = 20	20/25 = .80	80%
1050	20 + 3 = 23	23/25 = .92	92%
1070	23 + 2 = 25	25/25 = 1.00	100%

OGIVES

An *ogive* is a graph in which a point is plotted above each class boundary at a height equal to the cumulative frequency corresponding to that boundary. Ogives can also be constructed for a cumulative relative frequency distribution as well as a cumulative percentage distribution.

EXAMPLE 2.12 The ogive corresponding to the cumulative frequency distribution in Table 2.12 is shown in Fig. 2-7.

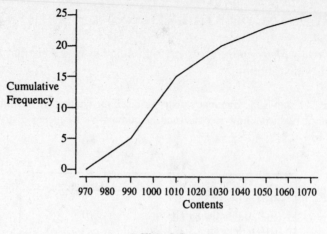

Fig. 2-7

STEM-AND-LEAF DISPLAYS

In a *stem-and-leaf display* each value is divided into a stem and a leaf. The leaves for each stem are shown separately. The stem-and-leaf diagram preserves the information on individual observations.

EXAMPLE 2.13 The following are the California Achievement Percentile Scores (CAT scores) for 30 seventh-grade students:

50	65	70	35	40	57	66	65	70	35
29	33	44	56	66	60	44	50	58	46
67	78	79	47	35	36	44	57	60	57

A stem-and-leaf diagram for these CAT scores is shown in Fig. 2-8.

Stem	Leaves
2	9
3	3 5 5 5 6
4	0 4 4 4 6 7
5	0 0 6 7 7 7 8
6	0 0 5 5 6 6 7
7	0 0 8 9

Fig. 2-8

Solved Problems

RAW DATA

2.1 Classify the following data as either qualitative data or quantitative data. In addition, classify the quantitative data as discrete or continuous.

(*a*) The number of times that a movement authority is sent to a train from a relay station is recorded for several trains over a two-week period. The movement authority, which is an electronic transmission, is sent repeatedly until a return signal is received from the train.

(b) A physician records the follow-up condition of patients with optic neuritis as improved, unchanged, or worse.

(c) A quality technician records the length of material in a roll product for several products selected from a production line.

(d) The *Bureau of Justice Statistics Sourcebook of Criminal Justice Statistics* in reporting on the daily use within the last 30 days of drugs among young adults lists the type of drug as marijuana, cocaine, or stimulants (adjusted).

(e) The number of aces in five-card poker hands is noted by a gambler over several weeks of gambling at a casino.

Ans. (a) The number of times that the moving authority must be sent is countable and therefore these data are quantitative and discrete.

(b) These data are categorical or qualitative.

(c) The length of material can be any number within an interval of values, and therefore these data are quantitative and continuous.

(d) These data are categorical or qualitative

(e) Each value in the data set would be one of the five numbers 0, 1, 2, 3, or 4. These data are quantitative and discrete.

FREQUENCY DISTRIBUTION FOR QUALITATIVE DATA

2.2   The following list gives the academic ranks of the 25 female faculty members at a small liberal arts college:

instructor	assistant professor	assistant professor	instructor
associate professor	assistant professor	associate professor	assistant professor
full professor	associate professor	instructor	assistant professor
full professor	associate professor	assistant professor	instructor
assistant professor	assistant professor	associate professor	assistant professor
full professor	assistant professor	assistant professor	assistant professor
associate professor			

Give a frequency distribution for these data.

Ans.   The academic ranks are tallied into the four possible categories and the results are shown in Table 2.13.

Table 2.13

Academic rank	Frequency
Full professor	3
Associate professor	6
Assistant professor	12
Instructor	4

RELATIVE FREQUENCY OF A CATEGORY AND PERCENTAGE

2.3   Give the relative frequencies and percentages for the categories shown in Table 2.13.

Ans. Each frequency in Table 2.13 is divided by 25 to obtain the relative frequencies for the categories. The relative frequencies are then multiplied by 100 to obtain percentages. The results are shown in Table 2.14.

Table 2.14

Academic rank	Relative frequency	Percentage
Full professor	.12	12%
Associate professor	.24	24%
Assistant professor	.48	48%
Instructor	.16	16%

2.4 Refer to Table 2.14 to answer the following.

(*a*) What percent of the female faculty have a rank of associate professor or higher?
(*b*) What percent of the female faculty are not full professors?
(*c*) What percent of the female faculty are assistant or associate professors?

Ans. (*a*) 24% + 12% = 36%   (*b*) 16% + 48% + 24% = 88%   (*c*) 48% + 24% = 72%

BAR GRAPHS AND PIE CHARTS

2.5 The subjects in an eating disorders research study were divided into one of three different groups. Table 2.15 gives the frequency distribution for these three groups. Construct a bar graph.

Table 2.15

Group	Frequency
Bulimic	30
Anorexic	50
Control	20

Ans. The bar graph for the distribution given in Table 2.15 is shown in Fig. 2-9.

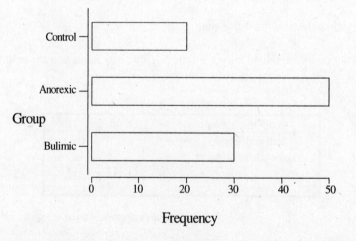

Fig. 2-9

2.6 Construct a pie chart for the frequency distribution given in Table 2.15.

Ans. Table 2.16 illustrates the determination of the angles for each sector of the pie chart.

Table 2.16

Group	Relative frequency	Angle size
Bulimic	.3	$360 \times .3 = 108°$
Anorexic	.5	$360 \times .5 = 180°$
Control	.2	$360 \times .2 = 72°$

Ans.   The pie chart for the distribution given in Table 2.15 is shown in Fig. 2-10.

Pie chart for eating disorder subjects

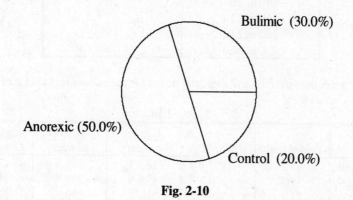

Fig. 2-10

2.7   A survey of 500 randomly chosen individuals is conducted. The individuals are asked to name their favorite sport. The pie chart in Fig. 2-11 summarizes the results of this survey.

Pie chart for favorite sport

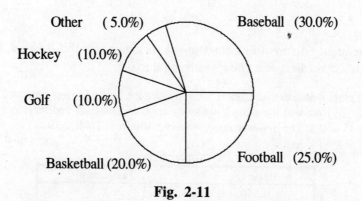

Fig. 2-11

(a)   How many individuals in the 500 gave baseball as their favorite sport?
(b)   How many gave a sport other than basketball as their favorite sport?
(c)   How many gave hockey or golf as their favorite sport?

Ans.   (a) $.3 \times 500 = 150$ (b) $.8 \times 500 = 400$ (c) $.2 \times 500 = 100$

FREQUENCY DISTRIBUTION FOR QUANTITATIVE DATA:
CLASS LIMITS, CLASS BOUNDARIES, CLASS MARKS, AND CLASS WIDTH

2.8 Table 2.17 gives the frequency distribution for the cholesterol values of 45 patients in a cardiac rehabilitation study. Give the lower and upper class limits and boundaries as well as the class marks for each class.

Table 2.17

Cholesterol value	Frequency
170 to 189	3
190 to 209	10
210 to 229	17
230 to 249	13
250 to 269	2

Ans. Table 2.18 gives the limits, boundaries, and marks for the classes in Table 2.17.

Table 2.18

Class	Lower limit	Upper limit	Lower boundary	Upper boundary	Class mark
170 to 189	170	189	169.5	189.5	179.5
190 to 209	190	209	189.5	209.5	199.5
210 to 229	210	229	209.5	229.5	219.5
230 to 249	230	249	229.5	249.5	239.5
250 to 269	250	269	249.5	269.5	259.5

2.9 The following data set gives the yearly food stamp expenditure in thousands of dollars for 25 households in Alcorn County:

2.3	1.9	1.1	3.2	2.7	1.5	0.7	2.5	2.5	3.1	2.5
2.0	2.7	1.9	2.2	1.2	1.3	1.7	2.9	3.0	3.2	1.7
2.2	2.7	2.0								

Construct a frequency distribution consisting of six classes for this data set. Use 0.5 as the lower limit for the first class and use a class width equal to 0.5.

Ans. The first class would extend from 0.5 to 0.9 since the desired lower limit is 0.5 and the desired class width is 0.5. Note that the class boundaries are 0.45 and 0.95 and therefore the class width equals 0.95 − 0.45 or 0.5. The frequency distribution is shown in Table 2.19.

Table 2.19

Expenditure	Frequency
0.5 to 0.9	1
1.0 to 1.4	3
1.5 to 1.9	5
2.0 to 2.4	5
2.5 to 2.9	7
3.0 to 3.4	4

2.10 Express the frequency distribution given in Table 2.19 using the "less than" form for the classes.

Ans. The answer is shown in Table 2.20.

Table 2.20

Expenditure	Frequency
0.5 to less than 1.0	1
1.0 to less than 1.5	3
1.5 to less than 2.0	5
2.0 to less than 2.5	5
2.5 to less than 3.0	7
3.0 to less than 3.5	4

2.11 The manager of a convenience store records the number of gallons of gasoline purchased for a sample of customers chosen over a one-week period. Table 2.21 lists the raw data. Construct a frequency distribution having five classes, each of width 4. Use 0.000 as the lower limit of the first class.

Table 2.21

12.357	19.900	17.500	12.000	8.000	16.000
15.500	18.500	10.000	16.500	6.000	14.675
13.345	13.450	12.500	13.345	5.500	11.234
17.790	19.000	13.456	17.680	15.000	17.678
12.345	4.458	4.000	18.900	12.000	13.200
1.000	14.400	7.500	6.650	17.890	19.500
14.350	16.678	5.500	14.000	3.600	14.000
17.789	13.567	2.000	13.500	15.000	7.500

Ans. When the data are tallied into the five classes, the frequency distribution shown in Table 2.22 is obtained.

Table 2.22

Gallons	Frequency
0.000 to 3.999	3
4.000 to 7.999	8
8.000 to 11.999	3
12.000 to 15.999	20
16.000 to 19.999	14

2.12 Using the data given in Table 2.21, form a frequency distribution consisting of the classes 0 to less than 4, 4 to less than 8, 8 to less than 12, 12 to less than 16, and 16 to less than 20. Compare this frequency distribution with the one given in Table 2.22.

Ans. The frequency distribution is given in Table 2.23. Table 2.23 may have more popular appeal than Table 2.22.

Table 2.23

Gallons	Frequency
0 to less than 4	3
4 to less than 8	8
8 to less than 12	3
12 to less than 16	20
16 to less than 20	14

SINGLE-VALUED CLASSES

2.13 The Food Guide Pyramid divides food into the following six groups: Fats, Oils, and Sweets Group; Milk, Yogurt, and Cheese Group; Vegetable Group; Bread, Cereal, Rice, and Pasta Group; Meat, Poultry, Fish, Dry Beans, Eggs, and Nuts Group; Fruit Group. One question in a nutrition study asked the individuals in the study to give the number of groups included in their daily meals. The results are given below:

6	4	5	4	4	3	4	5	5	5
6	5	4	3	6	6	6	5	2	3
4	5	6	4	5	5	5	6	5	6
5									

Give a frequency distribution for these data.

Ans.   A frequency distribution with single-valued classes is appropriate since only five unique values occur. The frequency distribution is shown in Table 2.24.

Table 2.24

Number of food groups	Frequency
2	1
3	3
4	7
5	12
6	8

2.14 A sociological study involving Mexican-American women utilized a 50-question survey. One question concerned the number of children living at home. The data for this question are given below:

5	2	3	0	4	6	2	1	1	2
3	3	4	4	5	5	5	3	3	3
3	4	4	4	5	5	2	3	4	4
5									

Give a frequency distribution for these data.

Ans.   Since only a small number of unique values occur, the classes will be chosen to be single valued. The frequency distribution is shown in Table 2.25.

Table 2.25

Number of children	Frequency
0	1
1	2
2	4
3	8
4	8
5	7
6	1

HISTOGRAMS

2.15 Construct a histogram for the frequency distribution shown in Table 2.23.

Ans.   The histogram for the frequency distribution in Table 2.23 is shown in Fig. 2-12.

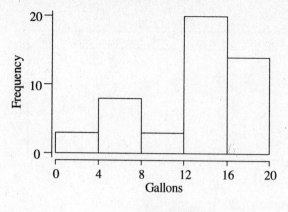

Fig. 2-12

2.16 Construct a histogram for the frequency distribution given in Table 2.24.

   Ans.   The histogram for the frequency distribution in Table 2.24 is shown in Fig. 2-13.

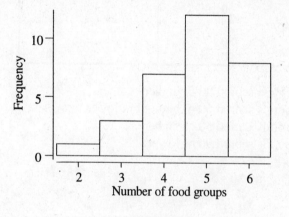

Fig. 2-13

CUMULATIVE FREQUENCY DISTRIBUTIONS

2.17 The Beckmann-Beal mathematics competency test is administered to 150 high school students for an educational study. The test consists of 48 questions and the frequency distribution for the scores is given in Table 2.26. Construct a cumulative frequency distribution for the scores.

Table 2.26

Beckmann-Beal score	Frequency
0–7	5
8–15	15
16–23	20
24–31	30
32–39	50
40–47	30

   Ans.   The cumulative frequency distribution is shown in Table 2.27.

Table 2.27

Scores less than	Cumulative frequency
0	0
8	5
16	20
24	40
32	70
40	120
48	150

2.18 Table 2.28 gives the cumulative frequency distribution of reading readiness scores for 35 kindergarten pupils.

Table 2.28

Scores less than	Cumulative frequency
50	0
60	5
70	15
80	30
90	35

(a) How many of the pupils scored 80 or higher?

(b) How many of the pupils scored 60 or higher but lower than 80?

(c) How many of the pupils scored 50 or higher?

(d) How many of the pupils scored 90 or lower?

Ans.   (a) $35 - 30 = 5$ (b) $30 - 5 = 25$ (c) 35   (d) 35

CUMULATIVE RELATIVE FREQUENCY DISTRIBUTIONS

2.19 Give the cumulative relative frequencies and the cumulative percentages for the reading readiness scores in Table 2.28.

Ans.   The cumulative relative frequencies and cumulative percentages are shown in Table 2.29.

Table 2.29

Scores less than	Cumulative relative frequency	Cumulative percentages
50	0.0	0.0%
60	0.143	14.3%
70	0.429	42.9%
80	0.857	85.7%
90	1.0	100.0%

OGIVES

2.20 Table 2.30 gives the cumulative frequency distribution for the daily breast-milk production in grams for 25 nursing mothers in a research study. Construct an ogive for this distribution.

Table 2.30

Daily production less than	Cumulative frequency
500	0
550	3
600	11
650	20
700	22
750	25

Ans.   The ogive curve for this distribution is shown in Fig. 2-14.

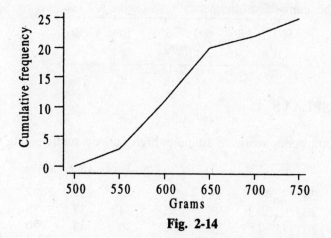

Fig. 2-14

2.21 Construct an ogive curve for the cumulative relative frequency distribution that corresponds to the cumulative frequency distribution in Table 2.30.

Ans.   Each of the cumulative frequencies in Table 2.30 is divided by 25 and the cumulative relative frequencies 0, .12, .44, .80, .88, and 1.00 are determined. Using these, the cumulative relative frequency distribution shown in Fig. 2-15 can be constructed.

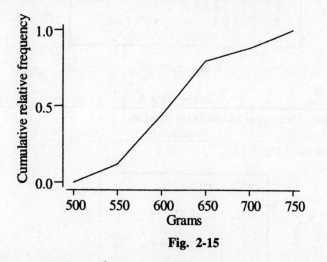

Fig. 2-15

2.22 Construct an ogive curve for the cumulative percentage distribution that corresponds to the cumulative frequency distribution in Table 2.30.

Ans. The cumulative percentages are obtained by multiplying the cumulative relative frequencies given in problem 2.21 by 100 to obtain 0%, 12%, 44%, 80%, 88%, and 100%. These percentages are then used to construct the ogive shown in Fig. 2-16.

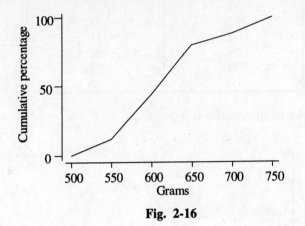

Fig. 2-16

STEM-AND-LEAF DISPLAYS

2.23 The mathematical competency scores of 30 junior high students participating in an educational study are as follows:

30	35	28	44	33	22	40	38	37	36
28	29	30	30	40	30	34	37	38	40
38	34	40	37	23	26	30	45	29	40

Construct a stem-and-leaf display for these data. Use 2, 3, and 4 as your stems.

Ans. The stem-and-leaf display is shown in Fig. 2-17.

Stem	Leaves
2	2 3 6 8 8 9 9
3	0 0 0 0 0 3 4 4 5 6 7 7 7 8 8 8
4	0 0 0 0 0 4 5

Fig. 2-17

2.24 Refine the display shown in Fig. 2-17 by separating the leaves into two groups, one consisting of 0, 1, 2, 3, and 4 and the other consisting of 5, 6, 7, 8, and 9.

Ans. The solution is shown in Fig. 2-18.

Stem	Leaves
2	2 3
2	6 8 8 9 9
3	0 0 0 0 0 3 4 4
3	5 6 7 7 7 8 8 8
4	0 0 0 0 0 4
4	5

Fig. 2-18

2.25 The stem-and-leaf display shown in Fig. 2-19 gives the savings in 15 randomly selected accounts. If the total amount in these 15 accounts equals $4340, find the value of the missing leaf, x.

Stem	Leaves
1	05 50 65 90
2	10 20 55 x 75
3	25 50 70
4	15 80
5	65

Fig. 2-19

Ans. Let A be the amount in the account with the missing leaf. Then the following equation must hold.

$$(105 + 150 + 165 + 190 + 210 + 220 + 255 + A + 275 + 325 + 350 + 370 + 415 + 480 + 565) = 4340$$

The solution to this equation is $A = 265$, which implies that $x = 65$.

Supplementary Problems

RAW DATA

2.26 Classify the data described in the following scenarios as qualitative or quantitative. Classify the quantitative data as either discrete or continuous.

(a) The individuals in a sociological study are classified into one of five income classes as follows: low, low to middle, middle, middle to upper, or upper.
(b) The fasting blood sugar readings are determined for several individuals in a study involving diabetics.
(c) The number of questions correctly answered on a 25-item test is recorded for each student in a computer science class.
(d) The number of attempts needed before successfully finding the path through a maze that leads to a reward is recorded for several rats in a psychological study.
(e) The race of each inmate is recorded for the individuals in a criminal justice study.

Ans. (a) qualitative　　　　(b) quantitative, continuous　　(c) quantitative, discrete
　　　(d) quantitative, discrete　(e) qualitative

FREQUENCY DISTRIBUTION FOR QUALITATIVE DATA

2.27 The following responses were obtained when 50 randomly selected residents of a small city were asked the question "How safe do you think your neighborhood is for kids?"

very	very	not sure	not at all	very	not very	not sure
very	not sure	somewhat	not very	very	not at all	not very
very	very	very	very	not very	somewhat	somewhat
very	very	not sure	not very	not at all	not very	very
very	not sure	very	very	not very	very	very
not very	somewhat	somewhat	very	somewhat	very	very
not very	not at all	very	very	very	somewhat	very
somewhat						

Give a frequency distribution for these data.

Ans.   See Table 2.31.

Table 2.31

Response	Frequency
Very	24
Somewhat	8
Not very	9
Not at all	4
Not sure	5

RELATIVE FREQUENCY OF A CATEGORY AND PERCENTAGE

2.28 Give the relative frequencies and percentages for the categories shown in Table 2.31.

Ans.   See Table 2.32.

Table 2.32

Response	Relative frequency	Percentage
Very	.48	48%
Somewhat	.16	16%
Not very	.18	18%
Not at all	.08	8%
Not sure	.10	10%

2.29 Refer to Table 2.32 to answer the following questions.
(a) What percent of the respondents have no opinion, i.e., responded not sure, on how safe the neighborhood is for children?
(b) What percent of the respondents think the neighborhood is very or somewhat safe for children?
(c) What percent of the respondents give a response other than very safe?

Ans.   (a) 10%   (b) 64%   (c) 52%

BAR GRAPHS AND PIE CHARTS

2.30 Construct a bar graph for the frequency distribution in Table 2.31.

2.31 Construct a pie chart for the frequency distribution given in Table 2.31.

2.32 The bar graph given Fig. 2-20 shows the distribution of responses of 300 individuals to the question "How do you prefer to spend stressful times?"
(a) What percent preferred to spend time alone?
(b) What percent gave a response other than "with friends"?
(c) How many individuals responded "with family," "with friends," or "other"?

Ans.   (a) 50%   (b) 83.3%   (c) 150

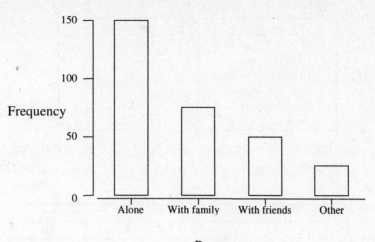

Fig. 2-20

FREQUENCY DISTRIBUTION FOR QUANTITATIVE DATA:
CLASS LIMITS, CLASS BOUNDARIES, CLASS MARKS, AND CLASS WIDTH

2.33 Table 2.33 gives the distribution of response times in minutes for 911 emergency calls classified as domestic disturbance calls. Give the lower and upper class limits and boundaries as well as the class marks for each class. What is the class width for the distribution?

Table 2.33

Response time	Frequency
5–9	3
10–14	7
15–19	25
20–24	19
25–29	14
30–34	2

Ans. See Table 2.34.

Table 2.34

Class	Lower limit	Upper limit	Lower boundary	Upper boundary	Class mark
5–9	5	9	4.5	9.5	7
10–14	10	14	9.5	14.5	12
15–19	15	19	14.5	19.5	17
20–24	20	24	19.5	24.5	22
25–29	25	29	24.5	29.5	27
30–34	30	34	29.5	34.5	32

The class width equals 5 minutes.

2.34 Express the distribution given in Table 2.33 in the "less than" form.

Ans. See Table 2.35.

Table 2.35

Response time	Frequency
5 to less than 10	3
10 to less than 15	7
15 to less than 20	25
20 to less than 25	19
25 to less than 30	14
30 to less than 35	2

2.35 Table 2.36 gives the response times in minutes for 50 randomly selected 911 emergency calls classified as robbery in progress calls. Group the data into five classes, using 1.0 to 2.9 as your first class.

Table 2.36

2.5	5.0	8.5	5.5	10.5
5.0	5.5	7.0	5.0	10.0
6.5	6.0	7.0	6.0	10.0
1.5	7.5	7.0	6.5	4.5
7.0	2.0	5.5	7.0	5.0
10.0	3.0	6.5	5.5	5.0
9.5	2.5	3.5	3.5	5.0
8.0	3.0	7.5	2.0	4.5
7.5	6.5	7.0	1.5	9.0
2.5	7.5	10.0	7.0	8.5

Ans.   See Table 2.37.

Table 2.37

Response time	Frequency
1.0 to 2.9	7
3.0 to 4.9	6
5.0 to 6.9	16
7.0 to 8.9	14
9.0 to 10.9	7

2.37 Refer to the frequency distribution of response times to 911 robbery in progress calls given in Table 2.37 to answer the following.
(*a*) What percent of the response times are less than seven minutes?
(*b*) What percent of the response times are equal to or greater than three minutes but less than seven minutes?
(*c*) What percent of the response times are nine or more minutes in length?

Ans.   (*a*) 58%   (*b*) 44%   (*c*) 14%

2.38 Refer to the frequency distribution given in Table 2.38 to find the following.
(*a*) The boundaries for the class c to d
(*b*) The class mark for the class e to f
(*c*) The width for the class g to i
(*d*) The lower class limit for the class g to i
(*e*) The total number of observations

Table 2.38

Class	Frequency
a to b	f_1
c to d	f_2
e to f	f_3
g to i	f_4
j to k	f_5

Ans.   *(a)* (b + c)/2 and (d + e)/2   *(b)* (e + f)/2   *(c)* (i + j)/2 − (f + g)/2
     (d) g   *(e)* $(f_1 + f_2 + f_3 + f_4 + f_5)$

SINGLE-VALUED CLASSES

2.39 A quality control technician records the number of defective items found in samples of size 50 for each of 30 days. The data are as follows:

0	2	3	0	0	0	0	1	2	1
1	2	0	0	1	2	2	2	0	0
1	1	1	0	0	0	3	2	0	1

Give a frequency distribution for these data.

Ans.   See Table 2.39

Table 2.39

Number of defectives	Frequency
0	13
1	8
2	7
3	2

2.40 The number of daily traffic citations issued over a 100-mile section of Interstate 80 is recorded for each day of September. The frequency distribution for these data is shown in Table 2.40. Find the value for x.

Table 2.40

Number of citations	Frequency
10	5
11	7
14	10
16	x
17	3
20	3

Ans.   x = 2

HISTOGRAMS

2.41 Construct a histogram for the response times frequency distribution given in Table 2.37.

2.42 Construct a histogram for the number of defectives frequency distribution given in Table 2.39.

CUMULATIVE FREQUENCY DISTRIBUTIONS

2.43 Give the cumulative frequency distribution for the frequency distribution shown in Table 2.37

Ans.   See Table 2.41.

Table 2.41

Response time less than	Cumulative frequency
1.0	0
3.0	7
5.0	13
7.0	29
9.0	43
11.0	50

2.44 Refer to Table 2.41 to answer the following questions.
  (*a*) How many of the response times are less than 5.0 minutes?
  (*b*) How many of the response times are 7.0 or more minutes?
  (*c*) How many of the response times are equal to or greater than 5.0 minutes but less than 9.0 minutes?

  Ans.   (*a*) 13   (*b*) 21   (*c*) 30

CUMULATIVE RELATIVE FREQUENCY DISTRIBUTIONS

2.45 Give the cumulative relative frequencies and the cumulative percentages for the cumulative frequency distribution shown in Table 2.41.

  Ans.   See Table 2.42.

Table 2.42

Response time less than	Cumulative relative frequency	Cumulative percentage
1.0	0.0	0%
3.0	0.14	14%
5.0	0.26	26%
7.0	0.58	58%
9.0	0.86	86%
11.0	1.0	100%

OGIVES

2.46 Construct an ogive for the cumulative frequency distribution given in Table 2.41.

2.47 Construct an ogive for the cumulative relative frequency distribution given in Table 2.42.

STEM-AND-LEAF DISPLAYS

2.48 The number of calls per 24-hour period to a 911 emergency number is recorded for 50 such periods. The results are shown in Table 2.43. Construct a stem-and-leaf display for these data.

Table 2.43

450	333	660	765	589
345	347	456	501	543
678	543	234	597	521
345	780	435	456	678
189	567	675	525	453
500	304	805	586	267
456	435	346	514	508
654	564	225	524	465
324	475	505	707	556
700	125	578	286	531

  Ans.   See Fig. 2-21.

Stem	Leaves
1	25 89
2	25 34 67 86
3	04 24 33 45 45 46 47
4	35 35 50 53 56 56 56 65 75
5	00 01 05 08 14 21 24 25 31 43 43 56 64 67 78 86 89 97
6	54 60 75 78 78
7	00 07 65 80
8	05

Fig. 2-21

2.49 The number of syringes used per month by the patients of a diabetic specialist is recorded and the results are given in Fig. 2-22. Answer the following questions by referring to Fig. 2-22.

(a) What is the minimum number of syringes used per month by these patients?

(b) What is the maximum number of syringes used per month by these patients?

(c) What is the total usage per month by these patients?

(d) What usage occurs most frequently?

Stem	Leaves
3	0 0 5 5 5 5
4	0 0 3 5 6 8 8
5	2 4 4 4 8 8 9 9 9
6	0 0 0 0 0 0 5 5 5 5 5 8 8 8
7	3 3 3 5 5
8	0 0 5
9	0 0

Fig. 2-22

Ans.   (a) 30   (b) 90   (c) 2700   (d) 60

2.50 Refine the stem-and-leaf display in Fig.2-22 by using either the leaves 0, 1, 2, 3, or 4 or the leaves 5, 6, 7, 8, or 9 on a particular row.

Ans.   See Fig. 2-23.

Stem	Leaves
3	0 0
3	5 5 5 5
4	0 0 3
4	5 6 8 8
5	2 4 4 4
5	8 8 9 9 9
6	0 0 0 0 0 0
6	5 5 5 5 5 8 8 8
7	3 3 3
7	5 5
8	0 0
8	5
9	0 0

Fig. 2-23

Chapter 3

Descriptive Measures

MEASURES OF CENTRAL TENDENCY

Chapter 2 gives several techniques for organizing data. Bar graphs, pie charts, frequency distributions, histograms, and stem-and-leaf plots are techniques for describing data. Often times, we are interested in a typical numerical value to help us describe a data set. This typical value is often called an *average value* or *a measure of central tendency*. We are looking for a single number that is in some sense representative of the complete data set.

EXAMPLE 3.1 The following are examples of measures of central tendency: median priced home, average cost of a new automobile, the average household income in the United States, and the modal number of televisions per household. Each of these examples is a single number, which is intended to be typical of the characteristic of interest.

MEAN, MEDIAN, AND MODE FOR UNGROUPED DATA

A data set consisting of the observations for some variable is referred to as raw data or *ungrouped data*. Data presented in the form of a frequency distribution are called *grouped data*. The measures of central tendency discussed in this chapter will be described for both grouped and ungrouped data since both forms of data occur frequently.

There are many different measures of central tendency. The three most widely used measures of central tendency are the *mean, median,* and *mode*. These measures are defined for both samples and populations.

The mean for a sample consisting of n observations is

$$\bar{x} = \frac{\sum x}{n} \tag{3.1}$$

and the mean for a population consisting of N observations is

$$\mu = \frac{\sum x}{N} \tag{3.2}$$

EXAMPLE 3.2 The number of 911 emergency calls classified as domestic disturbance calls in a large metropolitan location were sampled for thirty randomly selected 24 hour periods with the following results. Find the mean number of calls per 24-hour period.

$$
\begin{array}{cccccccccccccccc}
25 & 46 & 34 & 45 & 37 & 36 & 40 & 30 & 29 & 37 & 44 & 56 & 50 & 47 & 23 \\
40 & 30 & 27 & 38 & 47 & 58 & 22 & 29 & 56 & 40 & 46 & 38 & 19 & 49 & 50
\end{array}
$$

$$\bar{x} = \frac{\sum x}{n} = \frac{1168}{30} = 38.9$$

EXAMPLE 3.3 The total number of 911 emergency calls classified as domestic disturbance calls last year in a large metropolitan location was 14,950. Find the mean number of such calls per 24-hour period if last year was not a leap year.

$$\mu = \frac{\Sigma x}{N} = \frac{14,950}{365} = 41.0$$

The *median* of a set of data is a value that divides the bottom 50% of the data from the top 50% of the data. To find the median of a data set, first arrange the data in increasing order. If the number of observations is odd, the median is the number in the middle of the ordered list. If the number of observations is even, the median is the mean of the two values closest to the middle of the ordered list. There is no widely used symbol used to represent the median. Occasionally, the symbol $\tilde{x}$ is used to represent the sample median and the symbol $\tilde{\mu}$ is used to represent the population median.

EXAMPLE 3.4 To find the median number of domestic disturbance calls per 24-hour period for the data in Example 3.1, first arrange the data in increasing order.

19　22　23　25　27　29　29　30　30　34　36　37　37　38　38
40　40　40　44　45　46　46　47　47　49　50　50　56　56　58

The two values closest to the middle are 38 and 40. The median is the mean of these two values or 39.

EXAMPLE 3.5 A bank auditor selects 11 checking accounts and records the amount in each of the accounts. The 11 observations in increasing order are as follows:

150.25　175.35　195.00　200.00　235.00　240.45　250.55　256.00　275.50　290.10　300.55

The median is 240.45 since this is the middle value in the ordered list.

The *mode* is the value in a data set that occurs the most often. If no such value exists, we say that the data set has no mode. If two such values exist, we say the data set is *bimodal*. If three such values exist, we say the data set is *trimodal*. There is no symbol that is used to represent the mode.

EXAMPLE 3.6 Find the mode for the data given in Example 3.2. Often it is helpful to arrange the data in increasing order when finding the mode. The data, in increasing order, are given in Example 3.4. When the data are examined, it is seen that 40 occurs three times, and that no other value occurs that often. The mode is equal to 40.

For a large data set, as the number of classes is increased (and the width of the classes is decreased), the histogram becomes a smooth curve. Oftentimes, the smooth curve assumes a shape like that shown in Fig. 3-1. In this case, the data set is said to have a *bell-shaped distribution* or a *mound-shaped distribution*. For such a distribution, the mean, median, and mode are equal and they are located at the center of the curve.

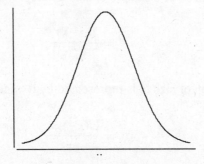

Fig. 3-1

For a data set having a skewed to the right distribution, the mode is usually less than the median which is usually less than the mean. For a data set having a skewed to the left distribution, the mean is usually less than the median which is usually less than the mode.

EXAMPLE 3.7 Find the mean, median, and mode for the following three data sets and confirm the above paragraph.

Data set 1: 10, 12, 15, 15, 18, 20   Data set 2: 2, 4, 6, 15, 15, 18   Data set 3: 12, 15, 15, 24, 26, 28

Table 3.1 gives the shape of the distribution, the mean, the median, and the mode for the three data sets.

Table 3.1

Data set	Distribution shape	Mean	Median	Mode
1	Bell-shaped	15	15	15
2	Left-skewed	10	10.5	15
3	Right-skewed	20.5	19.5	15

MEASURES OF DISPERSION

In addition to measures of central tendency, it is desirable to have numerical values to describe the spread or dispersion of a data set. Measures that describe the spread of a data set are called *measures of dispersion.*

EXAMPLE 3.8 Jon and Jack are two golfers who both average 85. However, Jon has shot as low as 75 and as high as 99 whereas Jack has never shot below 80 nor higher than 90. When we say that Jack is a more consistent golfer than Jon is, we mean that the spread in Jack's scores is less than the spread in Jon's scores. A measure of dispersion is a numerical value that illustrates the differences in the spread of their scores.

RANGE, VARIANCE, AND STANDARD DEVIATION FOR UNGROUPED DATA

The *range* for a data set is equal to the maximum value in the data set minus the minimum value in the data set. It is clear that the range is reflective of the spread in the data set since the difference between the largest and the smallest value is directly related to the spread in the data.

EXAMPLE 3.9 Compare the range in golf scores for Jon and Jack in Example 3.8. The range for Jon is 99 – 75 = 24 and the range for Jack is 90 – 80 = 10. The spread in Jon's scores, as measured by range, is over twice the spread in Jack's scores.

The *variance* and the *standard deviation* of a data set measures the spread of the data about the mean of the data set. The variance of a sample of size n is represented by s^2 and is given by

$$s^2 = \frac{\Sigma(x-\overline{x})^2}{n-1} \tag{3.3}$$

and the variance of a population of size N is represented by σ^2 and is given by

$$\sigma^2 = \frac{\Sigma(x-\mu)^2}{N} \tag{3.4}$$

The symbol σ is the lowercase sigma of the Greek alphabet and σ^2 is read as sigma squared.

EXAMPLE 3.10 The times required in minutes for five preschoolers to complete a task were 5, 10, 15, 3, and 7. The mean time for the five preschoolers is 8 minutes. Table 3.2 illustrates the computation indicated by formula (3.3). The first column lists the observations, x. The second column lists the *deviations from the mean*, $x - \overline{x}$. The third column lists the *squares of the deviations*. The sum at the bottom of the second column is called *the sum of the deviations*, and is always equal to zero for any data set. The sum at the bottom of the third column is referred to as the *sum of the squares of the deviations*. The sample variance is obtained by dividing the sum of the squares of the deviations by n – 1, or 5 –1 = 4. The sample variance equals 88 divided by 4 which is 22 minutes squared.

Table 3.2

x	$x - \overline{x}$	$(x - \overline{x})^2$
5	5 – 8 = –3	$(-3)^2 = 9$
10	10 – 8 = 2	$(2)^2 = 4$
15	15 – 8 = 7	$(7)^2 = 49$
3	3 – 8 = –5	$(-5)^2 = 25$
7	7 – 8 = –1	$(-1)^2 = 1$
	$\Sigma(x-\overline{x}) = 0$	$\Sigma(x-\overline{x})^2 = 88$

The variance of the data in Example 3.10 is 22 minutes squared. The units for the variance are minutes squared since the terms which are added in column 3 are minutes squared. The square root of the variance is called the *standard deviation* and the standard deviation is measured in the same units as the variable. The standard deviation of the times to complete the task is $\sqrt{22}$ or 4.7 minutes.

The *sample standard deviation* is

$$s = \sqrt{s^2} \qquad (3.5)$$

and the *population standard deviation* is

$$\sigma = \sqrt{\sigma^2} \qquad (3.6)$$

Shortcut formulas equivalent to formulas (3.3) and (3.4) are useful in computing variances and standard deviations. The *shortcut formulas* for computing sample and population variances are

$$s^2 = \frac{\Sigma x^2 - \frac{(\Sigma x)^2}{n}}{n-1} \qquad (3.7)$$

and

$$\sigma^2 = \frac{\Sigma x^2 - \frac{(\Sigma x)^2}{N}}{N} \qquad (3.8)$$

EXAMPLE 3.11 Formula (3.7) can be used to find the variance and standard deviation of the times given in Example 3.10. The term Σx^2 is called the *sum of the squares* and is found as follows:

$$\Sigma x^2 = 5^2 + 10^2 + 15^2 + 3^2 + 7^2 = 408$$

The term $(\Sigma x)^2$ is referred to as the *sum squared* and is found as follows:

$$(\Sigma x)^2 = (5 + 10 + 15 + 3 + 7)^2 = 1600$$

The variance is given as follows:

$$s^2 = \frac{408 - \frac{1600}{5}}{4} = \frac{408 - 320}{4} = 22$$

and the standard deviation is

$$s = \sqrt{22} = 4.7$$

These are the same values we found in Example 3.10 for the variance and standard deviation of the times.

Since most populations are large, the computation of σ^2 is rarely performed. In practice, the population variance (or standard deviation) is usually estimated by taking a sample from the population and using s^2 as a estimate of σ^2. The use of $n-1$ rather than n in the denominator of the formula for s^2 enhances the ability of s^2 to estimate σ^2.

For data sets having a symmetric mound-shaped distribution, the standard deviation is approximately equal to one-fourth of the range of the data set. This fact can be used to estimate s for bell-shaped distributions.

Statistical software packages are frequently used to compute the standard deviation as well as many other statistical measures.

EXAMPLE 3.12 The costs for scientific calculators with comparable built-in functions were recorded for 20 different sales locations. The results were as follows:

| 10.50 | 12.75 | 11.00 | 16.50 | 19.30 | 20.00 | 16.50 | 13.90 | 17.50 | 18.00 |
| 13.50 | 17.75 | 18.50 | 20.00 | 15.00 | 14.45 | 17.85 | 15.00 | 17.50 | 13.50 |

The analysis of these data using Minitab is as follows.

```
MTB > name c1 'cost'
MTB > set c1
DATA > 10.50  13.50  12.75  17.75  11.00  18.50  16.50  20.00  19.30
DATA > 15.00  20.00  14.45  16.50  17.85  13.90  15.00  17.50  17.50
DATA > 18.00  13.50
DATA > end
MTB > describe c1
```

Descriptive Statistics

Variable	N	Mean	Median	TrMean	StDev	SEMean
cost	20	15.950	16.500	16.028	2.826	0.632

Variable	Min	Max	Q1	Q3
cost	10.500	20.000	13.600	17.962

The cost data are set into column c1, and the command *describe c1* produces 10 different descriptive measures. The student is encouraged to confirm the values for the mean, median, standard deviation, minimum, and maximum. The other four measures are described elsewhere in the outline. Even though these data are not symmetric mound shaped, a "ballpark" approximation to the standard deviation is obtained by dividing the range by 4. One-fourth of the range is 2.375 and the value of the standard deviation is 2.826.

MEASURES OF CENTRAL TENDENCY AND DISPERSION FOR GROUPED DATA

Statistical data are often given in grouped form, i.e., in the form of a frequency distribution, and the raw data corresponding to the grouped data are not available or may be difficult to obtain. The articles that appear in newspapers and professional journals do not give the raw data, but give the results in grouped form. Table 3.3 gives the frequency distribution of the ages of 5000 shoplifters in a recent psychological study of these individuals.

Table 3.3

Age	Frequency
5–14	750
15–24	2005
25–34	1950
35–44	195
45–54	100

The techniques for finding the mean, median, mode, range, variance and standard deviation of grouped data will be illustrated by finding these measures for the data given in Table 3.3. The formulas and techniques given will be for sample data. Similar formulas and techniques are used for population data.

The *mean for grouped data* is given by

$$\overline{x} = \frac{\sum xf}{n} \tag{3.9}$$

where x represents the class marks, f represents the class frequencies, and n = Σf.

EXAMPLE 3.13 The class marks in Table 3.3 are $x_1 = 9.5$, $x_2 = 19.5$, $x_3 = 29.5$, $x_4 = 39.5$, $x_5 = 49.5$ and the frequencies are $f_1 = 750$, $f_2 = 2005$, $f_3 = 1950$, $f_4 = 195$, and $f_5 = 100$. The sample size n is 5000. The mean is

$$\overline{x} = \frac{9.5 \times 750 + 19.5 \times 2005 + 29.5 \times 1950 + 39.5 \times 195 + 49.5 \times 100}{5000} = \frac{116,400}{5000} = 23.3 \text{ years}$$

The *median for grouped data* is found by locating the value that divides the data into two equal parts. In finding the median for grouped data, it is assumed that the data in each class is uniformly spread across the class.

EXAMPLE 3.14 The median age for the data in Table 3.3 is a value such that 2500 ages are less than the value and 2500 are greater than the value. The median age must occur in the age group 15–24, since 750 are less than 15 and 2755 are 24 years or less. The class 15–24 is called the *median class* since the median must fall in this class. Since 750 are less than 15 years, there must be 1750 additional ages in the class 15– 24 that are less than the median. In other words, we need to go the fraction 1750/2005 across the class 15–24 to locate the median. We give the value 14.5 + (1750/2005) × 10 = 23.2 years as the median age. To summarize, 14.5 is the lower boundary of the median class, 1750/2005 is the fraction we must go across the median class to reach the median, and 10 is the class width for the median class.

The *modal class* is defined to be the class with the maximum frequency. The *mode for grouped data* is defined to be the class mark of the modal class.

EXAMPLE 3.15 The modal class for the distribution in Table 3.3 is the class 15–24. The mode is the class mark for this class that equals 19.5 years.

The *range for grouped data* is given by the difference between the upper boundary of the class having the largest values minus the lower boundary of the class having the smallest values.

EXAMPLE 3.16 The upper boundary for the class 45–54 is 54.5 and the lower boundary for the class 5–14 is 4.5, and the range is 54.5 – 4.5 = 50.0 years.

The *variance for grouped data* is given by

$$s^2 = \frac{\sum x^2 f - \frac{(\sum xf)^2}{n}}{n-1} \tag{3.10}$$

and the standard deviation is given by

$$s = \sqrt{s^2} \tag{3.11}$$

EXAMPLE 3.17 In order to find the variance and standard deviation for the distribution in Table 3.3 using formulas (*3.10*) and (*3.11*), we first evaluate $\sum x^2 f$ and $\frac{(\sum xf)^2}{n}$.

$$\sum x^2 f = 9.5^2 \times 750 + 19.5^2 \times 2005 + 29.5^2 \times 1950 + 39.5^2 \times 195 + 49.5^2 \times 100 = 3,076,350$$

From Example 3.13, we see that $\Sigma xf = 116,400$, and therefore $\frac{(\sum xf)^2}{n} = 2,709,792$. The variance is

$$s^2 = \frac{3,076,350 - 2,709,792}{4999} = 73.3$$

and the standard deviation is $\sqrt{73.3} = 8.6$ years.

CHEBYSHEV'S THEOREM

Chebyshev's theorem provides a useful interpretation of the standard deviation. Chebyshev's theorem states that the fraction of any data set lying within k standard deviations of the mean is at least $1 - \frac{1}{k^2}$, where k is a number greater than 1. The theorem applies to either a sample or a population. If k = 2, this theorem states that at least $1 - \frac{1}{2^2} = 1 - \frac{1}{4} = \frac{3}{4}$ or 75% of the data set will fall between $\bar{x} - 2s$ and $\bar{x} + 2s$. Similarly, for k = 3, the theorem states that at least $\frac{8}{9}$ or 89% of the data set will fall between $\bar{x} - 3s$ and $\bar{x} + 3s$.

EXAMPLE 3.18 The reading readiness scores for a group of 4 and 5 year old children have a mean of 73.5 and a standard deviation equal to 5.5. At least 75% of the scores are between $73.5 - 2 \times 5.5 = 62.5$ and $73.5 + 2 \times 5.5 = 84.5$. At least 89% of the scores are between $73.5 - 3 \times 5.5 = 57.0$ and $73.5 + 3 \times 5.5 = 90.0$.

EMPIRICAL RULE

The *empirical rule* states that for a data set having a bell-shaped distribution, approximately 68% of the observations lie within one standard deviation of the mean, approximately 95% of the observations lie within two standard deviations of the mean, and approximately 99.7% of the observations lie within three standard deviations of the mean. The empirical rule applies to either large samples or populations.

EXAMPLE 3.19 Assuming the incomes for all single parent households last year had a bell-shaped distribution with a mean equal to $23,500 and a standard deviation equal to $4,500, the following conclusions follow: 68% of the incomes lie between $19,000 and $28,000, 95% of the incomes lie between $14,500 and $32,500, and 99.7% of the incomes lie between $10,000 and $37,000. If the shape of the distribution is unknown, then Chebyshev's theorem does not give us any information about the percent of the distribution between $19,500 and $28,000. However, Chebyshev's theorem assures us that at least 75% of the incomes are between $14,500 and $32,500 and at least 89% of the incomes are between $10,000 and $37,000.

COEFFICIENT OF VARIATION

The *coefficient of variation* is equal to the standard deviation divided by the mean. The result is usually multiplied by 100 to express it as a percent. The coefficient of variation for a sample is given by

$$CV = \frac{s}{\bar{x}} \times 100\% \qquad (3.12)$$

and the coefficient of variation for a population is given by

$$CV = \frac{\mu}{\sigma} \times 100\% \qquad (3.13)$$

The coefficient of variation is a measure of relative variation, whereas the standard deviation is a measure of absolute variation.

EXAMPLE 3.20 A national sampling of prices for new and used cars found that the mean price for a new car is $20,100 and the standard deviation is $6,125 and that the mean price for a used car is $5,485 with a standard deviation equal to $2,730. In terms of absolute variation, the standard deviation of price for new cars is more than twice that of used cars. However, in terms of relative variation, there is more relative variation in the price of used cars than in new cars. The CV for used cars is $\frac{2,730}{5,485} \times 100 = 49.8\%$ and the CV for new cars is $\frac{6,125}{20,100} \times 100 = 30.5\%$.

Z SCORES

A *z score* is the number of standard deviations that a given observation, x, is below or above the mean. For sample data, the z score is

$$z = \frac{x - \bar{x}}{s} \qquad (3.14)$$

and for population data, the z score is

$$z = \frac{x - \mu}{\sigma} \qquad (3.15)$$

EXAMPLE 3.21 The mean salary for deputies in Douglas County is $27,500 and the standard deviation is $4,500. The mean salary for deputies in Hall County is $24,250 and the standard deviation is $2,750. A deputy who makes $30,000 in Douglas County makes $1,500 more than a deputy does in Hall County who makes $28,500. Which deputy has the higher salary relative to the county in which he works?

For the deputy in Douglas County who makes $30,000, the z score is

$$z = \frac{x-\mu}{\sigma} = \frac{30,000-27,500}{4,500} = 0.56$$

For the deputy in Hall County who makes \$28,500, the z score is

$$z = \frac{x-\mu}{\sigma} = \frac{28,500-24,250}{2,750} = 1.55$$

When the county of employment is taken into consideration, the \$28,500 salary is a higher relative salary than the \$30,000 salary.

MEASURES OF POSITION: PERCENTILES, DECILES, AND QUARTILES

Measures of position are used to describe the location of a particular observation in relation to the rest of the data set. *Percentiles* are values that divide the ranked data set into 100 equal parts. The *pth percentile* of a data set is a value such that at least p percent of the observations take on this value or less and at least (100 – p) percent of the observations take on this value or more. *Deciles* are values that divide the ranked data set into 10 equal parts. *Quartiles* are values that divide the ranked data set into four equal parts. The techniques for finding the various measures of position will be illustrated by using the data in Table 3.4. Table 3.4 contains the aortic diameters measured in centimeters for 45 patients. Notice that the data in Table 3.4 are already ranked. Raw data need to be ranked prior to finding measures of position.

Table 3.4

3.0	5.0	6.2	7.6	9.4
3.3	5.2	6.3	7.6	9.5
3.5	5.5	6.4	7.7	9.5
3.5	5.5	6.6	7.8	10.0
3.6	5.5	6.6	7.8	10.5
4.0	5.8	6.8	8.5	10.8
4.0	5.8	6.8	8.5	10.9
4.2	5.9	6.8	8.8	11.0
4.6	6.0	7.0	8.8	11.0

The *percentile for observation* x is found by dividing the number of observations less than x by the total number of observations and then multiplying this quantity by 100. This percent is then rounded to the nearest whole number to give the percentile for observation x.

EXAMPLE 3.22 The number of observations in Table 3.4 less than 5.5 is 11. Eleven divided by 45 is .244 and .244 multiplied by 100 is 24.4%. This percent rounds to 24. The diameter 5.5 is the 24th percentile and we express this as $P_{24} = 5.5$. The number of observations less than 5.0 is 9. Nine divided by 45 is .20 and .20 multiplied by 100 is 20%. $P_{20} = 5.0$. The number of observations less than 10.0 is 39. Thirty-nine divided by 45 is .867 and .867 multiplied by 100 is 86.7%. Since 86.7% rounds to 87%, we write $P_{87} = 10.0$

The *pth percentile* for a ranked data set consisting of n observations is found by a two-step procedure. The first step is to compute index $i = \frac{(p)(n)}{100}$. If i is not an integer, the next integer greater than i locates the position of the pth percentile in the ranked data set. If i is an integer, the pth percentile is the average of the observations in positions i and i + 1 in the ranked data set.

EXAMPLE 3.23 To find the tenth percentile for the data of Table 3.4, compute $i = \dfrac{(10)(45)}{100} = 4.5$. The next integer greater than 4.5 is 5. The observation in the fifth position in Table 3.4 is 3.6. Therefore, $P_{10} = 3.6$. Note that at least 10% of the data in Table 3.4 are 3.6 or less (the actual amount is 11.1%) and at least 90% of the data are 3.6 or more (the actual amount is 91.1%). For very large data sets, the percentage of observations equal to or less than P_{10} will be very close to 10% and the percentage of observations equal to or greater than P_{10} will be very close to 90%.

EXAMPLE 3.24 To find the fortieth percentile for the data in Table 3.4, compute $i = \dfrac{(40)(45)}{100} = 18$. The fortieth percentile is the average of the observations in the 18th and 19th positions in the ranked data set. The observation in the 18th position is 6.0 and the observation in the 19th position is 6.2. Therefore $P_{40} = \dfrac{6.0 + 6.2}{2} =$ 6.1. Note that 40% of the data in Table 3.4 are 6.1 or less and that 60% of the observations are 6.1 or more.

Deciles and quartiles are determined in the same manner as percentiles, since they may be expressed as percentiles. The deciles are represented as $D_1, D_2, \ldots, D_9$ and the quartiles are represented by Q_1, Q_2, and Q_3. The following equalities hold for deciles and percentiles:

$$D_1 = P_{10}, D_2 = P_{20}, D_3 = P_{30}, D_4 = P_{40}, D_5 = P_{50}, D_6 = P_{60}, D_7 = P_{70}, D_8 = P_{80}, D_9 = P_{90}$$

The following equalities hold for quartiles and percentiles:

$$Q_1 = P_{25}, Q_2 = P_{50}, Q_3 = P_{75}$$

From the above definitions of percentiles, deciles, and quartiles, the following equalities also hold:

$$\text{Median} = P_{50} = D_5 = Q_2$$

The techniques for finding percentiles, deciles, and quartiles differ somewhat from textbook to textbook, but the values obtained by the various techniques are usually very close to one another.

INTERQUARTILE RANGE

The *interquartile range*, designated by IQR, is defined as follows:

$$IQR = Q_3 - Q_1 \qquad\qquad (3.16)$$

The interquartile range shows the spread of the middle 50% of the data and is not affected by extremes in the data set.

EXAMPLE 3.25 The interquartile range for the aortic diameters in Table 3.4 is found by subtracting the value of Q_1 from Q_3. The first quartile is equal to the 25th percentile and is found by noting that $\dfrac{45 \times 25}{100} = 11.25$ and therefore $i = 12$. Q_1 is in the 12th position in Table 3.4 and $Q_1 = 5.5$. The third quartile is equal to the 75th percentile and is found by noting that $\dfrac{45 \times 75}{100} = 33.75$ and therefore $i = 34$. Q_3 is in the 34th position in Table 3.4 and $Q_3 = 8.5$. The IQR equals 8.5 − 5.5 or 3.0 cm.

BOX-AND-WHISKER PLOT

A *box-and-whisker plot*, sometimes simply called a *boxplot* is a graphical display in which a box extending from Q_1 to Q_3 is constructed and which contains the middle 50% of the data. Lines, called *whiskers*, are drawn from Q_1 to the smallest value and from Q_3 to the largest value. In addition, a vertical line is constructed inside the box corresponding to the median.

EXAMPLE 3.26 A Minitab generated boxplot for the data in Table 3.4 is shown in Fig. 3-2. For these data, the minimum diameter is 3.0 cm, the maximum diameter is 11.0 cm., $Q_1 = 5.5$ cm, $Q_2 = 6.6$ cm, and $Q_3 = 8.5$ cm. Because Minitab uses a slightly different technique for finding the first and third quartile, the box extends from 5.35 to 8.65, rather than from 5.5 to 8.5.

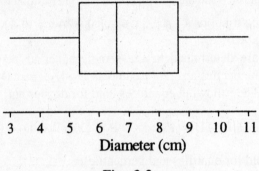

$$3 \quad 4 \quad 5 \quad 6 \quad 7 \quad 8 \quad 9 \quad 10 \quad 11$$

Diameter (cm)

Fig. 3-2

Another type boxplot, called a *modified boxplot,* is also sometimes constructed in which possible and probable outliers are identified. The modified boxplot is illustrated in problem 3.27.

Solved Problems

MEASURES OF CENTRAL TENDENCY: MEAN, MEDIAN, AND MODE FOR UNGROUPED DATA

Table 3.5

10.5	12.5	14.5	22.0	12.5
−2.5	20.2	3.5	7.5	14.5
14.0	17.5	14.0	12.0	17.0
20.3	27.5	22.5	10.5	40.0
5.5	12.7	35.5	38.0	10.5
4.0	−5.5	19.0	14.5	10.5

Table 3.5 gives the annual returns for 30 randomly selected mutual funds. Problems 3.1, 3.2, and 3.3 refer to this data set.

3.1 Find the mean for the annual returns in Table 3.5.

Ans. For the data in Table 3.5, $\Sigma x = 455.20$, $n = 30$, and $\bar{x} = 15.17$.

3.2 Find the median for the annual returns in Table 3.5.

Ans. The ranked annual returns are as follows:

–5.5	–2.5	3.5	4.0	5.5	7.5	10.5	10.5	10.5	10.5	12.0
12.5	12.5	12.7	14.0	14.0	14.5	14.5	14.5	17.0	17.5	19.0
20.2	20.3	22.0	22.5	27.5	35.5	38.0	40.0			

The median is the average of the 15th and 16th values in the ranked returns or the average of 14.0 and 14.0, which equals 14.0.

3.3   Find the mode for the annual returns in Table 3.5.

Ans. By considering the ranked annual returns in the solution to problem 3.2, we see that the observation 10.5 occurs more frequently than any other value and is therefore the mode for this data set.

3.4   Table 3.6 gives the distribution of the cause of death due to accidents or violence for white males during a recent year.

Table 3.6

Cause of death	Number
Motor vehicle accident	30,500
All other accidents	27,500
Suicide	20,234
Homicide	8,342

What is the modal cause of death due to accidents or violence for white males? Can the mean or median be calculated for the cause of death?

Ans. The modal cause of death due to accidents or violence is motor vehicle accident. Because this is nominal level data, the mean and the median have no meaning.

3.5   Table 3.7 gives the selling prices in tens of thousands of dollars for 20 homes sold during the past month. Find the mean, median, and mode. Which measure is most representative for the selling price of such homes?

Table 3.7

60.5	113.5	79.0	475.5
75.0	70.0	122.5	150.0
100.0	125.5	90.0	175.5
89.0	130.0	111.5	100.0
50.0	340.5	100.0	525.0

Ans. The ranked selling prices are:

50.0	60.5	70.0	75.0	79.0	89.0	90.0	100.0	100.0	100.0
111.5	113.5	122.5	125.5	130.0	150.0	175.5	340.5	475.5	525.0

The mean is 154.1, the median is 105.7, and the mode is 100.0. The median is the most representative measure. The three selling prices 340.5, 475.5, and 525.0 inflate the mean and make it less representative than the median. Generally, the median is the best measure of central tendency to use when the data are skewed.

MEASURES OF DISPERSION: RANGE, VARIANCE, AND STANDARD DEVIATION FOR UNGROUPED DATA

3.6 Find the range, variance, and standard deviation for the annual returns of the mutual funds in Table 3.5.

Ans. The range is $40.0 - (-5.5) = 45.5$.

The variance is given by $s^2 = \dfrac{\sum x^2 - \dfrac{(\sum x)^2}{n}}{n-1} = \dfrac{10{,}060 - \dfrac{(455.2)^2}{30}}{29} = 108.7275$ and the standard

deviation is $s = \sqrt{s^2} = \sqrt{108.7275} = 10.43$.

3.7 Find the range, variance, and standard deviation for the selling prices of homes in Table 3.7.

Ans. The range is $525 - 50 = 475$.

The variance is given by $s^2 = \dfrac{\sum x^2 - \dfrac{(\sum x)^2}{n}}{n-1} = \dfrac{812{,}884 - \dfrac{(3083)^2}{20}}{19} = 17{,}770.5026$ and the

standard deviation is $s = \sqrt{s^2} = \sqrt{17{,}770.5026} = 133.31$.

3.8 Compare the values of the standard deviations with $\dfrac{\text{range}}{4}$ for problems 3.6 and 3.7.

Ans. For problem 3.6, the values are 10.43 and $45.5/4 = 11.38$ and for problem 3.7, the values are 133.31 and $475/4 = 118.75$. Even though the distributions are skewed in both problems, the values are reasonably close. The approximation is closer for mound-shaped distributions than for other distributions.

3.9 What are the chief advantage and the chief disadvantage of the range as a measure of dispersion?

Ans. The chief advantage is the simplicity of computation of the range and the chief disadvantage is that it is insensitive to the values between the extremes.

3.10 The ages and incomes of the 10 employees at Computer Services Inc. are given in Table 3.8.

Table 3.8

Age	Income
25	23,500
30	25,000
40	30,000
53	47,500
29	32,000
45	37,500
40	32,000
55	50,500
35	40,000
47	43,750

Compute the standard deviation of ages and incomes for these employees. Assuming that all employees remain with the company 5 years and that each income is multiplied by 1.5 over that period, what will the standard deviation of ages and incomes equal 5 years in the future?

Ans. The current standard deviations are 10.21 years and $9,224.10. Five years from now, the standard deviations will equal 10.21 years and $13,836.15. In general, adding the same constant to each observation does not affect the standard deviation of the data set and multiplying each observation by the same constant multiplies the standard deviation by the constant.

MEASURES OF CENTRAL TENDENCY AND DISPERSION FOR GROUPED DATA

3.11 Table 3.9 gives the age distribution of individuals starting new companies. Find the mean, median, and mode for this distribution.

Table 3.9

Age	Frequency
20–29	11
30–39	25
40–49	14
50–59	7
60–69	3

Ans. The mean is found by dividing Σxf by n, where $\Sigma xf = 24.5 \times 11 + 34.5 \times 25 + 44.5 \times 14 + 54.5 \times 7 + 64.5 \times 3 = 2,330$ and n = 60. $\bar{x} = 38.8$. The median class is the class 30–39. In order to find the middle of the age distribution, that is, the age where 30 are younger than this age and 30 are older, we must proceed through the 11 individuals in the 20–29 age group and 19 in the 30–39 age group. This gives $29.5 + \dfrac{19}{25} \times 10 = 37.1$ as the median age. The modal class is the class 30–39, and the mode is the class mark for this class that equals 34.5.

3.12 Find the range, variance, and standard deviation for the distribution in Table 3.9.

Ans. The range is $69.5 - 19.5 = 50$.

The variance is given by $s^2 = \dfrac{\Sigma x^2 f - \dfrac{(\Sigma xf)^2}{n}}{n-1} = \dfrac{97,355 - \dfrac{(2,330)^2}{60}}{59} = 116.497$ and the standard deviation is $s = \sqrt{116.497} = 10.8$.

3.13 The raw data corresponding to the grouped data in Table 3.9 is given in Table 3.10. Find the mean, median, and mode for the raw data and compare the results with the mean, median, and mode for the grouped data found in problem 3.11.

Table 3.10

20	29	34	37	41	50
22	30	34	37	41	50
22	30	34	38	44	50
24	30	34	38	44	55
24	32	34	39	44	55
24	32	34	39	45	55
25	32	36	40	45	58
26	33	36	40	45	62
27	33	36	40	46	62
28	33	37	41	47	66

Ans. The sum of the raw data equals 2,299, and the mean is 38.2. This compares with 38.8 for the grouped data. The median is seen to be 37, and this compares with 37.1 for the grouped data. The mode is seen to be 34 and this compares with 34.5 for the grouped data. This problem illustrates that the measures of central tendency for a set of data in grouped and ungrouped form are relatively close.

3.14 Find the range, variance, and standard deviation for the ungrouped data in Table 3.10 and compare these with the same measures found for the grouped form in problem 3.12.

Ans. The range is $66 - 20 = 46$. $\Sigma x = 2,299$, $\Sigma x^2 = 94,685$ and $n = 60$.

The variance is $s^2 = \dfrac{\Sigma x^2 - \dfrac{(\Sigma x)^2}{n}}{n-1} = \dfrac{94,685 - \dfrac{(2,299)^2}{60}}{59} = 111.788$ and $s = 10.6$.

These measures of dispersion compare favorably with the measures for the grouped data given in problem 3.12.

CHEBYSHEV'S THEOREM AND THE EMPIRICAL RULE

3.15 The mean lifetime of rats used in many psychological experiments equals 3.5 years, and the standard deviation of lifetimes is 0.5 year. At least what percent will have lifetimes between 2.5 years and 4.5 years? At least what percent will have lifetimes between 2.0 years and 5.0 years?

Ans. The interval from 2.5 years to 4.5 years is a 2 standard deviation interval about the mean, i.e., $k = 2$ in Chebyshev's theorem. At least 75% of the rats will have lifetimes between 2.5 years and 4.5 years. The interval from 2.0 years to 5.0 years is a 3 standard deviation interval about the mean. At least 89% of the lifetimes fall within this interval.

3.16 The mean height of adult females is 66 inches and the standard deviation is 2.5 inches. The distribution of heights is mound-shaped. What percent have heights between: (*a*) 63.5 inches and 68.5 inches? (*b*) 61.0 inches and 71.0 inches? (*c*) 58.5 inches and 73.5 inches?

Ans. 63.5 to 68.5 is a one standard deviation interval about the mean, 61.0 to 71.0 is a two standard deviation interval about the mean, and 58.5 to 73.5 is a three standard deviation interval about the mean. According to the empirical rule, the percentages are: (*a*) 68%; (*b*) 95% and (*c*) 99.7%.

3.17 The mean length of service for Federal Bureau of Investigation (FBI) agents equals 9.5 years and the standard deviation is 2.5 years. At least what percent of the employees have between 2.0 years of service and 17.0 years of service? If the lengths of service have a bell-shaped distribution, what can you say about the percent having between 2.0 and 17.0 years of service?

Ans. The interval from 2.0 years to 17.0 years is a 3 standard deviation interval about the mean, i.e., k = 3 in Chebyshev's theorem. Therefore, at least 89% of the agents have lengths of service in this interval. If we know that the distribution is bell shaped, then 99.7% of the agents will have lengths of service between 2.0 and 17.0 years.

COEFFICIENT OF VARIATION

3.18 Find the coefficient of variation for the ages in Table 3.10.

Ans. From problem 3.13, the mean age is 38.2 years and from problem 3.14 the standard deviation is 10.6 years. Using formula (*3.12*), the coefficient of variation is $CV = \dfrac{s}{\overline{x}} \times 100\% = \dfrac{10.6}{38.2} \times 100 = 27.7\%$.

3.19 The mean yearly salary of all the employees at Pretty Printing is $42,500 and the standard deviation is $4,000. The mean number of years of education for the employees is 16 and the standard deviation is 2.5 years. Which of the two variables has the higher relative variation?

Ans. The coefficient of variation for salaries is $CV = \dfrac{4,000}{42,500} \times 100 = 9.4\%$ and the coefficient of variation for years of education is $CV = \dfrac{2.5}{16} \times 100 = 15.6\%$. Years of education has a higher relative variation.

Z SCORES

3.20 The mean daily intake of protein for a group of individuals is 80 grams and the standard deviation is 8 grams. Find the z scores for individuals with the following daily intakes of protein: (*a*) 95 grams; (*b*) 75 grams; (*c*) 80 grams.

Ans. (*a*) $z = \dfrac{95-80}{8} = 1.88$ (*b*) $z = \dfrac{75-80}{8} = -.63$ (*c*) $z = \dfrac{80-80}{8} = 0$

3.21 Three individuals were selected from the group described in problem 3.20 who have daily intakes with z scores equal to –1.4, 0.5, and 3.0. Find their daily intakes of protein.

Ans. If the equation $z = \dfrac{x - \overline{x}}{s}$ is solved for x, the result is $x = \overline{x} + zs$. The daily intake corresponding to a z score of –1.4 is x = 80 + (–1.4)(8) = 68.8 grams. For a z score equal to 0.5, x = 80 + (0.5)(8) = 84 grams. For a z score equal to 3.0, x = 80 + (3.0)(8) = 104 grams.

MEASURES OF POSITION: PERCENTILES, DECILES, AND QUARTILES

3.22 Find the percentiles for the ages 34, 45, and 55 in Table 3.10.

Ans.   The number of ages less than 34 is 20, and $\dfrac{20}{60} \times 100 = 33.3\%$, which rounds to 33%. The age 34 is the thirty-third percentile.

The number of ages less than 45 is 45, and $\dfrac{45}{60} \times 100 = 75\%$. The age 45 is the seventy-fifth percentile.

The number of ages less than 55 is 53, and $\dfrac{53}{60} \times 100 = 88.3\%$, which rounds to 88%. The age 55 is the eighty-eighth percentile.

3.23   Find the ninety-fifth percentile, the seventh decile, and the first quartile for the age distribution given in Table 3.10.

Ans.   To find P_{95}, compute $i = \dfrac{np}{100} = \dfrac{(60)(95)}{100} = 57$. P_{95} is the average of the observations in positions 57 and 58 in the ranked data set, or the average of 58 and 62 which is 60 years.

To find D_7, which is the same as P_{70}, compute $i = \dfrac{np}{100} = \dfrac{(60)(70)}{100} = 42$. D_7 is the average of the observations in positions 42 and 43 in the ranked data set, or the average of 41 and 44 which is 42.5 years.

To find Q_1, compute $i = \dfrac{np}{100} = \dfrac{(60)(25)}{100} = 15$. Q_1 is the average of the observations in positions 15 and 16 in the ranked data set, or the average of 32 and 32 which is 32 years.

INTERQUARTILE RANGE

3.24   Find the interquartile range for the annual returns of the mutual funds given in Table 3.5.

Ans.   The ranked annual returns are as follows.

−5.5	−2.5	3.5	4.0	5.5	7.5	10.5	10.5	10.5	10.5	12.0
12.5	12.5	12.7	*14.0*	*14.0*	14.5	14.5	14.5	17.0	17.5	19.0
20.2	20.3	22.0	22.5	27.5	35.5	38.0	40.0			

The first quartile, which is the same as P_{25}, is found by computing $i = \dfrac{np}{100} = \dfrac{(30)(25)}{100} = 7.5$.

The next integer greater than 7.5 is 8, and this locates the position of Q_1 in the ranked data set. $Q_1 = 10.5$.

The third quartile is found by computing $i = \dfrac{np}{100} = \dfrac{(30)(75)}{100} = 22.5$, and rounding this to 23.

The third quartile is found in position 23 in the ranked data set. $Q_3 = 20.2$
$IQR = Q_3 - Q_1 = 20.2 - 10.5 = 9.7$.

3.25   Find the interquartile range for the selling prices given in Table 3.7.

Ans.   The ranked selling prices are:

50.0	60.5	70.0	75.0	79.0	89.0	90.0	100.0	100.0	100.0
111.5	113.5	122.5	125.5	130.0	150.0	175.5	340.5	475.5	525.0

The first quartile, which is the same as P_{25} is found by computing $i = \dfrac{np}{100} = \dfrac{(20)(25)}{100} = 5$. The first quartile is the average of the observations in positions 5 and 6 in the ranked data set.

$$Q_1 = \frac{79.0 + 89.0}{2} = 84.0$$

The third quartile is found by computing $i = \dfrac{np}{100} = \dfrac{(20)(75)}{100} = 15$. The third quartile is the average of the observations in positions 15 and 16 in the ranked data set.

$$Q_3 = \frac{130.0 + 150.0}{2} = 140.0 \qquad IQR = Q_3 - Q_1 = 140.0 - 84.0 = 56.0$$

BOX-AND-WHISKER PLOT

3.26 Table 3.11 gives the number of days that 25 individuals spent in house arrest in a criminal justice study. Use Minitab to construct a boxplot for these data.

Table 3.11

35	25	90	60	45
40	58	90	90	55
60	55	80	90	60
55	60	85	75	60
56	55	75	80	90

Ans. The solution is shown in Fig. 3-3. It is seen that the minimum value is 25, the maximum value is 90, $Q_1 = 55$, median = 60, and $Q_3 = 83$.

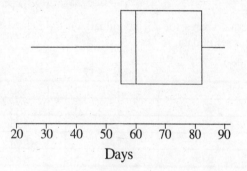

Days

Fig. 3-3

3.27 Construct a modified boxplot for the selling prices given in Table 3.7.

Ans. The ranked selling prices are:

50.0	60.5	70.0	75.0	79.0	89.0	90.0	100.0	100.0	100.0
111.5	113.5	122.5	125.5	130.0	150.0	175.5	340.5	475.5	525.0

In problem 3.25 it is shown that $Q_1 = 84.0$, $Q_3 = 140.0$, and IQR = 56.0.

A *lower inner fence* is defined to be $Q_1 - 1.5 \times IQR = 84.0 - 1.5 \times 56.0 = 0.0$
An *upper inner fence* is defined to be $Q_3 + 1.5 \times IQR = 140.0 + 1.5 \times 56.0 = 224.0$
A *lower outer fence* is defined to be $Q_1 - 3 \times IQR = 84.0 - 3 \times 56.0 = -84.0$
An *upper outer fence* is defined to be $Q_3 + 3 \times IQR = 140.0 + 3 \times 56.0 = 308.0$

The *adjacent values* are the most extreme values still lying within the inner fences. The adjacent values for the above data are 50.0 and 175.5, since they are the most extreme values between 0.0 and 224.0. In a modified boxplot, the whiskers extend only to the adjacent values.

Data values that lie between the inner and outer fences are *possible outliers* and data values that lie outside the outer fences are *probable outliers*. The observations 340.5, 475.5, and 525.0 lie outside the outer fences and are called probable outliers.

A Minitab printout for a boxplot of the data is given in Fig. 3-4. Each probable outlier is represented by an asterisk.

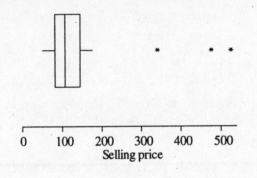

Fig. 3-4

Supplementary Problems

MEASURES OF CENTRAL TENDENCY: MEAN, MEDIAN, AND MODE FOR UNGROUPED DATA

3.28 Table 3.12 gives the verbal scores for 25 individuals on the Scholastic Aptitude Test (SAT). Find the mean verbal score for this set of data.

Table 3.12

340	375	450	580	565
445	675	545	380	630
500	350	635	495	640
467	560	680	605	590
635	625	440	420	400

Ans. $\Sigma x = 13{,}027$ $\quad \bar{x} = 521.1$

3.29 Find the median verbal score for the data in Table 3.12.

Ans. The ranked scores are shown below.

340 350 375 380 400 420 440 445 450 467 495 500 545
560 565 580 590 605 625 630 635 635 640 675 680

The median verbal score is 545.

3.30 Find the mode for the verbal scores in Table 3.12.

Ans. 635

3.31 Give the output produced by the Describe command of Minitab when the data in Table 3.12 are analyzed.

Ans. Descriptive Statistics

Variable	N	Mean	Median	TrMean	StDev	SEMean
SAT	25	521.1	545.0	522.0	108.1	21.6

Variable	Min	Max	Q1	Q3
SAT	340.0	680.0	430.0	627.5

3.32 Table 3.13 gives a stem-and-leaf display for the number of hours per week spent watching TV for a group of teenagers. Find the mean, median, and mode for this distribution. What is the shape of the distribution?

Table 3.13

stem	leaf
1	0 5 5 5
2	0 0 0 0 0 0 0 0 0 0 0 0 5 5 5
3	0

Ans. The mean, median, and mode are each equal to 20 hours. The distribution has a bell shape with center at 20.

MEASURES OF DISPERSION: RANGE, VARIANCE, AND STANDARD DEVIATION FOR UNGROUPED DATA

3.33 Find the range, variance, and standard deviation for the SAT scores in Table 3.12.

Ans. range = 340 $s^2 = 11,692.91$ $s = 108.13$

3.34 Find the range, variance, and standard deviation for the number of hours spent watching TV given in Table 3.13.

Ans. range = 20 $s^2 = 18.4213$ $s = 4.29$

3.35 What should your response be if you find that the variance of a data set equals −5.5?

Ans. Check your calculations, since the variance can never be negative

3.36 Consider a data set in which all observations are equal. Find the range, variance, and standard deviation for this data set.

Ans. The range, variance, and standard deviation will all equal zero.

3.37 A data set consisting of 10 observations has a mean equal to 0, and a variance equal to a. Express Σx^2 in terms of a.

Ans. $\Sigma x^2 = 9a$

MEASURES OF CENTRAL TENDENCY AND DISPERSION FOR GROUPED DATA

3.38 Table 3.14 gives the distribution of the words per minute for 60 individuals using a word processor. Find the mean, median, and mode for this distribution.

Table 3.14

Words per minute	Frequency
40–49	3
50–59	9
60–69	15
70–79	15
80–89	12
90–99	6

Ans.   mean = 71.5, median = 71.5, modes = 64.5 and 74.5

3.39   Find the range, variance, and standard deviation for the distribution in Table 3.14.

Ans.   range = 60 $s^2 = 184.0678$ $s = 13.57$

3.40   A quality control technician records the number of defective units found daily in samples of size 100 for the month of July.   The distribution of the number of defectives per 100 units is shown in Table 3.15. Find the mean, median, and mode for this distribution. Which of the three measures is the most representative for the distribution?

Table 3.15

Number of defectives	Frequency
0	12
1	7
8	1
27	1

Ans.   mean = 2   median = 0   mode = 0

The median or mode would be more representative than the mean. On most days, there were none or one defective in the sample. The two days on which the process was out of control inflated the mean.

3.41   The following formula is sometimes used for finding the median of grouped data.

$$\text{Median} = L + \frac{5n - c}{f} \times (U - L)$$

In this formula, L is the lower boundary of the median class, U is the upper boundary of the median class, n is the number of observations, f is the frequency of the median class, and c is the cumulative frequency of the class proceeding the median class. Give the values for L, U, n, f, c, and find the median for the distribution given in Table 3.14.

Ans.   L = 69.5   U = 79.5n = 60   f = 15   c = 27   median = 71.5

CHEBYSHEV'S THEOREM AND THE EMPIRICAL RULE

3.42   A sociological study of gang members in a large midwestern city found the mean age of the gang members in the study to be 14.5 years and the standard deviation to be 1.5 years. According to Chebyshev's theorem, at least what percent will be between 10.0 and 19.0 years of age?

Ans.   89%

3.43   A psychological study of Alcoholic Anonymous members found the mean number of years without drinking alcohol for individuals in the study to be 5.5 years and the standard deviation to be 1.5 years. The distribution of the number of years without drinking is bell-shaped. What percent of the distribution is between: (*a*) 4.0 and 7.0 years; (*b*) 2.5 and 8.5 years; (*c*) 1.0 and 10.0 years?

Ans.   (*a*) 68%   (*b*) 95%   (*c*) 99.7%

3.44   The ranked verbal SAT scores in Table 3.12 are:

340   350   375   380   400   420   440   445   450   467   495   500   545
560   565   580   590   605   625   630   635   635   640   675   680

The mean and standard deviation of these data are 521.1 and 108.1, respectively. According to Chebyshev's theorem, at least 75% of the observations are between 304.9 and 737.3. What is the actual percent of observations within two standard deviations of the mean?

Ans.   100%

COEFFICIENT OF VARIATION

3.45 The verbal scores on the SAT given in Table 3.12 have a mean equal to 521.1 and a standard deviation equal to 108.1. Find the coefficient of variation for these scores.

Ans.   20.7%

3.46 Fastners Inc. produces nuts and bolts. One of their bolts has a mean length of 2.00 inches with a standard deviation equal to 0.10 inch, and another type bolt has a mean length of 0.25 inch. What standard deviation would the second type bolt need to have in order that both types of bolts have the same coefficient of variation?

Ans.   0.0125 inch

Z SCORES

3.47 The low-density lipoprotein (LDL) cholesterol concentration for a group has a mean equal to 140 mg/dL and a standard deviation equal to 40 mg/dL. Find the z scores for individuals having LDL values of (*a*) 115; (*b*) 140; and (*c*) 200.

Ans.   (*a*) −0.63   (*b*) 0.0   (*c*) 1.50

3.48 Three individuals from the group described in problem 3.47 have z scores equal to (*a*) −1.75; (*b*) 0.5; and (*c*) 2.0. Find their LDL values.

Ans.   (*a*) 70   (*b*) 160   (*c*) 220

MEASURES OF POSITION: PERCENTILES, DECILES, AND QUARTILES

3.49 Table 3.16 gives the ages of commercial aircraft randomly selected from several airlines. Find the percentiles for the ages 10, 15, and 20.

Table 3.16

2	7	11	15	19
2	7	11	15	19
2	7	12	15	20
2	7	12	15	20
4	7	12	15	20
4	10	14	15	22
4	10	14	16	24
4	10	14	16	25
5	10	14	17	25
5	10	15	17	27

Ans.   The age 10 is the thirtieth percentile. The age 15 is the fifty-eighth percentile. The age 20 is the eighty-fourth percentile.

3.50 Find P_{90}, D_8, and Q_3 for the commercial aircraft ages in Table 3.16.

Ans. $P_{90} = 21$ $D_8 = 18$ $Q_3 = 16$

INTERQUARTILE RANGE

3.51   The first quartile for the salaries of county sheriffs in the United States is $37,500 and the third quartile is $50,500. What is the interquartile range for the salaries of county sheriffs?

Ans. $13,000

3.52   Find the interquartile range for the ages of commercial aircraft given in Table 3.16.

Ans.   9 years

BOX-AND-WHISKER PLOT

3.53   A Minitab produced boxplot for the weights of high school football players is shown in Fig. 3-5. Give the minimum, maximum, first quartile, median, and third quartile.

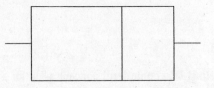

190 200 210 220 230 240 250 260 270
Weights

Fig. 3-5

Ans.   minimum = 195   maximum = 270 $Q_1 = 205$ $Q_2 = 240$ $Q_3 = 260$

3.54   Construct or use Minitab to construct a boxplot for the ages of the commercial aircraft in Table 3.16.

Ans.   The boxplot is shown in Fig. 3-6.

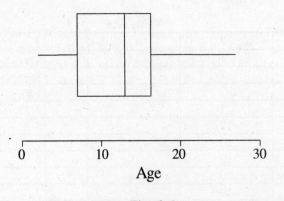

0   10   20   30
Age

Fig. 3-6

Chapter 4

Probability

EXPERIMENT, OUTCOMES, AND SAMPLE SPACE

An *experiment* is any operation or procedure whose outcomes cannot be predicted with certainty. The set of all possible *outcomes* for an experiment is called the *sample space* for the experiment.

EXAMPLE 4.1 Games of chance are examples of experiments. The single toss of a coin is an experiment whose outcomes cannot be predicted with certainty. The sample space consists of two outcomes, heads or tails. The letter S is used to represent the sample space and may be represented as S = {H, T}. The single toss of a die is an experiment resulting in one of six outcomes. S may be represented as {1, 2, 3, 4, 5, 6}. When a card is selected from a standard deck, 52 outcomes are possible. When a roulette wheel is spun, the outcome cannot be predicted with certainty.

EXAMPLE 4.2 When a quality control technician selects an item for inspection from a production line, it may be classified as defective or nondefective. The sample space may be represented by S = {D, N}. When the blood type of a patient is determined, the sample space may be represented as S = {A, AB, B, O}. When the Myers-Briggs personality type indicator is administered to an individual, the sample space consists of 16 possible outcomes.

The experiments discussed in Examples 4.1 and 4.2 are rather simple experiments and the descriptions of the sample spaces are straightforward. More complicated experiments are discussed in the following section and techniques such as tree diagrams are utilized to describe the sample space for these experiments.

TREE DIAGRAMS AND THE COUNTING RULE

In a *tree diagram*, each outcome of an experiment is represented as a branch of a geometric figure called a tree.

EXAMPLE 4.3 Figure 4-1 shows a tree diagram for the experiment of tossing a coin twice. The tree has four branches. Each branch is an outcome for the experiment. If the experiment is expanded to three tosses, the branches are simply continued with H or T added to the end of each branch shown in Fig. 4-1. This would result in the eight outcomes: HHH, HHT, HTH, HTT, THH, THT, TTH, and TTT. This technique could be continued systematically to give the outcomes for n tosses of a coin. Notice that 2 tosses has 4 outcomes and 3 tosses has 8 outcomes. N tosses has 2^N possible outcomes.

The *counting rule* for a two-step experiment states that if the first step can result in any one of n_1 outcomes, and the second step in any one of n_2 outcomes, then the experiment can result in $(n_1)(n_2)$ outcomes. If a third step is added with n_3 outcomes, then the experiment can result in $(n_1)(n_2)(n_3)$ outcomes. The counting rule applies to an experiment consisting of any number of steps. If the counting rule is applied to Example 4.3, we see that for two tosses of a coin, $n_1 = 2$, $n_2 = 2$, and the number of outcomes for the experiment is $2 \times 2 = 4$. For three tosses, there are $2 \times 2 \times 2 = 8$ outcomes and so forth. The counting rule may be used to figure the number of outcomes of an experiment and then a tree diagram may be used to actually represent the outcomes.

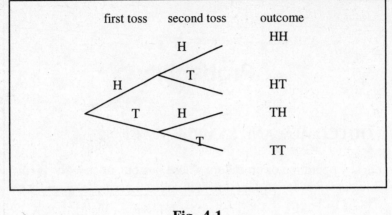

Fig. 4-1

EXAMPLE 4.4 For the experiment of rolling a pair of dice, the first die may be any of six numbers and the second die may be any one of six numbers. According to the counting rule, there are $6 \times 6 = 36$ outcomes. The outcomes may be represented by a tree having 36 branches. The sample space may also be represented by a two-dimensional plot as shown in Fig. 4-2.

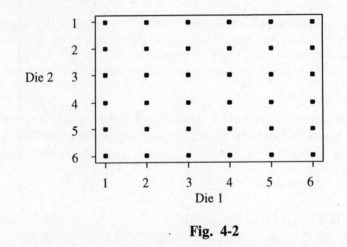

Fig. 4-2

EXAMPLE 4.5 An experiment consists of observing the blood types for five randomly selected individuals. Each of the five will have one of four blood types A, B, AB, or O. Using the counting rule, we see that the experiment has $4 \times 4 \times 4 \times 4 \times 4 = 1,024$ possible outcomes. In this case constructing a tree diagram would be difficult.

EVENTS, SIMPLE EVENTS, AND COMPOUND EVENTS

An *event* is a subset of the sample space consisting of at least one outcome from the sample space. If the event consists of exactly one outcome, it is called a *simple event*. If an event consists of more than one outcome, it is called a *compound event*.

EXAMPLE 4.6 A quality control technician selects two computer mother boards and classifies each as defective or nondefective. The sample space may be represented as S = {NN, ND, DN, DD}, where D represents a defective unit and N represents a nondefective unit. Let A represent the event that neither unit is defective and let B represent the event that at least one of the units is defective. A = {NN} is a simple event and B = {ND, DN, DD} is a compound event. Figure 4-3 is a *Venn Diagram* representation of the sample space S and the events A and B. In a Venn diagram, the sample space is usually represented by a rectangle and events are represented by circles within the rectangle.

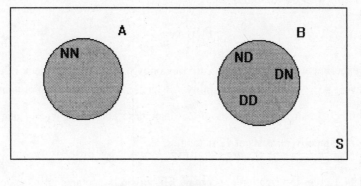

Fig. 4-3

EXAMPLE 4.7 For the experiment described in Example 4.5, there are 1,024 different outcomes for the blood types of the five individuals. The compound event that all five have the same blood type is composed of the following four outcomes: (A, A, A, A, A), (B, B, B, B, B), (AB, AB, AB, AB, AB), and (O, O, O, O, O). The simple event that all five have blood type O would be the outcome (O, O, O, O, O).

PROBABILITY

Probability is a measure of the likelihood of the occurrence of some event. There are several different definitions of probability. Three definitions are discussed in the next section. The particular definition that is utilized depends upon the nature of the event under consideration. However, all the definitions satisfy the following two specific properties and obey the rules of probability developed later in this chapter.

The probability of any event E is represented by the symbol P(E) and the symbol is read as "P of E" or as "the probability of event E." P(E) is a real number between zero and one as indicated in the following inequality:

$$0 \leq P(E) \leq 1 \qquad (4.1)$$

The sum of the probabilities for all the simple events of an experiment must equal one. That is, if E_1, E_2, . . . , E_n are the simple events for an experiment, then the following equality must be true:

$$P(E_1) + P(E_2) + \ldots + P(E_n) = 1 \qquad (4.2)$$

Equality (*4.2*) is also sometimes expressed as in formula (*4.3*):

$$P(S) = 1 \qquad (4.3)$$

Equation (*4.3*) states that the probability that some outcome in the sample space will occur is one.

CLASSICAL, RELATIVE FREQUENCY, AND SUBJECTIVE PROBABILITY DEFINITIONS

The *classical definition of probability* is appropriate when all outcomes of an experiment are equally likely. For an experiment consisting of n outcomes, the classical definition of probability assigns probability $\frac{1}{n}$ to each outcome or simple event. For an event E consisting of k outcomes, the probability of event E is given by formula (4.4)

$$P(E) = \frac{k}{n} \qquad\qquad (4.4)$$

EXAMPLE 4.8 The experiment of selecting one card randomly from a standard deck of cards has 52 equally likely outcomes. The event A_1 = {club} has probability $\frac{13}{52}$, since A_1 consists of 13 outcomes. The event A_2 = {red card} has probability $\frac{26}{52}$, since A_2 consists of 26 outcomes. The event A_3 = {face card (Jack, Queen, King)} has probability $\frac{12}{52}$, since A_3 consists of 12 outcomes.

EXAMPLE 4.9 Table 4.1 gives information concerning fifty organ transplants in the state of Nebraska during a recent year. Each patient represented in Table 4.1 had only one transplant. If one of the 50 patient records is randomly selected, the probability that the patient had a heart transplant is $\frac{15}{50}$ = .30, since 15 of the patients had heart transplants. The probability that a randomly selected patient had to wait one year or more for the transplant is $\frac{20}{50}$ = .40, since 20 of the patients had to wait one year or more. The display in Table 4.1 is called a *two-way table*. It displays two different variables concerning the patients.

Table 4.1

	Waiting Time for Transplant	
Type of transplant	Less than one year	One year or more
Heart	10	5
Kidney	7	3
Liver	5	5
Pancreas	3	2
Eyes	5	5

EXAMPLE 4.10 To find the probability of the event A that the sum of the numbers on the faces of a pair of dice equals seven when a pair of dice is rolled, consider the sample space shown in Fig. 4-4. The event A is shown as a rectangular box in the sample space. The outcomes in A are as follows A = {(1, 6), (2, 5), (3, 4), (4, 3), (5, 2), (6, 1)}. Since A contains six of the thirty-six equally likely outcomes for the experiment, the probability of event A is $\frac{6}{36}$.

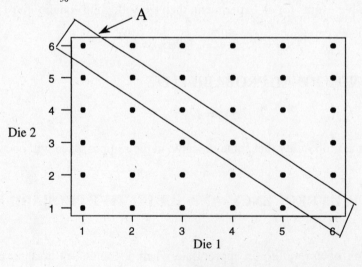

Fig. 4-4

The classical definition of probability is not always appropriate in computing probabilities of events. If a coin is bent, heads and tails are not equally likely outcomes. If a die has been loaded, each of the six faces do not have probability of occurrence equal to $\frac{1}{6}$. For experiments not having equally likely outcomes, the *relative frequency definition of probability* is appropriate. The relative frequency definition of probability states that if an experiment is performed n times, and if event E occurs f times, then the probability of event E is given by formula (*4.5*).

$$P(E) = \frac{f}{n} \qquad\qquad\qquad\qquad (4.5)$$

EXAMPLE 4.11 A bent coin is tossed 50 times and a head appears on 35 of the tosses. The relative frequency definition of probability assigns the probability $\frac{35}{50}$ = .70 to the event that a head occurs when this coin is tossed. A loaded die is tossed 75 times and the face "6" appears 15 times in the 75 tosses. The relative frequency definition of probability assigns the probability $\frac{15}{75}$ = .20 to the event that the face "6" will appear when this die is tossed.

EXAMPLE 4.12 A study by the state of Tennessee found that when 750 drivers were randomly stopped 471 were found to be wearing seat belts. The relative frequency probability that a driver wears a seat belt in Tennessee is $\frac{471}{750}$ = 0.63.

There are many circumstances where neither the classical definition nor the relative frequency definition of probability is applicable. The *subjective definition of probability* utilizes intuition, experience, and collective wisdom to assign a degree of belief that an event will occur. This method of assigning probabilities allows for several different assignments of probability to a given event. The different assignments must satisfy formulas (*4.1*) and (*4.2*).

EXAMPLE 4.13 A military planner states that the probability of nuclear war in the next year is 1%. The individual is assigning a subjective probability of .01 to the probability of the event "nuclear war in the next year." This event does not lend itself to either the classical definition or the relative frequency definition of probability.

EXAMPLE 4.14 A medical doctor tells a patient with a newly diagnosed cancer that the probability of successfully treating the cancer is 90%. The doctor is assigning a subjective probability of .90 to the event that the cancer can be successfully treated. The probability for this event cannot be determined by either the classical definition or the relative frequency definition of probability.

MARGINAL AND CONDITIONAL PROBABILITIES

Table 4.2 classifies the 500 members of a police department according to their minority status as well as their promotional status during the past year. One hundred of the individuals were classified as being a minority and seventy were promoted during the past year. The probability that a randomly selected individual from the police department is a minority is $\frac{100}{500}$ = .20 and the probability that a randomly selected person was promoted during the past year is $\frac{70}{500}$ = .14. Table 4.3 is obtained by dividing each entry in Table 4.2 by 500.

Table 4.2

	Minority		
Promoted	No	Yes	Total
No	350	80	430
Yes	50	20	70
Total	400	100	500

The four probabilities in the center of Table 4.3, .70, .16, .10, and .04, are called *joint probabilities*. The four probabilities in the margin of the table, .80, .20, .86, and .14, are called *marginal probabilities*.

Table 4.3

	Minority		
Promoted	No	Yes	Total
No	.70	.16	.86
Yes	.10	.04	.14
Total	.80	.20	1.00

The joint probabilities concerning the selected police officer may be described as follows:

.70 = the probability that the selected officer is not a minority and was not promoted
.16 = the probability that the selected officer is a minority and was not promoted
.10 = the probability that the selected officer is not a minority and was promoted
.04 = the probability that the selected officer is a minority and was promoted

The marginal probabilities concerning the selected police officer may be described as follows:

.80 = the probability that the selected officer is not a minority
.20 = the probability that the selected officer is a minority
.86 = the probability that the selected officer was not promoted during the last year
.14 = the probability that the selected officer was promoted during the last year

In addition to the joint and marginal probabilities discussed above, another important concept is that of a *conditional probability*. If it is known that the selected police officer is a minority, then the conditional probability of promotion during the past year is $\frac{20}{100}$ = .20, since 100 of the police officers in Table 4.2 were classified as minority and 20 of those were promoted. This same probability may be obtained from Table 4.3 by using the ratio $\frac{.04}{.20}$ = .20.

The formula for the conditional probability of the occurrence of event A given that event B is known to have occurred for some experiment is represented by P(A | B) and is the ratio of the joint probability of A and B divided by the probability of B. The following formula is used to compute a conditional probability.

$$P(A \mid B) = \frac{P(A \text{ and } B)}{P(B)} \qquad (4.6)$$

The following example summarizes the above discussion and the newly introduced notation.

EXAMPLE 4.15 For the experiment of selecting one police officer at random from those described in Table 4.2, define event A to be the event that the individual was promoted last year and define event B to be the event that the individual is a minority. The joint probability of A and B is expressed as P(A and B) = .04. The marginal probabilities of A and B are expressed as P(A) = .14 and P(B) = .20. The conditional probability of A given B is $P(A \mid B) = \dfrac{P(A \text{ and } B)}{P(B)} = \dfrac{.04}{.20} = .20$.

MUTUALLY EXCLUSIVE EVENTS

Two or more events are said to be *mutually exclusive* if the events do not have any outcomes in common. They are events that cannot occur together. If A and B are mutually exclusive events then the joint probability of A and B equals zero, that is, P(A and B) = 0. A Venn diagram representation of two mutually exclusive events is shown in Fig. 4-5.

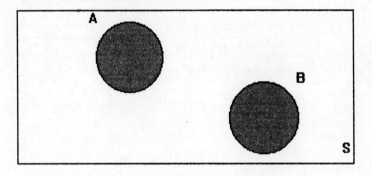

Fig. 4-5

EXAMPLE 4.16 An experiment consists in observing the gender of two randomly selected individuals. The event, A, that both individuals are male and the event, B, that both individuals are female are mutually exclusive since if both are male, then both cannot be female and P(A and B) = 0.

EXAMPLE 4.17 Let event A be the event that an employee at a large company is a white collar worker and let B be the event that an employee is a blue collar worker. Then A and B are mutually exclusive since an employee cannot be both a blue collar worker and a white collar worker and P(A and B) = 0.

DEPENDENT AND INDEPENDENT EVENTS

If the knowledge that some event B has occurred influences the probability of the occurrence of another event A, then A and B are said to be *dependent events*. If knowing that event B has occurred does not affect the probability of the occurrence of event A, then A and B are said to be *independent events*. Two events are independent if the following equation is satisfied. Otherwise the events are dependent.

$$P(A \mid B) = P(A) \tag{4.7}$$

The event of having a criminal record and the event of not having a father in the home are dependent events. The events of being a diabetic and having a family history of diabetes are dependent events, since diabetes is an inheritable disease. The events of having 10 letters in your last name and being a sociology major are independent events. However, many times it is not obvious whether two events are dependent or independent. In such cases, formula (*4.7*) is used to determine whether the events are independent or not.

EXAMPLE 4.18 For the experiment of drawing one card from a standard deck of 52 cards, let A be the event that a club is selected, let B be the event that a face card (jack, queen, or king) is drawn, and let C be the event that a jack is drawn. Then A and B are independent events since $P(A) = \frac{13}{52} = .25$ and $P(A \mid B) = \frac{3}{12} = .25$. $P(A \mid B) = \frac{3}{12} = .25$, since there are 12 face cards and 3 of them are clubs. The events B and C are dependent events since $P(C) = \frac{4}{52} = .077$ and $P(C \mid B) = \frac{4}{12} = .333$. $P(C \mid B) = \frac{4}{12} = .333$, since there are 12 face cards and 4 of them are jacks.

EXAMPLE 4.19 Suppose one patient record is selected from the 125 represented in Table 4.4. The event that a patient has a history of heart disease, A, and the event that a patient is a smoker, B, are dependent events, since $P(A) = \frac{15}{125} = .12$ and $P(A \mid B) = \frac{10}{45} = .22$. For this group of patients, knowing that an individual is a smoker almost doubles the probability that the individual has a history of heart disease.

Table 4.4

Smoker	History of Heart Disease		
	No	Yes	Total
No	75	5	80
Yes	35	10	45
Total	110	15	125

COMPLEMENTARY EVENTS

To every event A, there corresponds another event A^c, called the *complement of A* and consisting of all other outcomes in the sample space not in event A. The word *not* is used to describe the complement of an event. The complement of selecting a red card is not selecting a red card. The complement of being a smoker is not being a smoker. Since an event and its complement must account for all the outcomes of an experiment, their probabilities must add up to one. If A and A^c are complementary events then the following equation must be true.

$$P(A) + P(A^c) = 1 \qquad\qquad (4.8)$$

EXAMPLE 4.20 Approximately 2% of the American population is diabetic. The probability that a randomly chosen American is not diabetic is .98, since $P(A) = .02$, where A is the event of being diabetic, and $.02 + P(A^c) = 1$. Solving for $P(A^c)$ we get $P(A^c) = 1 - .02 = .98$.

EXAMPLE 4.21 Find the probability that on a given roll of a pair of dice that "snake eyes" are not rolled. Snake eyes means that a one was observed on each of the dice. Let A be the event of rolling snake eyes. Then $P(A) = \frac{1}{36} = .028$. The event that snake eyes are not rolled is A^c. Then using formula (4.8), $.028 + P(A^c) = 1$, and solving for $P(A^c)$, it follows that $P(A^c) = 1 - .028 = .972$.

Complementary events are always mutually exclusive events but mutually exclusive events are not always complementary events. The events of drawing a club and drawing a diamond from a standard deck of cards are mutually exclusive, but they are not complementary events.

MULTIPLICATION RULE FOR THE INTERSECTION OF EVENTS

The *intersection* of two events A and B consists of all those outcomes which are common to both A and B. The intersection of the two events is represented as A *and* B. The intersection of two events is also represented by the symbol A $\cap$ B, read as A *intersect* B. A Venn diagram representation of the intersection of two events is shown in Fig. 4-6.

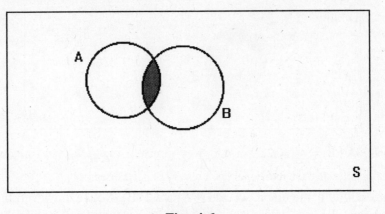

Fig. 4-6

The probability of the intersection of two events is given by the *multiplication rule*. The multiplication rule is obtained from the formula for conditional probabilities, and is given by formula *(4.9)*.

$$P(A \text{ and } B) = P(A) \, P(B \mid A) \tag{4.9}$$

EXAMPLE 4.22 A small hospital has 40 physicians on staff of which 5 are cardiologists. The probability that two randomly selected physicians are both cardiologists is determined as follows. Let A be the event that the first selected physician is a cardiologist, and B the event that the second selected physician is a cardiologist. Then $P(A) = \frac{5}{40} = .125$, $P(B \mid A) = \frac{4}{39} = .103$ and $P(A \text{ and } B) = .125 \times .103 = .013$. If two physicians were selected from a group of 40,000 of which 5,000 were cardiologists, then $P(A) = \frac{5,000}{40,000} = .125$, $P(B \mid A) = \frac{4,999}{39,999}$ = .125, and $P(A \text{ and } B) = (.125)^2 = .016$. Notice that when the selection is from a large group, the probability of selecting a cardiologist on the second selection is approximately the same as selecting one on the first selection.

Following this line of reasoning, suppose it is known that 12.5% of all physicians are cardiologists. If three physicians are selected randomly, the probability that all three are cardiologists equals $(.125)^3 = .002$. The probability that none of the three are cardiologists is $(.875)^3 = .670$.

If events A and B are independent events, then $P(A \mid B) = P(A)$ and $P(B \mid A) = P(B)$. When $P(B \mid A)$ is replaced by $P(B)$, formula *(4.9)* simplifies to

$$P(A \text{ and } B) = P(A) \, P(B) \tag{4.10}$$

EXAMPLE 4.23 Ten percent of a particular population have hypertension and 40 percent of the same population have a home computer. Assuming that having hypertension and owning a home computer are independent events, the probability that an individual from this population has hypertension and owns a home computer is $.10 \times .40 = .04$. Another way of stating this result is that 4 percent have hypertension and own a home computer.

ADDITION RULE FOR THE UNION OF EVENTS

The *union* of two events A and B consists of all those outcomes that belong to A or B or both A and B. The union of events A and B is represented as A $\cup$ B or simply as A *or* B. A Venn diagram representation of the union of two events is shown in Fig. 4-7. The darker part of the shaded union of the two events corresponds to overlap and corresponds to the outcomes in both A and B.

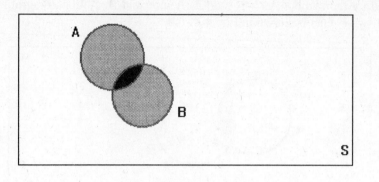

Fig. 4-7

To find the probability in the union, we add P(A) and P(B). Notice, however, that the darker part indicates that P(A *and* B) gets added twice and must be subtracted out to obtain the correct probability. The resultant equation is called the *addition rule* for probabilities and is given by formula (*4.11*).

$$P(A \ or \ B) = P(A) + P(B) - P(A \ and \ B) \qquad (4.11)$$

If A and B are mutually exclusive events, then P(A *and* B) = 0 and the formula (*4.11*) simplifies to the following.

$$P(A \ or \ B) = P(A) + P(B)$$

EXAMPLE 4.24 Forty percent of the employees at Computec, Inc. have a college degree, 30 percent have been with Computec for at least three years, and 15 percent have both a college degree and have been with the company for at least three years. If A is the event that a randomly selected employee has a college degree and B is the event that a randomly selected employee has been with the company at least three years, then A *or* B is the event that an employee has a college degree or has been with the company at least three years. The probability of A *or* B is .40 + .30 − .15 = .55. Another way of stating the result is that 55 percent of the employees have a college degree or have been with Computec for at least three years.

EXAMPLE 4.25 A hospital employs 25 medical-surgical nurses, 10 intensive care nurses, 15 emergency room nurses, and 50 floor care nurses. If a nurse is selected at random, the probability that the nurse is a medical-surgical nurse or an emergency room nurse is .25 + .15 = .40. Since the events of being a medical-surgical nurse and an emergency room nurse are mutually exclusive, the probability is simply the sum of probabilities of the two events.

BAYES' THEOREM

A computer disk manufacturer has three locations that produce computer disks. The Omaha plant produces 30% of the disks, of which 0.5% are defective. The Memphis plant produces 50% of the disks, of which 0.75% are defective. The Kansas City plant produces the remaining 20%, of which 0.25% are defective. If a disk is purchased at a store and found to be defective, what is the probability that it was manufactured by the Omaha plant? This type of problem can be solved using Bayes' theorem. To formalize our approach, let A_1 be the event that the disk was manufactured by the Omaha plant, let A_2 be the event that the disk was manufactured by the Memphis plant, and let A_3 be the event that it was manufactured by the Kansas City plant. Let B be the event that the disk is defective. We are asked to find $P(A_1 \mid B)$. This probability is obtained by dividing $P(A_1$ and B) by $P(B)$.

The event that a disk is defective occurs if the disk is manufactured by the Omaha plant and is defective or if the disk is manufactured by the Memphis plant and is defective or if the disk is manufactured by the Kansas City plant and is defective. This is expressed as follows

$$B = (A_1 \ and \ B) \ or \ (A_2 \ and \ B) \ or \ (A_3 \ and \ B) \qquad (4.12)$$

Because the three events which are connected by or's in formula (4.12) are mutually exclusive, $P(B)$ may be expressed as

$$P(B) = P(A_1 \ and \ B) + P(A_2 \ and \ B) + P(A_3 \ and \ B) \qquad (4.13)$$

By using the multiplication rule, formula (4.13) may be expressed as

$$P(B) = P(B \mid A_1) \ P(A_1) + P(B \mid A_2) \ P(A_2) + P(B \mid A_3) \ P(A_3) \qquad (4.14)$$

Using formula (4.14), $P(B) = .005 \times .3 + .0075 \times .5 + .0025 \times .2 = .00575$. That is, 0.575% of the disks manufactured by all three plants are defective. The probability $P(A_1 \ and \ B)$ equals $P(B \mid A_1) \ P(A_1) = .005 \times .3 = .0015$. The probability we are seeking is equal to $\dfrac{.0015}{.00575} = .261$. Summarizing, if a defective disk is found, the probability that it was manufactured by the Omaha plant is .261.

In using Bayes' theorem to find $P(A_1 \mid B)$ use the following steps:

Step 1: Compute $P(A_1 \ and \ B)$ by using the equation $P(A_1 \ and \ B) = P(B \mid A_1) \ P(A_1)$.
Step 2: Compute $P(B)$ by using formula (4.14).
Step 3: Divide the result in step 1 by the result in step 2 to obtain $P(A_1 \mid B)$.

These same steps may be used to find $P(A_2 \mid B)$ and $P(A_3 \mid B)$.

Events like A_1, A_2, and A_3 are called *collectively exhaustive*. They are mutually exclusive and their union equals the sample space. Bayes' theorem is applicable to any number of collectively exhaustive events.

EXAMPLE 4.26 Using the three-step procedure given above, the probability that a defective disk was manufactured by the Memphis plant is found as follows.

Step 1: $P(A_2 \ and \ B) = P(B \mid A_2) \ P(A_2) = .0075 \times .5 = .00375$.
Step 2: $P(B) = .005 \times .3 + .0075 \times .5 + .0025 \times .2 = .00575$.
Step 3: $P(A_2 \mid B) = \dfrac{.00375}{.00575} = .652$.

The probability that a defective disk was manufactured by the Kansas City plant is found as follows.

Step 1: $P(A_3 \ and \ B) = P(B \mid A_3) \ P(A_3) = .0025 \times .2 = .0005$.
Step 2: $P(B) = .005 \times .3 + .0075 \times .5 + .0025 \times .2 = .00575$.
Step 3: $P(A_2 \mid B) = \dfrac{.0005}{.00575} = .087$.

PERMUTATIONS AND COMBINATIONS

Many of the experiments in statistics involve the selection of a subset of items from a larger group of items. The experiment of selecting two letters from the four letters a, b, c, and d is such an experiment. The following pairs are possible: (a, b), (a, c), (a, d), (b, c), (b, d), and (c, d). We say that when selecting two items from four distinct items that there are six possible *combinations*. The number of combinations possible when selecting n from N items is represented by the symbol C_n^N and is given by

$$C_n^N = \frac{N!}{n!(N-n)!} \qquad (4.15)$$

$_NC_n$, $C(N, n)$, and $\binom{N}{n}$ are three other notations that are used for the number of combinations in addition to the symbol C_n^N.

The symbol n!, read as "n factorial," is equal to $n \times (n-1) \times (n-2) \times \ldots \times 1$. For example, $3! = 3 \times 2 \times 1 = 6$, and $4! = 4 \times 3 \times 2 \times 1 = 24$. The values for n! become very large even for small values of n. The value of 10! is 3,628,800, for example.

In the context of selecting two letters from four, $N! = 4! = 4 \times 3 \times 2 \times 1 = 24$, $n! = 2! = 2 \times 1 = 2$ and $(N-n)! = 2! = 2$. The number of combinations possible when selecting two items from four is given by $C_2^4 = \frac{4!}{2!2!} = \frac{24}{2 \times 2} = 6$, the same number obtained when we listed all possibilities above. When the number of items is larger than four or five, it is difficult to enumerate all of the possibilities.

EXAMPLE 4.27 The number of five card poker hands that can be dealt from a deck of 52 cards is given by $C_5^{52} = \frac{52!}{5!47!} = \frac{52 \times 51 \times 50 \times 49 \times 48 \times 47!}{120 \times 47!} = \frac{52 \times 51 \times 50 \times 49 \times 48}{120} = 2,598,960$. Notice that by expressing 52! as $52 \times 51 \times 50 \times 49 \times 48 \times 47!$, we are able to divide 47! out because it is a common factor in both the numerator and the denominator.

If the order of selection of items is important, then we are interested in the number of *permutations* possible when selecting n items from N items. The number of permutations possible when selecting n objects from N objects is represented by the symbol P_n^N, and given by

$$P_n^N = \frac{N!}{(N-n)!} \qquad (4.16)$$

$_NP_n$, $P(N, n)$ and $(N)_n$ are other symbols used to represent the number of permutations.

EXAMPLE 4.28 The number of permutations possible when selecting two letters from the four letters a, b, c, and d is $P_2^4 = \frac{4!}{(4-2)!} = \frac{4!}{2!} = \frac{24}{2} = 12$. In this case, the 12 permutations are easy to list. They are ab, ba, ac, ca, ad, da, bc, cb, bd, db, cd, and dc. There are always more permutations than combinations when selecting n items from N, because each different ordering is a different permutation but not a different combination.

EXAMPLE 4.29 A president, vice president, and treasurer are to be selected from a group of 10 individuals. How many different choices are possible? In this case, the order of listing of the three individuals for the three offices is important because a slate of Jim, Joe, and Jane for president, vice president, and treasurer is different from Joe, Jim, and Jane for president, vice president, and treasurer, for example. The number of permutations is $P_3^{10} = \frac{10!}{7!} = 10 \times 9 \times 8 = 720$. That is, there are 720 different sets of size three that could serve as president, vice president, and treasurer.

USING PERMUTATIONS AND COMBINATIONS TO SOLVE PROBABILITY PROBLEMS

EXAMPLE 4.30 For a lotto contest in which six numbers are selected from the numbers 01 through 45, the number of combinations possible for the six numbers selected is $C_6^{45} = \dfrac{45!}{6!39!} = 8{,}145{,}060$. The probability that you select the correct six numbers in order to win this lotto is $\dfrac{1}{8{,}145{,}060} = .000000123$. The probability of winning the lotto can be reduced by requiring that the six numbers be selected in the correct order. The number of permutations possible when six numbers are selected from the numbers 01 through 45 is given by $P_6^{45} = \dfrac{45!}{39!}$ $= 45 \times 44 \times 43 \times 42 \times 41 \times 40 = 5{,}864{,}443{,}200$. The probability of winning the lotto is $\dfrac{1}{5{,}864{,}443{,}200} = .000000000171$, when the order of selection of the six numbers is important.

EXAMPLE 4.31 A Royal Flush is a five-card hand consisting of the ace, king, queen, jack, and ten of the same suit. The probability of a Royal Flush is equal to $\dfrac{4}{2{,}598{,}960} = .00000154$, since from Example 4.27, there are 2,598,960 five-card hands possible and four of them are Royal Flushes.

Solved Problems

EXPERIMENT, OUTCOMES, AND SAMPLE SPACE

4.1   An experiment consists of flipping a coin, followed by tossing a die. Give the sample space for this experiment.

   Ans.   One of many possible representations of the sample space is S = {H1, H2, H3, H4, H5, H6, T1, T2, T3, T4, T5, T6}.

4.2   Give the sample space for observing a patient's Rh blood type.

   Ans.   One of many possible representations of the sample space is S = {Rh$^-$, Rh$^+$}.

TREE DIAGRAMS AND THE COUNTING RULE

4.3   Use a tree diagram to illustrate the sample space for the experiment of observing the sex of the children in families consisting of three children.

   Ans.   The tree diagram representation for the sex distribution of the three children is shown in Fig. 4-8, where, for example, the branch or outcome mfm represents the outcome that the first born was a male, the second born was a female, and the last born was a male.

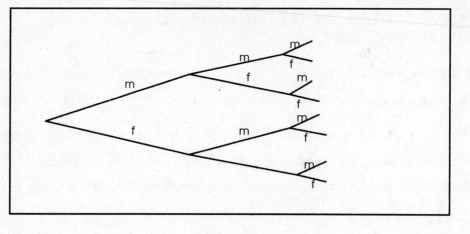

Fig. 4-8

4.4 A sociological study consists of recording the marital status, religion, race, and income of an individual. If marital status is classified into one of four categories, religion into one of three categories, race into one of five categories, and income into one of five categories, how many outcomes are possible for the experiment of recording the information of one of these individuals?

Ans. Using the counting rule, we see that there are $4 \times 3 \times 5 \times 5 = 300$ outcomes possible.

EVENTS, SIMPLE EVENTS, AND COMPOUND EVENTS

4.5 For the sample space given in Fig. 4-8, give the outcomes associated with the following events and classify each as a simple event or a compound event.
(a) At least one of the children is a girl.
(b) All the children are of the same sex.
(c) None of the children are boys.
(d) All of the children are boys.

Ans. The event that at least one of the children is a girl means that either one of the three was a girl, or two of the three were girls, or all three were girls. The event that all were of the same sex means that all three were boys or all three were girls. The event that none were boys means that all three were girls. The outcomes for these events are as follows:
(a) mmf, mfm, mff, fmm, fmf, ffm, fff; compound event
(b) mmm, fff; compound event
(c) fff; simple event
(d) mmm; simple event

4.6 In the game of *Yahtzee*, five dice are thrown simultaneously. How many outcomes are there for this experiment? Give the outcomes that correspond to the event that the same number appeared on all five dice.

Ans. By the counting rule, there are $6 \times 6 \times 6 \times 6 \times 6 = 7,776$ outcomes possible. Six of these 7,776 outcomes correspond to the event that the same number appeared on all five dice. These six outcomes are as follows: (1, 1, 1, 1, 1), (2, 2, 2, 2, 2), (3, 3, 3, 3, 3), (4, 4, 4, 4, 4), (5, 5, 5, 5, 5), (6, 6, 6, 6, 6).

PROBABILITY

4.7 Which of the following are permissible values for the probability of the event E?

(a) $P(E) = .75$ (b) $P(E) = -.25$ (c) $P(E) = 1.50$

(d) $P(E) = 1$ (e) $P(E) = .01$

Ans.   The probabilities given in (a), (d), and (e) are permissible since they are between 0 and 1 inclusive. The probability in (b) is not permissible, since probability measure can never be negative. The probability in (c) is not permissible, since probability measure can never exceed one.

4.8 An experiment is made up of five simple events designated A, B, C, D, and E. Given that $P(A) = .1$, $P(B) = .2$, $P(C) = .3$, and $P(E) = .2$, find $P(D)$.

Ans.   The sum of the probabilities for all simple events in an experiment must equal one. This implies that $.1 + .2 + .3 + P(D) + .2 = 1$, and solving for $P(D)$, we find that $P(D) = .2$. Note that this experiment does not have equally likely outcomes.

CLASSICAL, RELATIVE FREQUENCY, AND SUBJECTIVE PROBABILITY DEFINITIONS

4.9 A container has 5 red balls, 10 white balls, and 35 blue balls. One of the balls is selected randomly. Find the probability that the selected ball is (a) red; (b) white; (c) blue.

Ans.   This experiment has 50 equally likely outcomes. The event that the ball is red consists of five outcomes, the event that the ball is white consists of 10 outcomes, and the event that the ball is blue consists of 35 outcomes. Using the classical definition of probability, the following probabilities are obtained: (a) $\frac{5}{50} = .10$; (b) $\frac{10}{50} = .20$; (c) $\frac{35}{50} = .70$.

4.10 A store manager notes that for 250 randomly selected customers, 75 use coupons in their purchase. What definition of probability should the manager use to compute the probability that a customer will use coupons in their store purchase? What probability should be assigned to this event?

Ans.   The relative frequency definition of probability should be used. The probability of using coupons in store purchases is approximately $\frac{75}{250} = .30$.

4.11 Statements such as "the probability of snow tonight is 70%," "the probability that it will rain today is 20%," and "the probability that a new computer software package will be successful is 99%" are examples of what type of probability assignment?

Ans.   Since all three of the statements are based on professional judgment and experience, they are subjective probability assignments.

MARGINAL AND CONDITIONAL PROBABILITIES

4.12 Financial Planning Consultants Inc. keeps track of 500 stocks. Table 4.5 classifies the stocks according to two criteria. Three hundred are from the New York exchange and 200 are from the American exchange. Two hundred are up, 100 are unchanged, and 200 are down.

Table 4.5

	Up	Unchanged	Down	Total
NYSE	50	75	175	300
AMEX	150	25	25	200
Total	200	100	200	500

If one of these stocks is randomly selected find the following.

(a) The joint probability that the selected stock is from AMEX and unchanged.

(b) The marginal probability that the selected stock is from NYSE.

(c) The conditional probability that the stock is unchanged given that it is from AMEX.

Ans. (a) There are 25 from the AMEX and unchanged. The joint probability is $\frac{25}{500} = .05$.

   (b) There are 300 from NYSE. The marginal probability is $\frac{300}{500} = .60$.

   (c) There are 200 from the AMEX. Of these 200, 25 are unchanged. The conditional probability is $\frac{25}{200} = .125$.

4.13 Twenty percent of a particular age group has hypertension. Five percent of this age group has hypertension and diabetes. Given that an individual from this age group has hypertension, what is the probability that the individual also has diabetes?

Ans. Let A be the event that an individual from this age group has hypertension and let B be the event that an individual from this age group has diabetes. We are given that P(A) = .20 and P(A and B) = .05. We are asked to find P(B | A). $P(B \mid A) = \frac{P(A \text{ and } B)}{P(A)} = \frac{.05}{.20} = .25$.

MUTUALLY EXCLUSIVE EVENTS

4.14 For the experiment of drawing a card from a standard deck of 52, the following events are defined: A is the event that the card is a face card, B is the event that the card is an Ace, C is the event that the card is a heart, and D is the event that the card is black. List the six pairs of events and determine which are mutually exclusive.

Ans.

Pair	Mutually exclusive	Common outcomes
A, B	Yes	None
A, C	No	Jack, queen, and king of hearts
A, D	No	Jack, queen, and king of spades and clubs
B, C	No	Ace of hearts
B, D	No	Ace of clubs and ace of spades
C, D	Yes	None

4.15 Three items are selected from a production process and each is classified as defective or non-defective. Give the outcomes in the following events and check each pair to see if the pair is mutually exclusive. Event A is the event that the first item is defective, B is the event that there is exactly one defective in the three, and C is the event that all three items are defective.

Ans. A consists of the outcomes DDD, DDN, DND, and DNN. B consists of the outcomes DNN, NDN, and NND. C consists of the outcome DDD. (D represents a defective, and N represents a nondefective.)

Pair	Mutually exclusive	Common outcomes
A, B	No	DNN
A, C	No	DDD
B, C	Yes	None

DEPENDENT AND INDEPENDENT EVENTS

4.16 African-American males have a higher rate of hypertension than the general population. Let A represent the event that an individual is hypertensive and let B represent the event that an individual is an African-American male. Are A and B independent or dependent events?

  Ans. To say that African-American males have a higher rate of hypertension than the general population means that $P(A \mid B) > P(A)$. Since $P(A \mid B) \neq P(A)$, A and B are dependent events.

4.17 Table 4.6 gives the number of defective and nondefective items in samples from two different machines. Is the event of a defective item being produced by the machines dependent upon which machine produced it?

Table 4.6

	Number defective	Number nondefective
Machine 1	5	195
Machine 2	15	585

  Ans. Let D be the event that a defective item is produced by the machines, Let M_1 be the event that the item is produced by machine 1, and let M_2 be the event that the item is produced by machine 2. $P(D) = \frac{20}{800} = .025$, $P(D \mid M_1) = \frac{5}{200} = .025$, and $P(D \mid M_2) = \frac{15}{600} = .025$, and the event of producing a defective item is independent of which machine produces it.

COMPLEMENTARY EVENTS

4.18 The probability that a machine does not produce a defective item during a particular shift is .90. What is the complement of the event that a machine does not produce a defective item during that particular shift and what is the probability of that complementary event?

  Ans. The complementary event is that the machine produces at least one defective item during the shift, and the probability that the machine produces at least one defective item during the shift is $1 - .90 = .10$.

4.19 Events E_1, E_2, and E_3 have the following probabilities of occurrence: $P(E_1) = .05$, $P(E_2) = .50$, and $P(E_3) = .99$. Find the probabilities of the complements of these events.

  Ans. $P(E_1{}^c) = 1 - P(E_1) = 1 - .05 = .95$ $P(E_2{}^c) = 1 - P(E_2) = 1 - .50 = .50$
  $P(E_3{}^c) = 1 - P(E_3) = 1 - .99 = .01$

MULTIPLICATION RULE FOR THE INTERSECTION OF EVENTS

4.20 If one card is drawn from a standard deck, what is the probability that the card is a face card? If two cards are drawn, without replacement, what is the probability that both are face cards? If five cards are drawn, without replacement, what is the probability that all five are face cards?

Ans. Let E_1 be the event that the first card is a face card, let E_2 be the event that the second drawn card is a face card, and so on until E_5 represents the event that the fifth card is a face card. The probability that the first card is a face card is $P(E_1) = \frac{12}{52} = .231$. The event that both are face cards is the event E_1 *and* E_2, and $P(E_1$ *and* $E_2) = P(E_1) \, P(E_2 \mid E_1) = \frac{12}{52} \times \frac{11}{51} = \frac{132}{2652} = .050$. The event that all five are face cards is the event E_1 *and* E_2 *and* E_3 *and* E_4 *and* E_5. The probability of this event is given by $P(E_1) \, P(E_2 \mid E_1) \, P(E_3 \mid E_1, E_2) \, P(E_4 \mid E_1, E_2, E_3) \, P(E_5 \mid E_1, E_2, E_3, E_4)$, or
$$\frac{12}{52} \times \frac{11}{51} \times \frac{10}{50} \times \frac{9}{49} \times \frac{8}{48} = \frac{95,040}{311,875,200} = .000305.$$

4.21 If 60 percent of all Americans own a handgun, find the probability that all five in a sample of five randomly selected Americans own a handgun. Find the probability that none of the five own a handgun.

Ans. Let E_1 be the event that the first individual owns a handgun, E_2 be the event that the second individual owns a handgun, E_3 be the event that the third individual owns a handgun, E_4 be the event that the fourth individual owns a handgun, and E_5 be the event that the fifth individual owns a handgun. The probability that all five own a handgun is $P(E_1$ *and* E_2 *and* E_3 *and* E_4 *and* $E_5)$. Because of the large group from which the individuals are selected, the events E_1 through E_5 are independent and the probability is given by $P(E_1) \, P(E_2) \, P(E_3) \, P(E_4) \, P(E_5) = (.6)^5 = .078$. Similarly, the probability that none of the five own a handgun is $(.4)^5 = .010$.

ADDITION RULE FOR THE UNION OF EVENTS

4.22 Table 4.7 gives the IQ rating as well as the creativity rating of 250 individuals in a psychological study. Find the probability that a randomly selected individual from this study will be classified as having a high IQ or as having high creativity.

Table 4.7

	Low IQ	High IQ
Low creativity	75	30
High creativity	20	125

Ans. Let A be the event that the selected individual has a high IQ, and let B be the event that the individual has high creativity. Then $P(A) = \frac{155}{250} = .62$, $P(B) = \frac{145}{250} = .58$, $P(A$ *and* $B) = \frac{125}{250} = .50$, and $P(A$ *or* $B) = P(A) + P(B) - P(A$ *and* $B) = .62 + .58 - .50 = .70$.

4.23 The probability of event A is .25, the probability of event B is .10, and A and B are independent events. What is the probability of the event A *or* B?

Ans. Since A and B are independent events, $P(A$ *and* $B) = P(A) \, P(B) = .25 \times .10 = .025$. $P(A$ *or* $B) = P(A) + P(B) - P(A$ *and* $B) = .25 + .10 - .025 = .325$.

BAYES' THEOREM

4.24 Box 1 contains 30 red and 70 white balls, box 2 contains 50 red and 50 white balls, and box 3 contains 75 red and 25 white balls. The three boxes are all emptied into a large box, and a ball is selected at random. If the selected ball is red, what is the probability that it came from (*a*) box 1; (*b*) box 2; (*c*) box 3?

Ans. Let B_1 be the event that the selected ball came from box 1, let B_2 be the event that the selected ball came from box 2, let B_3 be the event that the ball came from box 3, and let R be the event that the selected ball is red.

(a) We are asked to find $P(B_1 \mid R)$. The three-step procedure is as follows:

Step 1: $P(R \text{ and } B_1) = P(R \mid B_1)\, P(B_1) = \frac{30}{100} \times \frac{1}{3} = .10$.

Step 2: $P(R) = P(R \mid B_1)\, P(B_1) + P(R \mid B_2)\, P(B_2) + P(R \mid B_3)\, P(B_3)$

$P(R) = \frac{155}{300} = .517$

Step 3: $P(B_1 \mid R) = \dfrac{.10}{.517} = .193$

(b) We are asked to find $P(B_2 \mid R)$.

Step 1: $P(R \text{ and } B_2) = P(R \mid B_2)\, P(B_2) = \frac{50}{100} \times \frac{1}{3} = .167$

Step 2: $P(R) = .517$

Step 3: $P(B_2 \mid R) = \dfrac{.167}{.517} = .323$

(c) We are asked to find $P(B_3 \mid R)$.

Step 1: $P(R \text{ and } B_3) = P(R \mid B_3)\, P(B_2) = \frac{75}{100} \times \frac{1}{3} = .250$

Step 2: $P(R) = .517$

Step 3: $P(B_3 \mid R) = \dfrac{.250}{.517} = .484$

4.25 Table 4.8 gives the percentage of the U.S. population in four regions of the United States, as well as the percentage of social security recipients within each region. For the population of all social security recipients, what percent live in each of the four regions?

Table 4.8

Region	Percentage of U.S. population	Percentage of social security recipients in the region
Northeast	20	15
Midwest	25	10
South	35	12
West	20	11

Ans. Let B_1 be the event that an individual lives in the Northeast region, B_2 be the event that an individual lives in the Midwest region, B_3 be the event that an individual lives in the South, and B_4 be the event that an individual lives in the West. Let S be the event that an individual is a social security recipient. We are given that $P(B_1) = .20$, $P(B_2) = .25$, $P(B_3) = .35$, $P(B_4) = .20$, $P(S \mid B_1) = .15$, $P(S \mid B_2) = .10$, $P(S \mid B_3) = .12$, and $P(S \mid B_4) = .11$. We are to find $P(B_1 \mid S)$, $P(B_2 \mid S)$, $P(B_3 \mid S)$, and $P(B_4 \mid S)$. $P(S)$ is needed to find each of the four probabilities.

$P(S) = P(S \mid B_1)\, P(B_1) + P(S \mid B_2)\, P(B_2) + P(S \mid B_3)\, P(B_3) + P(S \mid B_4)\, P(B_4)$
$P(S) = .15 \times .20 + .10 \times .25 + .12 \times .35 + .11 \times .20 = .119$.
This means that 11.9% of the population are social security recipients.

The three-step procedure to find $P(B_1 \mid S)$ is as follows:
Step 1: $P(B_1 \text{ and } S) = P(S \mid B_1)\, P(B_1) = .15 \times .20 = .03$
Step 2: $P(S) = .119$

Step 3: $P(B_1 \mid S) = \dfrac{.03}{.119} = .252$

The three-step procedure to find $P(B_2 \mid S)$ is as follows:

Step 1: $P(B_2 \text{ and } S) = P(S \mid B_2)\, P(B_2) = .10 \times .25 = .025$

Step 2: $P(S) = .119$

Step 3: $P(B_2 \mid S) = \dfrac{.025}{.119} = .210$

The three-step procedure to find $P(B_3 \mid S)$ is as follows:

Step 1: $P(B_3 \text{ and } S) = P(S \mid B_3)\, P(B_3) = .12 \times .35 = .042$

Step 2: $P(S) = .119$

Step 3: $P(B_3 \mid S) = \dfrac{.042}{.119} = .353$

The three-step procedure to find $P(B_4 \mid S)$ is as follows:

Step 1: $P(B_4 \text{ and } S) = P(S \mid B_4)\, P(B_4) = .11 \times .20 = .022$

Step 2: $P(S) = .119$

Step 3: $P(B_4 \mid S) = \dfrac{.022}{.119} = .185$

We can conclude that 25.2% of the social security recipients are from the Northeast, 21.0% are from the Midwest, 35.3% are from the South, and 18.5% are from the West.

PERMUTATIONS AND COMBINATIONS

4.26 Evaluate the following: (a) C_0^n; (b) C_n^n; (c) P_0^n; (d) P_n^n.

Ans.   Each of the four parts uses the fact that $0! = 1$.

(a) $C_0^n = \dfrac{n!}{0!(n-0)!} = \dfrac{n!}{1 \times n!} = 1$, since the n! in the numerator and denominator divide out.

(b) $C_n^n = \dfrac{n!}{n!(n-n)!} = \dfrac{n!}{n!0!} = 1$, since 0! is equal to one and the n! divides out of top and bottom.

(c) $P_0^n = \dfrac{n!}{(n-0)!} = \dfrac{n!}{n!} = 1.$

(d) $P_n^n = \dfrac{n!}{(n-n)!} = \dfrac{n!}{0!} = \dfrac{n!}{1} = n!$

4.27 An *exacta wager* at the racetrack is a bet where the bettor picks the horses that finish first and second. A *trifecta wager* is a bet where the bettor picks the three horses that finish first, second, and third. (a) In a 12-horse race, how many exactas are possible? (b) In a 12-horse race, how many trifectas are possible?

Ans.   Since the finish order of the horse is important, we use permutations to count the number of possible selections.

(a) The number of ordered ways you can select two horses from twelve is

$$P_2^{12} = \frac{12!}{(12-2)!} = \frac{12!}{10!} = \frac{12 \times 11 \times 10!}{10!} = 12 \times 11 = 132$$

(b) The number of ordered ways you can select three horses from twelve is

$$P_3^{12} = \frac{12!}{(12-3)!} = \frac{12!}{9!} = \frac{12 \times 11 \times 10 \times 9!}{9!} = 12 \times 11 \times 10 = 1320$$

4.28 A committee of five senators is to be selected from the U.S. Senate. How many different committees are possible?

Ans.   Since the order of the five senators is not important, the proper counting technique is combinations.

$$C_5^{100} = \frac{100!}{5!(100-5)!} = \frac{100 \times 99 \times 98 \times 97 \times 96 \times 95!}{120 \times 95!} = \frac{100 \times 99 \times 98 \times 97 \times 96}{120} = 75,287,520.$$

There are 75,287,520 different committees possible.

USING PERMUTATIONS AND COMBINATIONS TO SOLVE PROBABILITY PROBLEMS

4.29 Twelve individuals are to be selected to serve on a jury from a group consisting of 10 females and 15 males. If the selection is done in a random fashion, what is the probability that all 12 are males?

Ans.   Twelve individuals can be selected from 25 in the following number of ways:

$$C_{12}^{25} = \frac{25!}{12!13!} = \frac{25 \times 24 \times 23 \times 22 \times 21 \times 20 \times 19 \times 18 \times 17 \times 16 \times 15 \times 14 \times 13!}{12!13!}$$

After dividing out the common factor, 13!, we obtain the following.

$$C_{12}^{25} = \frac{25 \times 24 \times 23 \times 22 \times 21 \times 20 \times 19 \times 18 \times 17 \times 16 \times 15 \times 14}{12 \times 11 \times 10 \times 9 \times 8 \times 7 \times 6 \times 5 \times 4 \times 3 \times 2 \times 1} = 5,200,300$$

The above fraction may need to be evaluated in a zigzag fashion. That is, rather than multiply the 12 terms on top and then the 12 terms on bottom and then divide, do a multiplication, followed by a division, followed by a multiplication, and so on until all terms on top and bottom are accounted for. The jury can consist of all males in $C_{12}^{15} = \frac{15!}{12!3!} = \frac{15 \times 14 \times 13 \times 12!}{12!3!} = \frac{15 \times 14 \times 13}{6} = 455$ ways. The probability of an all-male jury is $\frac{455}{5,200,300} = .000087$. That is, there are about 9 chances out of 100,000 that an all-male jury would be chosen at random.

4.30 The five teams in the western division of the American conference of the National Football League are: Kansas City, Oakland, Denver, San Diego, and Seattle. Suppose the five teams are equally balanced. (*a*) What is the probability that Kansas City, Seattle, and Denver finish the season in first, second, and third place respectively? (*b*) What is the probability that the top three finishers are Kansas City, Seattle, and Denver?

Ans.   (*a*) Since the order of finish is specified, permutations are used to solve the problem. There are $P_3^5 = \frac{5!}{2!} = 60$ different ordered ways that three of the five teams could finish the season in first, second, and third place in the conference. The probability that Kansas City will finish first, Seattle will finish second, and Denver will finish third is $\frac{1}{60} = .017$.

(b)   Since the order of finish for the three teams is not specified, combinations are used to solve

the problem. There are $C_3^5 = \dfrac{5!}{3!2!} = 10$ combinations of three teams that could finish in the top

three. The probability that the top three finishers are Kansas City, Seattle, and Denver is $\dfrac{1}{10} = .10$.

Supplementary Problems

EXPERIMENT, OUTCOMES, AND SAMPLE SPACE

4.31   An experiment consists of using a 25-question test instrument to classify an individual as having either a type A or a type B personality. Give the sample space for this experiment. Suppose two individuals are classified as to personality type. Give the sample space. Give the sample space for three individuals.

Ans.   For one individual, S = {A, B}, where A means the individual has a type A personality, and B means the individual has a type B personality.

For two individuals, S = {AA, AB, BA, BB}, where AB, for example, is the outcome that the first individual has a type A personality and the second individual has a type B personality.

For three individuals, S = {AAA, AAB, ABA, ABB, BAA, BAB, BBA, BBB}, where ABA is the outcome that the first individual has a type A personality, the second has a type B personality, and the third has a type A.

4.32   At a roadblock, state troopers classify drivers as either driving while intoxicated, driving while impaired, or sober. Give the sample space for the classification of one driver. Give the sample space for two drivers. How many outcomes are possible for three drivers?

Ans.   Let A be the event that a driver is classified as driving while intoxicated, let B be the event that a driver is classified as driving while impaired, and let C be the event that a driver is classified as sober.

The sample space for one driver is S = {A, B, C}.

The sample space for two drivers is S = {AA, AB, AC, BA, BB, BC, CA, CB, CC}.

The sample space for three drivers has 27 possible outcomes.

TREE DIAGRAMS AND THE COUNTING RULE

4.33   An experiment consists of inspecting four items selected from a production line and classifying each one as defective, D, or nondefective, N. How many branches would a tree diagram for this experiment have? Give the branches that have exactly one defective. Give the branches that have exactly one nondefective.

Ans.   The tree would have $2^4 = 16$ branches which would represent the possible outcomes for the experiment.

The branches that have exactly one defective are DNNN, NDNN, NNDN, and NNND.

The branches that have exactly one nondefective are NDDD, DNDD, DDND, and DDDN.

4.34   An experiment consists of selecting one card from a standard deck, tossing a pair of dice, and then flipping a coin. How many outcomes are possible for this experiment?

Ans.   According to the counting rule, there are $52 \times 36 \times 2 = 3,744$ possible outcomes.

EVENTS, SIMPLE EVENTS, AND COMPOUND EVENTS

4.35 An experiment consists of rolling a single die. What are the simple events for this experiment?

> *Ans.* The simple events are the outcomes $\{1\}$, $\{2\}$, $\{3\}$, $\{4\}$, $\{5\}$, and $\{6\}$, where the number in braces represents the number on the turned up face after the die is rolled.

4.36 Suppose we consider a baseball game between the New York Yankees and the Detroit Tigers as an experiment. Is the event that the Tigers beat the Yankees one to nothing a simple event or a compound event? Is the event that the Tigers shut out the Yankees a simple event or a compound event? (A shutout is a game in which one of the teams scores no runs.)

> *Ans.* The event that the Tigers shut out the Yankees one to nothing is a simple event because it represents a single outcome. The event that the Tigers shut out the Yankees is a compound event because it could be a one to nothing shutout, or a two to nothing shutout, or a three to nothing shutout, etc.

PROBABILITY

4.37 Which of the following are permissible values for the probability of the event E?

(a) $\dfrac{3}{4}$ (b) $\dfrac{3}{2}$ (c) 0.0   (d) $\dfrac{-5}{7}$

> *Ans.* (a) and (c) are permissible values, since they are between 0 and 1 inclusive. (b) is not permissible because it exceeds one. (d) is not permissible because it is negative.

4.38 An experiment is made up of three simple events A, B, and C. If P(A) = x, P(B) = y, and P(C) = z, and x + y + z = 1, can you be sure that a valid assignment of probabilities has been made?

> *Ans.* No. Suppose x = .75, y = .75, and z = –.5, for example. Then x + y + z = 1, but this is not a valid assignment of probabilities.

CLASSICAL, RELATIVE FREQUENCY, AND SUBJECTIVE PROBABILITY DEFINITIONS

4.39 If a U.S. senator is chosen at random, what is the probability that he/she is from one of the 48 contiguous states?

> *Ans.* The are 96 senators from the 48 contiguous states and a total of 100 from the 50 states. The probability is $\dfrac{96}{100}$ = .96.

4.40 In an actuarial study, 9,875 females out of 10,000 females who are age 20 live to be 30 years old. What is the probability that a 20-year-old female will live to be 30 years old?

> *Ans.* Using the relative frequency definition of probability, the probability is $\dfrac{9,875}{10,000}$ = .988.

4.41 Casino odds for sporting events such as football games, fights, etc. are examples of which probability definition?

> *Ans.* subjective definition of probability

MARGINAL AND CONDITIONAL PROBABILITIES

4.42 Table 4.9 gives the joint probability distribution for a group of individuals in a sociological study.

Table 4.9

High school graduate	Welfare recipient	
	No	Yes
No	.15	.25
Yes	.55	.05

(a) Find the marginal probability that an individual in this study is a welfare recipient.

(b) Find the marginal probability that an individual in this study is not a high school graduate?

(c) If an individual is a welfare recipient, what is the probability that he/she is not a high school graduate?

(d) If an individual is a high school graduate, what is the probability that he/she is a welfare recipient?

Ans.   (a) $.25 + .05 = .30$ (c) $\dfrac{.25}{.30} = .83$

           (b) $.15 + .25 = .40$ (d) $\dfrac{.05}{.60} = .083$

4.43   Sixty percent of the registered voters in Douglas County are Republicans. Fifteen percent of the registered voters in Douglas County are Republican and have incomes above \$250,000 per year. What percent of the Republicans who are registered voters in Douglas County have incomes above \$250,000 per year?

      Ans. $\dfrac{.15}{.60} = .25$. Twenty-five percent of the Republicans have incomes above \$250,000.

MUTUALLY EXCLUSIVE EVENTS

4.44   With reference to the sociological study described in problem 4.42, are the events {high school graduate} and {welfare recipient} mutually exclusive?

      Ans.   No, since 5% of the individuals in the study satisfy both events.

4.45   Do mutually exclusive events cover all the possibilities in an experiment?

      Ans.   No. This is true only when the mutually exclusive events are also complementary.

DEPENDENT AND INDEPENDENT EVENTS

4.46   If two events are mutually exclusive, are they dependent or independent?

      Ans.   If A and B are two nontrivial events (that is, they have a nonzero probability) and if they are mutually exclusive, then $P(A \mid B) = 0$, since if B occurs, then A cannot occur. But since $P(A)$ is positive, $P(A \mid B) \neq P(A)$, and the events must be dependent.

4.47   Seventy-five percent of all Americans live in a metropolitan area. Eighty percent of all Americans consider themselves happy. Sixty percent of all Americans live in a metropolitan area and consider themselves happy. Are the events {lives in a metropolitan area} and {considers themselves happy} independent or dependent events?

      Ans.   Let A be the event {live in a metropolitan area} and let B be the event {consider themselves happy}. $P(A \mid B) = \dfrac{P(A \text{ and } B)}{P(B)} = \dfrac{.60}{.80} = .75 = P(A)$, and hence A and B are independent events.

COMPLEMENTARY EVENTS

4.48   What is the sum of the probabilities of two complementary events?

Ans. one

4.49 The probability that a machine used to make computer chips is out of control is .001. What is the complement of the event that the machine is out of control and what is the probability of this event?

    Ans. The complementary event is that the machine is in control. The probability of this event is .999.

MULTIPLICATION RULE FOR THE INTERSECTION OF EVENTS

4.50 In a particular state, 20 percent of the residents are classified as senior citizens. Sixty percent of the senior citizens of this state are receiving social security payments. What percent of the residents are senior citizens who are receiving social security payments?

    Ans. Let A be the event that a resident is a senior citizen and let B be the event that a resident is receiving social security payments. We are given that $P(A) = .20$ and $P(B \mid A) = .60$. $P(A$ and $B) = P(A) P(B \mid A) = .20 \times .60 = .12$. Twelve percent of the residents are senior citizens who are receiving social security payments.

4.51 If E_1, E_2, . . . , E_n are n independent events, then $P(E_1$ and E_2 and E_3 . . . and $E_n)$ is equal to the product of the probabilities of the n events. Use this probability rule to answer the following.
  (*a*) Find the probability of tossing five heads in a row with a coin.
  (*b*) Find the probability that the face 6 turns up every time in four rolls of a die.
  (*c*) If 43 percent of the population approve of the president's performance, what is the probability that all 10 individuals in a telephone poll disapprove of his performance.

    Ans. (*a*) $(.5)^5 = .03125$
          (*b*) $(\frac{1}{6})^4 = .00077$
          (*c*) $(.57)^{10} = .00362$

ADDITION RULE FOR THE UNION OF EVENTS

4.52 Events A and B are mutually exclusive and $P(A) = .25$ and $P(B) = .35$. Find $P(A$ and $B)$ and $P(A$ or $B)$.

    Ans. Since A and B are mutually exclusive, $P(A$ and $B) = 0$. Also $P(A$ or $B) = .25 + .35 = .60$.

4.53 Fifty percent of a particular market own a VCR or have a cell phone. Forty percent of this market own a VCR. Thirty percent of this market have a cell phone. What percent own a VCR and have a cell phone?

    Ans. $P(A$ and $B) = P(A) + P(B) - P(A$ or $B) = .3 + .4 - .5 = .2$, and therefore 20% own a VCR and a cell phone.

BAYES' THEOREM

4.54 In Arkansas, 30 percent of all cars emit excessive amounts of pollutants. The probability is 0.95 that a car emitting excessive amounts of pollutants will fail the state's vehicular emission test, and the probability is 0.15 that a car not emitting excessive amounts of pollutants will also fail the test. If a car fails the emission test, what is the probability that it actually emits excessive amounts of emissions?

    Ans. Let A be the event that a car emits excessive amounts of pollutants, and B the event that a car fails the emission test. Then, $P(A) = .30$, $P(A^c) = .70$, $P(B \mid A) = .95$, and $P(B \mid A^c) = .15$. We are asked to find $P(A \mid B)$. The three-step procedure results in $P(A \mid B) = .73$.

4.55 In a particular community, 15 percent of all adults over 50 have hypertension. The health service in this community correctly diagnoses 99 percent of all such persons with hypertension. The health service incorrectly diagnoses 5 percent who do not have hypertension as having hypertension.
    (a) Find the probability that the health service will diagnose an adult over 50 as having hypertension.
    (b) Find the probability that an individual over 50 who is diagnosed as having hypertension actually has hypertension.

    Ans. Let A be the event that an individual over 50 in this community has hypertension, and let B be the event that the health service diagnoses an individual over 50 as having hypertension. Then, $P(A) = .15$, $P(A^c) = .85$, $P(B \mid A) = .99$, and $P(B \mid A^c) = .05$.
        (a) $P(B) = .15 \times .99 + .85 \times .05 = .19$
        (b) Using the three-step procedure, $P(A \mid B) = .78$

PERMUTATIONS AND COMBINATIONS

4.56 How many ways can three letters be selected from the English alphabet if:
    (a) The order of selection of the three letters is considered important, i.e., abc is different from cba, for example.
    (b) The order of selection of the three letters is not important?

    Ans. (a) 15,600   (b) 2,600

4.57 The following teams comprise the Atlantic division of the Eastern conference of the National Hockey League: Florida, Philadelphia, N.Y. Rangers, New Jersey, Washington, Tampa Bay, and N.Y. Islanders.
    (a) Assuming no teams are tied at the end of the season, how many different final standings are possible for the seven teams?
    (b) Assuming no ties, how many different first-, second-, and third-place finishers are possible?

    Ans. (a) 5,040   (b) 210

4.58 A criminologist selects five prison inmates from 30 volunteers for more intensive study. How many such groups of five are possible when selected from the 30?

    Ans. 142,506

USING PERMUTATIONS AND COMBINATIONS TO SOLVE PROBABILITY PROBLEMS

4.59 Three individuals are to be randomly selected from the 10 members of a club to serve as president, vice president, and treasurer. What is the probability that Lana is selected for president, Larry for vice president, and Johnny for treasurer?

    Ans. $\dfrac{1}{P_3^{10}} = \dfrac{1}{720} = .00138$

4.60 A sample of size 3 is selected from a box which contains two defective items and 18 nondefective items. What is the probability that the sample contains one defective item?

    Ans. $\dfrac{C_1^2 \times C_2^{18}}{C_3^{20}} = \dfrac{2 \times 153}{1,140} = \dfrac{306}{1,140} = .268$

Chapter 5

Discrete Random Variables

RANDOM VARIABLE

A *random variable* associates a numerical value with each outcome of an experiment. A random variable is defined mathematically as a real-valued function defined on a sample space, and is represented as a letter such as X or Y.

EXAMPLE 5.1 For the experiment of flipping a coin twice, the random variable X is defined to be the number of tails to appear when the experiment is performed. The random variable Y is defined to be the number of heads minus the number of tails when the experiment is conducted. Table 5.1 shows the outcomes and the numerical value each random variable assigns to the outcome. These are two of the many random variables possible for this experiment.

Table 5.1

Outcome	Value of X	Value of Y
HH	0	2
HT	1	0
TH	1	0
TT	2	–2

EXAMPLE 5.2 An experimental study involving diabetics measured the following random variables: fasting blood sugar, hemoglobin, blood pressure, and triglecerides. These random variables assign numerical values to each of the individuals in the study. The numerical values range over different intervals for the different random variables.

DISCRETE RANDOM VARIABLE

A random variable is a *discrete random variable* if it has either a finite number of values or infinitely many values that can be arranged in a sequence. We say that a discrete random variable may assume a countable number of values. Discrete random variables usually arise from an experiment that involves counting. The random variables given in Example 5.1 are discrete, since they have a finite number of different values. Both of the variables are associated with counting.

EXAMPLE 5.3 An experiment consists of observing 100 individuals who get a flu shot and counting the number X who have a reaction. The variable X may assume 101 different values from 0 to 100. Another experiment consists of counting the number of individuals W who get a flu shot until an individual gets a flu shot and has a reaction. The variable W may assume the values 1, 2, 3, ... The variable W can assume a countably infinite number of values.

CONTINUOUS RANDOM VARIABLE

A random variable is a *continuous random variable* if it is capable of assuming all the values in an interval or in several intervals. Because of the limited accuracy of measuring devices, no random variables are truly continuous. However, we may treat random variables abstractly as being continuous.

EXAMPLE 5.4 The following random variables are considered continuous random variables: survival time of cancer patients, the time between release from prison and conviction for another crime, the daily milk yield of Holstein cows, weight loss during a dietary routine, and the household incomes for single-parent households in a sociological study.

PROBABILITY DISTRIBUTION

The *probability distribution* of a discrete random variable X is a list or table of the distinct numerical values of X and the probabilities associated with those values. The probability distribution is usually given in tabular form or in the form of an equation.

EXAMPLE 5.5 Table 5.2 lists the outcomes and the values of X, the sum of the up-turned faces for the experiment of rolling a pair of dice. Table 5.2 is used to build the probability distribution of the random variable X. This table lists the 36 possible outcomes. Only one outcome gives a value of 2 for X. The probability that X = 2 is 1 divided by 36 or .028 when rounded to three decimal places. We write this as P(2) = .028. The probability that X = 3, P(3), is equal to 2 divided by 36 or .056. The probability distribution for X is given in Table 5.3.

Table 5.2

Outcome	Value of X	Outcome	Value of X	Outcome	Value of X
(1, 1)	2	(3, 1)	4	(5, 1)	6
(1, 2)	3	(3, 2)	5	(5, 2)	7
(1, 3)	4	(3, 3)	6	(5, 3)	8
(1, 4)	5	(3, 4)	7	(5, 4)	9
(1, 5)	6	(3, 5)	8	(5, 5)	10
(1, 6)	7	(3, 6)	9	(5, 6)	11
(2, 1)	3	(4, 1)	5	(6, 1)	7
(2, 2)	4	(4, 2)	6	(6, 2)	8
(2, 3)	5	(4, 3)	7	(6, 3)	9
(2, 4)	6	(4, 4)	8	(6, 4)	10
(2, 5)	7	(4, 5)	9	(6, 5)	11
(2, 6)	8	(4, 6)	10	(6, 6)	12

Table 5.3

x	2	3	4	5	6	7	8	9	10	11	12
P(x)	.028	.056	.083	.111	.139	.167	.139	.111	.083	.056	.028

The probability distribution, $P(x) = P(X = x)$ satisfies formulas (*5.1*) and (*5.2*).

$$P(x) \geq 0 \quad \text{for each value x of X} \qquad (5.1)$$

$$\Sigma P(x) = 1 \quad \text{where the sum is over all values of X} \qquad (5.2)$$

Notice that the values for P(x) in Table 5.3 are all positive, which satisfies formula (*5.1*), and that the sum equals 1 except for rounding errors.

EXAMPLE 5.6 $P(x) = \dfrac{x}{10}$, x = 1, 2, 3, 4 is a probability distribution since P(1) = .1, P(2) = .2, P(3) = .3, and P(4) = .4 and (*5.1*) and (*5.2*) are both satisfied.

EXAMPLE 5.7 It is known from census data that for a particular income group that 10% of households have no children, 25% have one child, 50% have two children, 10% have three children, and 5% have four children. If X represents the number of children per household for this income group, then the probability distribution of X is given in Table 5.4.

Table 5.4

x	0	1	2	3	4
P(x)	.10	.25	.50	.10	.05

The event $X \geq 2$ is the event that a household in this income group has at least two children and means that $X = 2$, or $X = 3$, or $X = 4$. The probability that $X \geq 2$ is given by

$$P(X \geq 2) = P(X = 2) + P(X = 3) + P(X = 4) = .50 + .10 + .05 = .65$$

The event $X \leq 1$ is the event that a household in this income group has at most one child and is equivalent to $X = 0$, or $X = 1$. The probability that $X \leq 1$ is given by

$$P(X \leq 1) = P(X = 0) + P(X = 1) = .10 + .25 = .35$$

The event $1 \leq X \leq 3$ is the event that a household has between one and three children inclusive and is equivalent to $X = 1$, or $X = 2$, or $X = 3$. The probability that $1 \leq X \leq 3$ is given by

$$P(1 \leq X \leq 3) = P(X = 1) + P(X = 2) + P(X = 3) = .25 + .50 + .10 = .85$$

The above discussion may be summarized by stating that 65% of the households have at least two children, 35% have at most one child, and 85% have between one and three children inclusive.

MEAN OF A DISCRETE RANDOM VARIABLE

The *mean of a discrete random variable* is given by

$$\mu = \Sigma x\, P(x) \quad \text{where the sum is over all values of X} \tag{5.3}$$

The mean of a discrete random variable is also called the *expected value*, and is represented by E(X). The mean or expected value will also often be referred to as the *population mean*. Regardless of whether it is called the mean, the expected value, or the population mean, the numerical value is given by formula (*5.3*).

EXAMPLE 5.8 The mean of the random variable in Example 5.5 is found as follows.

x	2	3	4	5	6	7	8	9	10	11	12
P(x)	.028	.056	.083	.111	.139	.167	.139	.111	.083	.056	.028
xP(x)	.056	.168	.332	.555	.834	1.169	1.112	.999	.830	.616	.336

The sum of the row labeled xP(x) is the mean of X and is equal to 7.007. If fractions are used in place of decimals for P(x), the value will equal 7 exactly. In other words E(x) = 7. The long-term average value for the sum on the dice is 7. The mean value of the sum on the dice for the population of all possible rolls of the dice equals 7. If you were to record all the rolls of the dice at Las Vegas, the average value would equal 7.

EXAMPLE 5.9 The mean number of children per household for the distribution given in Example 5.7 is found as follows.

x	0	1	2	3	4
P(x)	.10	.25	.50	.10	.05
xP(x)	0	.25	1.00	.30	.20

The sum of the row labeled xP(x), 1.75, is the mean of the distribution. For this population of households, the mean number of children per household is 1.75. Notice that the mean does not have to equal a value assumed by the random variable.

STANDARD DEVIATION OF A DISCRETE RANDOM VARIABLE

The *variance of a discrete random variable* is represented by σ^2 and is defined by

$$\sigma^2 = \Sigma\,(x - \mu)^2\,P(x) \tag{5.4}$$

The variance is also represented by Var(X) and may be calculated by the alternative formula given by

$$\text{Var}(X) = \sigma^2 = \Sigma\,x^2\,P(x) - \mu^2 \tag{5.5}$$

The *standard deviation of a discrete random variable* is represented by σ or sd(X) and is given by

$$\sigma = \text{sd}\,(X) = \sqrt{\text{Var}(X)} \tag{5.6}$$

EXAMPLE 5.10 A social researcher is interested in studying the family dynamics caused by gender makeup. The distribution of the number of girls in families consisting of four children is as follows: 6.25% of such families have no girls, 25% have one girl, 37.5% have two girls, 25% have three girls, and 6.25% have four girls. Table 5.5 illustrates the computation of the variance of X, the number of girls in a family having four children. The standard deviation is the square root of the variance and equals one.

Table 5.5

x	P(x)	xP(x)	$x - \mu$	$(x - \mu)^2$	$(x - \mu)^2 P(x)$
0	0.0625	0.0	−2	4	0.25
1	0.25	0.25	−1	1	0.25
2	0.375	0.75	0	0	0.0
3	0.25	0.75	1	1	0.25
4	0.0625	0.25	2	4	0.25
Total	1.0	$\mu = 2$	$\Sigma(x - \mu) = 0$		Var(x) = 1

Table 5.6 illustrates the computation of the components needed when using the alternative formula (5.5) to compute the variance of X. Using formula (5.5), the variance is $\text{Var}(x) = \Sigma\,x^2\,P(x) - \mu^2 = 5 - 4 = 1$, and the standard deviation is also equal to one. We see that formulas (5.4) and (5.5) give the same results for the standard deviation. The mean number of girls in families of four children equals 2 and the standard deviation is equal to 1.

Table 5.6

x	P(x)	xP(x)	$x^2 P(x)$
0	0.0625	0.0	0.0
1	0.25	0.25	0.25
2	0.375	0.75	1.5
3	0.25	0.75	2.25
4	0.0625	0.25	1.0
Total	1.0	$\mu = 2.0$	$\Sigma\,x^2 P(x) = 5$

BINOMIAL RANDOM VARIABLE

A *binomial random variable* is a discrete random variable that is defined when the conditions of a *binomial experiment* are satisfied. The conditions of a binomial experiment are given in Table 5.7.

Table 5.7

Conditions of a Binomial Experiment
1. There are n identical trials.
2. Each trial has only two possible outcomes.
3. The probabilities of the two outcomes remain constant for each trial.
4. The trials are independent.

The two outcomes possible on each trial are called *success* and *failure*. The probability associated with success is represented by the letter p and the probability associated with failure is represented by the letter q, and since one or the other of success or failure must occur on each trial, p + q must equal one, i.e., p + q = 1. When the conditions of the binomial experiment are satisfied, the binomial random variable X is defined to equal the number of successes to occur in the n trials. The random variable X may assume any one of the whole numbers from zero to n.

EXAMPLE 5.11 A balanced coin is tossed 10 times, and the number of times a head occurs is represented by X. The conditions of a binomial experiment are satisfied. There are n = 10 identical trials. Each trial has two possible outcomes, head or tail. Since we are interested in the occurrence of a head on a trial, we equate the occurrence of a head with success, and the occurrence of a tail with failure. We see that p = .5 and q = .5. Also, it is clear that the trials are independent since the occurrence of a head on a given toss is independent of what occurred on previous tosses. The number of heads to occur in the 10 tosses, X, can equal any whole number between 0 and 10. X is a binomial random variable with n = 10 and p = .5.

EXAMPLE 5.12 A balanced die is tossed five times, and the number of times that the face with six spots on it faces up is counted. The conditions of a binomial experiment are satisfied. There are five identical trials. Each trial has two possible outcomes since the face 6 turns up or a face other than 6 turns up. Since we are interested in the face 6, we equate the face 6 with success and any other face with failure. We see that $p = \frac{1}{6}$ and $q = \frac{5}{6}$.

Also, the outcomes from toss to toss are independent of one another. The number of times the face 6 turns up, X, can equal 0, 1, 2, 3, 4, or 5. X is a binomial random variable with n = 5 and p = .167.

EXAMPLE 5.13 A manufacturer uses an injection mold process to produce disposable razors. One-half of one percent of the razors are defective. That is, on average, 500 out of every 100,000 razors are defective. A quality control technician chooses a daily sample of 100 randomly selected razors and records the number of defectives found in the sample in order to monitor the process. The conditions of a binomial experiment are satisfied. There are 100 identical trials. Each trial has two possible outcomes since the razor is either defective or non-defective. Since we are recording the number of defectives, we equate the occurrence of a defective with success and the occurrence of a nondefective with failure. We see that p = .005 and q = .995. The number of defectives in the 100, X, can equal any whole number between 0 and 100. X is a binomial random variable with n = 100 and p = .005.

BINOMIAL PROBABILITY FORMULA

The *binomial probability formula* is used to compute probabilities for binomial random variables. The binomial probability formula is given in

$$P(x) = \binom{n}{x} p^x q^{(n-x)} = \frac{n!}{x!(n-x)!} p^x q^{(n-x)} \text{ for } x = 0, 1, \ldots, n \qquad (5.7)$$

The symbol $\begin{pmatrix} n \\ x \end{pmatrix}$ is discussed in Chapter 4 and represents the number of combinations possible when

x items are selected from n.

EXAMPLE 5.14 A die is rolled three times and the random variable X is defined to be the number of times the face 6 turns up in the three tosses. X is a binomial random variable that assumes one of the values 0, 1, 2, or 3. In this binomial experiment, success occurs on a given trial if the face 6 turns up and failure occurs if any other face turns up. The probability of success is $p = \frac{1}{6} = .167$ and the probability of failure is $q = \frac{5}{6} = .833$. In order to help understand why the binomial probability formula works, the binomial probabilities will be computed using the basic principles of probability first and then by using formula (5.7).

To find P(0), note that X = 0 means that no successes occurred, that is, three failures occurred. Because of the independence of trials, the probability of three failures is $.833 \times .833 \times .833 = (.833)^3 = .578$. That is, the probability that the face 6 does not turn up on any of the three tosses is .578.

To find P(1), note that X = 1 means that one success and two failures occurred. One success and two failures occur if the sequence SFF, or the sequence FSF, or the sequence FFS occurs. The probability of SFF is $.167 \times .833 \times .833 = .1159$. The probability of FSF is $.833 \times .167 \times .833 = .1159$. The probability of FFS is $.833 \times .833 \times .167 = .1159$. The three probabilities for SFF, FSF, and FFS are added because of the addition rule for mutually exclusive events. Therefore, $P(X = 1) = P(1) = 3 \times .1159 = .348$. The probability that the face 6 turns up on one of the three tosses is .348.

To find P(2), note that X = 2 means that two successes and one failure occurred. Two successes and one failure occur if the sequence FSS or the sequence SFS or the sequence SSF occurs. The probability of FSS is $.833 \times .167 \times .167 = .0232$. The probability of SFS is $.167 \times .833 \times .167 = .0232$. The probability of SSF is $.167 \times .167 \times .833 = .0232$. The probabilities for FSS, SFS, and SSF are added because of the addition rule for mutually exclusive events. Therefore, $P(X = 2) = P(2) = 3 \times .0232 = .070$. The probability that the face 6 turns up on two of the three tosses is .070.

To find P(3), note that X = 3 means that three successes occurred. The probability of three consecutive successes is $.167 \times .167 \times .167 = (.167)^3 = .005$. There are five chances in a thousand of the face 6 turning up on each of the three tosses.

Using the binomial probability formula, we find the four probabilities as follows:

$$P(0) = \frac{3!}{0!3!}(.167)^0 (.833)^3 = (.833)^3 = .578$$

$$P(1) = \frac{3!}{1!2!}(.167)^1 (.833)^2 = 3 \times .167 \times (.833)^2 = .348$$

$$P(2) = \frac{3!}{2!1!}(.167)^2 (.833)^1 = .070$$

$$P(3) = \frac{3!}{3!0!}(.167)^3 (.833)^0 = .005$$

This example illustrates how much work the binomial probability formula saves us when solving problems involving the binomial distribution. The distribution for the variable in Example 5.14 is given in Table 5.8.

Table 5.8

x	0	1	2	3
P(x)	.578	.348	.070	.005

In formula (5.7), The term $p^x q^{(n-x)}$ gives the probability of x successes and (n − x) failures. The term $\frac{n!}{x!(n-x)!}$ counts the number of different arrangements which are possible for x successes and (n − x) failures. The role of these terms is illustrated in Example 5.14.

EXAMPLE 5.15 Fifty-seven percent of companies in the U.S. use networking to recruit workers. The probability that in a survey of ten companies exactly half of them use networking to recruit workers is

$$P(5) = \frac{10!}{5!5!} (.57)^5 (.43)^5 = .223$$

TABLES OF THE BINOMIAL DISTRIBUTION

Appendix 1 contains the *table of binomial probabilities*. This table lists the probabilities of x for n = 1 to n = 25 for selected values of p.

EXAMPLE 5.16 If X represents the number of girls in families having four children, then X is a binomial random variable with n = 4 and p = .5. Using formula (5.7), the distribution of X is determined as follows:

$$P(0) = \frac{4!}{0!4!} (.5)^0 (.5)^4 = (.5)^4 = .0625 \qquad P(1) = \frac{4!}{1!3!} (.5)^1 (.5)^3 = .25 \qquad P(2) = \frac{4!}{2!2!} (.5)^2 (.5)^2 = .375$$

$$P(3) = \frac{4!}{3!1!} (.5)^3 (.5)^1 = .25 \qquad P(4) = \frac{4!}{4!0!} (.5)^4 (.5)^0 = .0625$$

Table 5.9 contains a portion of the table of binomial probabilities found in Appendix 1. The numbers in bold print indicates the portion of the table from which the binomial probability distribution for X is obtained. The probabilities given are the same ones obtained by using formula (5.7).

Table 5.9

			p			
n	x	. . .	.40	**.50**	.60	. . .
4	0	. . .	.1296	**.0625**	.0256	. . .
	1	. . .	.3456	**.2500**	.1536	. . .
	2	. . .	.3456	**.3750**	.3456	. . .
	3	. . .	.1536	**.2500**	.3456	. . .
	4	. . .	.0256	**.0625**	.1296	. . .

EXAMPLE 5.17 Eighty percent of the residents in a large city feel that the government should allow more than one company to provide local telephone service. Using the table of binomial probabilities, the probability that at least five in a sample of ten residents feel that the government should allow more than one company to provide local telephone service is found as follows. The event "at least five" means five or more and is equivalent to X = 5 or X = 6 or X = 7 or X = 8 or X = 9 or X = 10. The probabilities are added because of the addition law for mutually exclusive events.

$$P(X \geq 5) = P(5) + P(6) + P(7) + P(8) + P(9) + P(10)$$

$$P(X \geq 5) = .0264 + .0881 + .2013 + .3020 + .2684 + .1074 = .9936$$

Statistical software is used to perform binomial probability computations and to some extent has rendered binomial probability tables obsolete. Minitab contains routines for computing binomial probabilities. Example 5.18 illustrates how to use Minitab to compute the probability for a single value or the total distribution for a binomial random variable.

EXAMPLE 5.18 The following Minitab output shows the binomial probability computations given in Examples 5.15 and 5.16. The binomial probabilities are shown in bold type.

MTB > # Minitab computation for Example 5.15 #
MTB > pdf 5;
SUBC > binomial n = 10 p = .57.

Probability Density Function
Binomial with n = 10 and p = 0.570000

X	P(X = x)
5.00	**0.2229**

MTB > # Minitab computation for Example 5.16 #
MTB > pdf;
SUBC > binomial n = 4 and p = .5.

Probability Density Function
Binomial with n = 4 and p = 0.500000

x	P(X = x)
0	**0.0625**
1	**0.2500**
2	**0.3750**
3	**0.2500**
4	**0.0625**

MEAN AND STANDARD DEVIATION OF A BINOMIAL RANDOM VARIABLE

The mean and variance for a binomial random variable may be found by using formulas (*5.3*) and (*5.4*). However, in the case of a binomial random variable, shortcut formulas exist for computing the mean and standard deviation of a binomial random variable. The mean of a binomial random variable is given by

$$\mu = np \qquad (5.8)$$

The variance of a binomial random variable is given by

$$\sigma^2 = npq \qquad (5.9)$$

EXAMPLE 5.19 The mean for the binomial distribution given in Example 5.18 using formula (*5.3*) is

$$\mu = \Sigma\, xP(x) = 0 \times .0625 + 1 \times .25 + 2 \times .375 + 3 \times .25 + 4 \times .0625 = 2$$

The mean using the shortcut formula (*5.8*) is

$$\mu = np = 4 \times .5 = 2$$

The variance for the binomial distribution given in Example 5.18 using formula (*5.4*) is

$$\sigma^2 = \Sigma\, x^2\, P(x) - \mu^2 = 0 \times .0625 + 1 \times .25 + 4 \times .375 + 9 \times .25 + 16 \times .0625 - 4 = 1$$

The variance using the shortcut formula (*5.9*) is

$$\sigma^2 = npq = 4 \times (.5) \times (.5) = 1$$

EXAMPLE 5.20 Chemotherapy provides a 5-year survival rate of 80% for a particular type of cancer. In a group of 20 cancer patients receiving chemotherapy for this type of cancer, the mean number surviving after 5 years is $\mu = 20 \times .8 = 16$ and the standard deviation is $\sigma = \sqrt{20 \times .8 \times .2} = 1.8$. On the average, 16 patients will survive, and typically the number will vary by no more than two from this figure.

POISSON RANDOM VARIABLE

The binomial random variable is applicable when counting the number of occurrences of an event called success in a finite number of trials. When the number of trials is large or potentially infinite, another random variable called the *Poisson random variable* may be appropriate. The Poisson probability distribution is applied to experiments with *random* and *independent* occurrences of an event. The occurrences are considered with respect to a time interval, a length interval, a fixed area or a particular volume.

EXAMPLE 5.21 The number of calls to arrive per hour at the reservation desk for Regional Airlines is a Poisson random variable. The calls arrive randomly and independently of one another. The random variable X is defined to be the number of calls arriving during a time interval equal to one hour and may be any number from 0 to some very large value.

EXAMPLE 5.22 The number of defects in a 10-foot coil of wire is a Poisson random variable. The defects occur randomly and independently of one another. The random variable X is defined to be the number of defects in a 10-foot coil of wire and may be any number between 0 and some very large value. The interval in this example is a length interval of 10 feet.

EXAMPLE 5.23 The number of pinholes in 1-yd^2 pieces of plastic is a Poisson random variable. The pinholes occur randomly and independently of one another. The random variable X is defined to be the number of pinholes per square yard piece and can assume any number between 0 and a very large value. The interval in this example is an area of 1 yd^2.

POISSON PROBABILITY FORMULA

The probability of x occurrences of some event in an interval where the Poisson assumptions of randomness and independence are satisfied is given by formula (*5.10*), where λ is the mean number of occurrences of the event in the interval and the value of e is approximately 2.71828. The value of e is found on most calculators and powers of e are easily evaluated. Tables of Poisson probabilities are found in many statistical texts. However, with the wide spread availability of calculators and statistical software, they are being less widely utilized.

$$P(x) = \frac{\lambda^x e^{-\lambda}}{x!} \quad \text{for } x = 0, 1, 2, \ldots \qquad (5.10)$$

The mean of a Poisson random variable is given by

$$\mu = \lambda \qquad (5.11)$$

The variance of a Poisson random variable is given by

$$\sigma^2 = \lambda \qquad (5.12)$$

EXAMPLE 5.24 The number of small pinholes in sheets of plastic are of concern to a manufacturer. If the number of pinholes is too large, the plastic is unusable. The mean number per square yard is equal to 2.5. The 1-yd^2 sheets are unusable if the number of pinholes exceeds 6. The probability of interest is $P(X > 6)$, where X represents the number of pinholes in a 1-yd^2 sheet. This probability is found by finding the probability of the complementary event and subtracting it from 1.

$$P(X > 6) = 1 - P(X \le 6)$$

The command cdf of Minitab can be used to find $P(X \le 6)$. The following Minitab output illustrates how this is accomplished.

MTB > cdf 6;
SUBC > poisson mean = 2.5.

Cumulative Distribution Function
Poisson with mu = 2.50000

    x   P(X <= x)
  6.00   **0.9858**

$$P(X > 6) = 1 - P(X \le 6) = 1 - .9858 = .0142$$

By multiplying .0142 by 100 we see that 1.42% of the plastic sheets are unusable.

EXAMPLE 5.25 The Poisson distribution approximates the binomial distribution closely when $n \ge 20$ and $p \le$ .05. A machine produces items of which 1% are defective. In a sample of 150 items selected from the output of this machine, the probability of two or fewer defectives in the sample is $P(X \le 2)$ and is found by the command cdf 2; when using Minitab. The following Minitab output gives the binomial probability that $X \le 2$.

MTB > cdf 2;
SUBC > binomial n = 150 p = .01.

Cumulative Distribution Function
Binomial with n = 150 and p = 0.0100000

    x   P(X <= x)
  2.00   **0.8095**

The mean number of defectives in a sample of 150 is $np = 150 \times .01 = 1.5$. The Poisson probability that $X \le 2$ where $\lambda = 1.5$ is given by the following Minitab output.

MTB > cdf 2;
SUBC > poisson mean = 1.5.

Cumulative Distribution Function
Poisson with mu = 1.50000

    x   P(X <= x)
  2.00   **0.8088**

The binomial probability of the event $X \le 2$ is .8095 and the Poisson approximation is .8088. Notice that the Poisson approximation is very close to the binomial probability.

HYPERGEOMETRIC RANDOM VARIABLE

The *hypergeometric random variable* is used in situations where success or failure is possible on each trial but where there is not independence from trial to trial. The lack of independence from trial to trial distinguishes the hypergeometric distribution from the binomial distribution. The hypergeometric random variable applies in situations where there are N items, of which k are classified as successes and N – k are classified as failures. A sample of size n ≤ k is selected from the N items and X is defined to equal the number of successes in the n items selected. X is a hypergeometric random variable which can equal any whole number from 0 to n.

EXAMPLE 5.26 A sociologist randomly selects 5 individuals from a group consisting of 10 male single parents and 15 female single parents. The random variable X is defined to equal the number of male single parents in the 5 selected individuals. In this example, N = 25, k = 10, N – k = 15, and n = 5. The number of male single parents in the 5 selected is a hypergeometric random variable. If this hypergeometric random variable is represented by X, then X may assume any one of the values 0, 1, 2, 3, 4, or 5.

EXAMPLE 5.27 A box contains 5 defective and 25 acceptable computer monitors. Three of the monitors are randomly selected and X is defined to be the number of defective monitors in the three. X is a hypergeometric random variable with N = 30, k = 5, N – k = 25, and n = 3. X may assume any one of the values 0, 1, 2, or 3.

HYPERGEOMETRIC PROBABILITY FORMULA

When n items are selected from N items of which k are successes and N – k are failures, the random variable X, defined to equal the number of successes in the n selected items, is a hypergeometric random variable. The probability distribution of X is given by the *hypergeometric probability formula* shown in formula (5.13).

$$P(x) = \frac{\binom{k}{x} \times \binom{N-k}{n-x}}{\binom{N}{n}} = \frac{\dfrac{k!}{x!(k-x)!}\dfrac{(N-k)!}{(n-x)!(N-k-n+x)!}}{\dfrac{N!}{n!(N-n)!}} \quad \text{for } x = 0, 1, \ldots, n \quad (5.13)$$

EXAMPLE 5.28 A police department consists of 25 officers of whom 5 are minorities. Three officers are randomly selected to meet with the mayor. Let X be the number of minorities in the three selected to meet with the mayor. X is a hypergeometric random variable with N = 25, k = 5, N – k = 20, and n = 3. The probability distribution of X is derived as follows:

$$P(0) = \frac{\binom{5}{0} \times \binom{20}{3}}{\binom{25}{3}} = \frac{1 \times 1140}{2300} = .496 \qquad P(1) = \frac{\binom{5}{1} \times \binom{20}{2}}{\binom{25}{3}} = \frac{5 \times 190}{2300} = .413$$

$$P(2) = \frac{\binom{5}{2} \times \binom{20}{1}}{\binom{25}{3}} = \frac{10 \times 20}{2300} = .087 \qquad P(3) = \frac{\binom{5}{3} \times \binom{20}{0}}{\binom{25}{3}} = \frac{10 \times 1}{2300} = .004$$

The probability distribution of X is given in Table 5.10. It is highly likely that at most one of the three will be a minority. The probability that X ≤ 1 is .909.

Table 5.10

x	0	1	2	3
P(x)	.496	.413	.087	.004

EXAMPLE 5.29 The binomial distribution approximates the hypergeometric distribution whenever n ≤ .05N. A box contains 200 computer chips, of which 7 are defective. The probability of finding one defective in a sample of 5 randomly selected chips is given by the following hypergeometric probability computation.

$$P(1) = \frac{\binom{7}{1} \times \binom{193}{4}}{\binom{200}{5}} = \frac{7 \times 56,031,760}{2,535,650,040} = .155$$

Since $n \le .05 \times 200 = 10$, the probability may be approximated by using the binomial distribution. The five selections of the computer chips may be viewed as $n = 5$ trials. The probability of success, selecting a defective chip, is $p = \frac{7}{200} = .035$, and $q = .965$. The binomial probability of one defective in the five chips is given as follows:

$$P(1) = \frac{5!}{1!4!} (.035)^1 (.965)^4 = .152$$

The approximation is very good when $n \le .05N$, as shown in this example.

Solved Problems

RANDOM VARIABLE

5.1 Let X represent the number of boys in families having three children. List all possible birth order permutations for families having three children and give the value of X for each outcome.

Ans. Table 5.11 gives the outcomes and the value X assigns to each outcome.

Table 5.11

Outcome	Value of X
BBB	3
BBG	2
BGB	2
BGG	1
GBB	2
GBG	1
GGB	1
GGG	0

5.2 For the experiment of rolling three dice, X is defined to be the sum of the three dice. What are the unique values assumed by X?

Ans. The values for X range from 3, corresponding to the outcome (1, 1, 1) to 18, corresponding to the outcome (6, 6, 6). The unique values are 3, 4, 5, 6, 7, 8, 9, 10, 11, 12, 13, 14, 15, 16, 17, and 18.

DISCRETE RANDOM VARIABLE

5.3 A telemarketing company administers an aptitude test consisting of 25 problems to potential employees. A variable of interest to the company is X, the number of problems worked correctly. How many different values are possible for X?

Ans. 26, since an individual can get from 0 to 25 correct.

5.4 A random variable assigns a single number to each outcome of an experiment. Is it true that each value of a random variable corresponds to a single outcome?

Ans. No. Consider the outcomes and values of X given in Table 5.11. The value X = 1 corresponds to the three outcomes BGG, GBG, and GGB for example.

CONTINUOUS RANDOM VARIABLE

5.5 Identify the continuous random variables in parts (*a*) through (*e*).
(*a*) The time that an individual is logged onto the internet during a given week
(*b*) The number of domestic violence calls responded to per day by the Chicago police department
(*c*) The mortality rate for women who used estrogen therapy for at least a year starting in 1969
(*d*) The daily room rate for luxury/upscale hotels in the U.S.
(*e*) The number of executions per state since 1975

Ans. Parts (*a*) and (*d*) are continuous random variables. Time and money are almost always considered continuous even though in practice they are probably discrete.

5.6 Is it possible to give a probability value to each individual value of a continuous random variable?

Ans. No. It is not possible to give an individual probability to each value of a continuous variable since the variable may assume an uncountably infinite number of different values. Instead, probabilities are assigned to an interval of values for a continuous random variable.

PROBABILITY DISTRIBUTION

5.7 According to the registrar's office at the University of Nebraska at Omaha (UNO), during the current semester, 9% of the students are registered for 3 credit hours, 13% are registered for 6 credit hours, 16% are registered for 9 credit hours, 21% are registered for 12 credit hours, 26% are registered for 15 credit hours, 13% are registered for 18 credit hours, and 2% are registered for 21 credit hours. If the random variable X represents the number of credit hours per student at UNO, give the probability distribution for X.

Ans. The probability distribution for X is given in Table 5.12.

Table 5.12

x	3	6	9	12	15	18	21
P(x)	.09	.13	.16	.21	.26	.13	.02

5.8 An experiment consists of rolling a die and flipping a coin. The coin has the number 1 stamped on one side and the number 2 stamped on the other side. The random variable Y is defined to equal the sum of the number showing on the coin plus the number showing on the die after the experiment is conducted. Give the probability distribution for Y.

Ans. Table 5.13 gives the outcomes for the experiment as well as the values the random variable Y assigns to the outcomes. From Table 5.13, the probability distribution given in Table 5.14 is determined.

Table 5.13

Outcome	Value of Y
(coin = 1, die = 1)	2
(coin = 1, die = 2)	3
(coin = 1, die = 3)	4
(coin = 1, die = 4)	5
(coin = 1, die = 5)	6
(coin = 1, die = 6)	7
(coin = 2, die = 1)	3
(coin = 2, die = 2)	4
(coin = 2, die = 3)	5
(coin = 2, die = 4)	6
(coin = 2, die = 5)	7
(coin = 2, die = 6)	8

Y = 2 corresponds to one of twelve equally likely outcomes. Therefore P(2) equals $\frac{1}{12}$ = .083. Y = 3 corresponds to two of twelve equally likely outcomes. Therefore P(3) equals $\frac{2}{12}$ = .167. The other probabilities are found in a similar fashion.

Table 5.14

y	2	3	4	5	6	7	8
P(y)	.083	.167	.167	.167	.167	.167	.083

MEAN OF A DISCRETE RANDOM VARIABLE

5.9 A roulette wheel has 18 red, 18 black, and 2 green slots. You bet $10 on red. If red comes up, you get your $10 back plus $10 more. If red does not come up, you lose your $10. Let random variable P represent your profit when playing this roulette wheel. Find the mean value of P.

Ans. The distribution of P is given in table 5.15. The probability of a $10 loss, i.e., P = −10, is $\frac{20}{38}$ = .526, and the probability of a $10 gain is $\frac{18}{38}$ = .474

Table 5.15

p	−10	10
P(p)	.526	.474

The mean profit is $\mu = \Sigma\, xP(x) = -10 \times .526 + 10 \times .474 = -.52$. Your average loss is 52 cents per play of the roulette wheel.

5.10 The distribution of the number of children per household for households receiving Aid to Dependent Children (ADC) in a large eastern city is as follows: Five percent of the ADC households have one child, 35% have 2 children, 30% have 3 children, 20% have 4 children, and 10% have 5 children. Find the mean number of children per ADC household in this city.

Ans. The mean is $\mu = \Sigma xP(x) = 1 \times .05 + 2 \times .35 + 3 \times .30 + 4 \times .20 + 5 \times .10 = 2.95$. The mean is about 3 per household.

STANDARD DEVIATION OF A DISCRETE RANDOM VARIABLE

5.11 Find the standard deviation of the profit when playing the color red on the roulette wheel described in problem 5.9.

Ans. The variance is given by $\Sigma\, x^2 P(x) - \mu^2 = 100 \times .526 + 100 \times .474 - .2704 = 99.7296$, and the standard deviation is $\sqrt{99.7296} = \$9.97$.

5.12 Find the standard deviation of the number of children per ADC household for the distribution given in problem 5.10.

Ans. The variance is given by $\Sigma\, x^2 P(x) - \mu^2 = 1 \times .05 + 4 \times .35 + 9 \times .30 + 16 \times .20 + 25 \times .10 - 2.95^2 = 1.1475$ and the standard deviation is $\sqrt{1.1475} = 1.07$.

BINOMIAL RANDOM VARIABLE

5.13 Ninety percent of the residents of Stanford, California, twenty-five years of age or older have at least a bachelor's degree. Three hundred residents of Stanford twenty-five years or older are selected for a poll concerning higher education. If X represents the number in the 300 who have at least a bachelor's degree, give the conditions necessary for X to be a binomial random variable and identify n, p, and q.

Ans. The main condition we need to be concerned about is that the 300 residents be selected independently. There are 300 identical trials and each trial has only two possible outcomes. We may identify success with having at least a bachelor's degree and failure with having less than a bachelor's degree. If the respondents are chosen randomly and independently, then the probabilities of success and failure should remain constant. Pollsters use standard techniques to ensure independence of individuals selected. The values of n, p, and q are 300, .90, and .10 respectively.

5.14 A box contains 20 items, of which 25% are defective. Three items are randomly selected and X, the number of defectives in the three selected items, is determined. Explain why X is not a binomial random variable with $n = 3$ and $p = .25$.

Ans. The probabilities p and q do not remain constant from trial to trial. Suppose we are interested in the probability that $X = 3$; that is, we are interested in the probability of getting three consecutive defectives. The probability that the first one is defective is .25. The probability that the second is defective after selecting a defective on the first selection is $\frac{4}{19} = .21$, and the probability that the third is defective after selecting defectives on the first two selections is $\frac{3}{18} = .17$. The binomial model assumes that the probability p remains constant at $p = .25$. In this experiment, p does not remain constant, but changes from .25 to .21 to .17.

BINOMIAL PROBABILITY FORMULA

5.15 Approximately 12% of the U.S. population is composed of African-Americans. Assuming that the same percentage is true for telephone ownership, what is the probability that when 25 phone numbers are selected at random for a small survey, that 5 of the numbers belong to an African-American family?

Ans. Let X represent the number of phone numbers in the 25 belonging to African-Americans. Then, X has a binomial distribution with n = 25 and p = .12. The probability P(X = 5) is given as follows:

$$P(X = 3) = \frac{25!}{5!20!}\,(.12)^5\,(.88)^{20} = .1025$$

5.16 It is estimated that 42% of women ages 45 to 54 are overweight. If 20 females between 45 and 54 are randomly selected, what is the probability that one-half of them are overweight?

Ans. Let X represent the number of women in the 20 who are overweight. Then, X has a binomial distribution with n = 20 and p = .42. The probability P(X = 10) is given as follows:

$$P(X = 10) = \frac{20!}{10!10!}\,(.42)^{10}\,(.58)^{10} = .1359$$

TABLES OF THE BINOMIAL DISTRIBUTION

5.17 Sixty percent of teenagers who drink alcohol do so because of peer pressure. Use the table of binomial probabilities to find the probability that in a sample of 15 teenagers who drink, 5 or fewer do so because of peer pressure.

Ans. Table 5.16 shows the portion of the table needed to compute P(X ≤ 5).

$$P(X \leq 5) = .0000 + .0000 + .0003 + .0016 + .0074 + .0212 = .0305$$

Table 5.16

n	x	p = .60
15	0	.0000
	1	.0000
	2	.0003
	3	.0016
	4	.0074
	5	.0212

5.18 A domestic homicide is one in which the victim and the killer are relatives or involved in a relationship. Suppose 40% of all murders are domestic homicides. A criminal justice study randomly selects 10 murder cases for investigation. Use the table of binomial probabilities to find the probability that between one and four inclusive of the murder cases will be domestic homicide cases.

Ans. Table 5.17 shows the portion of the table needed to compute P(1 ≤ X ≤ 4).

$$P(1 \leq X \leq 4) = .0403 + .1209 + .2150 + .2508 = .6270$$

Table 5.17

n	x	p = .40
10	1	.0403
	2	.1209
	3	.2150
	4	.2508

MEAN AND STANDARD DEVIATION OF A BINOMIAL RANDOM VARIABLE

5.19 Seventy-five percent of employed women say their income is essential to support their family. Let X be the number in a sample of 200 employed women who will say their income is essential to support their family. What is the mean and standard deviation of X?

Ans. X is a binomial random variable with n = 200 and p = .75. The mean is $\mu = np = 200 \times .75 = 150$, and the standard deviation is $\sigma = \sqrt{npq} = \sqrt{37.5} = 6.12$.

5.20 A binomial distribution has a mean equal to 8 and a standard deviation equal to 2. Find the values for n and p.

Ans. The following equations must hold: 8 = np and 4 = npq. Substituting 8 for np in the second equation gives 4 = 8q, which gives q = .5. Since p + q = 1, p = 1 − .5 = .5. Substituting .5 for p in the first equation gives n(.5) = 8, and it follows that n = 16.

POISSON RANDOM VARIABLE

5.21 Consider the number of customers arriving at the Grover street branch of Industrial and Federal Savings and Loan during a one-hour interval. What assumptions concerning the arrivals of customers are necessary in order that X, the number of customers arriving in a one hour interval, be a Poisson random variable?

Ans. The assumptions necessary are that the arrivals be random and independent of one another.

5.22 Why are the arrival of patients at a physician's office and the arrival of commercial airplanes at an airport not Poisson random variables?

Ans. Because of appointment times to see a physician, the arrivals are not random. Because of scheduled arrivals of commercial airplanes, the arrivals are not random.

POISSON PROBABILITY FORMULA

5.23 The mean number of patients arriving at the emergency room of University Hospital on Saturday nights between 10:00 and 12:00 is 6.5. Assuming that the patients arrive randomly and independently, what is the probability that on a given Saturday night, 5 or fewer patients arrive at the emergency room between 10:00 and 12:00?

Ans. Let X represent the number of patients to arrive at the emergency room of University Hospital between 10:00 and 12:00. The probability formula for X is $P(x) = \dfrac{6.5^x e^{-6.5}}{x!}$. The probability of the event that $X \le 5$ is P(0) + P(1) + P(2) + P(3) + P(4) + P(5). Each of these probabilities contain the common term $e^{-6.5}$, which may be factored out to give the following as the probability of the event of interest.

$$P(X \le 5) = e^{-6.5}\left(\frac{6.5^0}{0!} + \frac{6.5^1}{1!} + \frac{6.5^2}{2!} + \frac{6.5^3}{3!} + \frac{6.5^4}{4!} + \frac{6.5^5}{5!}\right)$$

$$P(X \le 5) = .0015(1 + 6.5 + 21.125 + 45.7708 + 74.7760 + 96.6809) = .369$$

The solution using Minitab is as follows.

MTB > cdf 5;
SUBC > Poisson 6.5.
Cumulative Distribution Function
Poisson with mu = 6.50000

    x   P(X <= x)
   5.00   **0.3690**

5.24 When working problems involving the Poisson random variable, it is important to remember that the interval for the mean number of occurrences and the interval for X must be equal. If they are not, the mean should be redefined to make them equal. This problem illustrates this important point.

The mean number of patients arriving at the emergency room of University Hospital on Saturday nights between 10:00 and 12:00 is 6.5. Assuming that the patients arrive randomly and independently, what is the probability that on a given Saturday night, 2 or fewer patients arrive at the emergency room between 11:00 and 12:00?

Ans. Let X be the number of patients to arrive at the emergency room of University Hospital on Saturday nights between 11:00 and 12:00. The mean for X is $\frac{6.5}{2} = 3.25$, and the event of interest is $X \le 2$.

$$P(X \le 2) = \frac{e^{-3.25}3.25^0}{0!} + \frac{e^{-3.25}3.25^1}{1!} + \frac{e^{-3.25}3.25^2}{2!} = .369$$

HYPERGEOMETRIC RANDOM VARIABLE

5.25 A box contains 10 red marbles and 90 blue marbles. Five marbles are selected randomly from the 100 in the box. Let X be the number of blue marbles in the five selected marbles. Identify the values for N, k, N – k, and n in the hypergeometric distribution which corresponds to X.

Ans. The total number of marbles is N = 100. The number of blue marbles (successes) is k = 90. The number of red marbles (failures) is N – k = 10. The sample size is n = 5.

5.26 In problem 5.25, consider the event X = 2. This is the event that five marbles are selected and 2 are blue and 3 are red.
(*a*) How many ways may 5 marbles be selected from 100 marbles?
(*b*) How many ways may 2 blue marbles be selected from 90 blue marbles?
(*c*) How many ways may 3 red marbles be selected from 10 red marbles?
(*d*) How many ways may 3 red marbles be selected from 10 red marbles and 2 blue marbles be selected from 90 blue marbles?

Ans. (*a*) $\binom{100}{5} = 75,287,520$ (*b*) $\binom{90}{2} = 4,005$ (*c*) $\binom{10}{3} = 120$

(*d*) $120 \times 4,005 = 480,600$

HYPERGEOMETRIC PROBABILITY FORMULA

5.27 Computec, Inc. manufactures personal computers. There are 40 employees at the Omaha plant and 15 employees at the Lincoln plant. Five employees of Computec, Inc. are randomly selected to fill out a benefits questionnaire.

(*a*) What is the probability that none of the five selected are from the Lincoln plant?

(*b*) What is the probability that all five of the selected employees are from the Lincoln plant?

Ans. (*a*) $\dfrac{\binom{15}{0}\binom{40}{5}}{\binom{55}{5}} = \dfrac{1 \times 658,008}{3,478,761} = .189$ (*b*) $\dfrac{\binom{15}{5}\binom{40}{0}}{\binom{55}{5}} = \dfrac{3,003 \times 1}{3,478,761} = .000863$

5.28 What is the probability of getting 5 face cards when 5 cards are selected from a deck of 52?

Ans. A deck of cards consists of 12 face cards and 40 nonface cards. Let X represent the number of face cards in the 5 cards selected. The event in which we are interested is $X = 5$. The probability is

$$P(X = 5) = \dfrac{\binom{12}{5}\binom{40}{0}}{\binom{52}{5}} = \dfrac{792 \times 1}{2,598,960} = .000305$$

Supplementary Problems

RANDOM VARIABLE

5.29 A taste test is conducted involving 35 individuals. Random variable X is the number in the 35 who prefer a locally produced nonalcoholic beer to a national brand. What are the possible values for X?

Ans. The whole numbers 0 through 35

5.30 A psychological experiment was conducted in which the time to traverse a maze was recorded for each of five dogs. The times were 4, 6, 8, 9, and 12 minutes. Two of the times were randomly selected and the difference X = largest of the pair − smallest of the pair was recorded. Give all possible pairs of possible selections, and the value of X for each outcome.

Ans. See Table 5.18.

Table 5.18

Outcome	Value of X
(4, 6)	2
(4, 8)	4
(4, 9)	5
(4, 12)	8
(6, 8)	2
(6, 9)	3
(6, 12)	6
(8, 9)	1
(8, 12)	4
(9, 12)	3

DISCRETE RANDOM VARIABLE

5.31 A die is tossed until the face 6 turns up. Let X be the number of tosses needed until the face 6 first turns up. Give the possible values for the variable X.

   Ans.   The possible values are the positive integers. That is, the possible values are 1, 2, 3, ...

5.32 Identify the discrete random variables in parts (*a*) through (*e*).
   (*a*) The number of arrests during a 10-day period during which the police apply a Zero-tolerance strategy
   (*b*) The time workers have spent with their current employer
   (*c*) The number of nurse practitioners per state
   (*d*) The career lifetime of major league baseball players
   (*e*) The number of executions of death row inmates per year in the U.S.

   Ans.   (*a*), (*c*), and (*e*)

CONTINUOUS RANDOM VARIABLE

5.33 Identify the continuous random variables in the following list.
   (*a*) Weight of individuals in kg
   (*b*) Serum cholesterol level in mg/dl
   (*c*) Length of intravenous therapy in hours
   (*d*) Body mass index in kg/m^2
   (*e*) Cardiac output in liters/minute

   Ans.   All five are continuous.

5.34 What is the primary difference between a discrete random variable and a continuous random variable?

   Ans.   There are values between the possible values of a discrete random variable which are not possible values for the random variable. This is not generally true for a continuous random variable.

PROBABILITY DISTRIBUTION

5.35 Suppose Table 5.19 gives the number in thousands of students in grades 9 through 12 for public schools in the United States. Let X represent the grade level. Give the probability distribution for X.

Table 5.19

Grade	9	10	11	12
Frequency	3,525	3,475	3,050	2,950

   Ans.   The distribution is given in Table 5.20.

Table 5.20

x	9	10	11	12
P(x)	.271	.267	.245	.227

5.36 Which of the following are probability distributions? For those which are not, tell why they are not.
   (*a*)

x	−4	−1	0	3	8
P(x)	.1	.2	.4	.3	.2

(b)

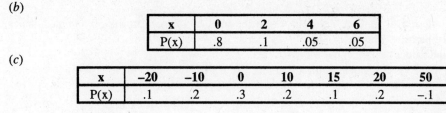

x	0	2	4	6
P(x)	.8	.1	.05	.05

(c)

x	–20	–10	0	10	15	20	50
P(x)	.1	.2	.3	.2	.1	.2	–.1

(d) $P(x) = .2x^2$, x = 1, 2

Ans.   Parts (b) and (d) are probability distributions. Part (a) is not because $\Sigma P(x) = 1.2$. Part (c) is not $P(50) < 0$.

MEAN OF A DISCRETE RANDOM VARIABLE

5.37 The number of personal computers per household in the United States has the probability distribution shown in Table 5.21. Find the mean number of personal computers per household.

Table 5.21

x	0	1	2
P(x)	.30	.55	.15

Ans. $\mu = 0.85$

5.38 Based on a large survey, the distribution in Table 5.22 was found for the number of pounds individuals desired to lose. Find the mean number of pounds they desired to lose.

Table 5.22

pounds	0	5	15	25	50
percent	33	21	11	25	10

Ans. $\mu = 13.95$

STANDARD DEVIATION OF A DISCRETE RANDOM VARIABLE

5.39 Find the standard deviation of the number of personal computers per household for the distribution given in Table 5.21.

Ans. $\sigma = 0.65$

5.40 Find the standard deviation of the number of pounds desired to be lost.

Ans. $\sigma = 15.55$

BINOMIAL RANDOM VARIABLE

5.41 Thirty percent of the trees in a national forest are infested with a parasite. Fifty trees are randomly selected from this forest and X is defined to equal the number of trees in the 50 sampled that are infested with the parasite. The infestation is uniformly spread throughout the forest. Identify the values for n, p, and q.

Ans.   n = 50, p = .30, and q = .70

5.42 Suppose in problem 5.41 that we define Y to be the number of trees in the 50 sampled that are not infested with the parasite. Then Y is a binomial random variable.
(a) What are the values of n, p, and q for Y?
(b) The event X = 20 is equivalent to the event that Y = a. Find the value for a.

Ans.   (a) n = 50, p = .70, and q = .30
         (b) a = 30

BINOMIAL PROBABILITY FORMULA

5.43 The Dallas–Fort Worth Airport claims that 85% of their flights are on time. If the claim is correct, what is the probability that in a sample of 20 flights at the Dallas–Fort Worth Airport that 15 or more of the sample flights are on time?

Ans.   .9327

5.44 A psychological study involving the troops in the Bosnia peacekeeping force was conducted. If 12 percent of the 21,496 troops are females, what is the probability that in a sample of 50 randomly selected individuals that five or fewer are female?

Ans.   .4353

TABLES OF THE BINOMIAL DISTRIBUTION

5.45 There are approximately 3,000 inmates on death row. Forty percent of the death row inmates are African-American. Twenty of the death row inmates are randomly selected for a sociological study. Use the table of binomial probabilities to find the probability that most of the selected inmates are African-American.

Ans.   .1275

5.46 Ten percent of the Rentwheels car-rental fleet are equipped with cellular phones. If five of the cars are randomly selected, what is the probability that none are equipped with a cellular phone?

Ans.   .5905

MEAN AND STANDARD DEVIATION OF A BINOMIAL RANDOM VARIABLE

5.47 It is conjectured that 60% of the deaths from melanoma can be prevented by a skin self-exam. If this conjecture is correct, how many of the 7,000 deaths due to this skin cancer would be prevented per year on the average? What is the standard deviation associated with the number of deaths prevented?

Ans.   4,200   41

5.48 Fifteen percent of the machinery and equipment at businesses is more than 10 years old. In a randomly selected sample of 35 businesses, how many would you expect to have machinery or equipment that is more than 10 years old? What standard deviation is associated with this expected number?

Ans.   5.25   2.11

POISSON RANDOM VARIABLE

5.49 Give four examples of a Poisson random variable.

   Ans. 1. The number of telephone calls received per hour by an office
2. The number of keyboard errors per page made by an individual using a word processor
3. The number of bacteria in a given culture
4. The number of imperfections per yard in a roll of fabric

5.50 What are the potential values of a Poisson random variable?

   Ans. 0, 1, 2, . . .

POISSON PROBABILITY FORMULA

5.51 Suppose the mean number of earthquakes per year is 13. What is the probability of 20 or more earthquakes in a given year?

   Ans. .0427

5.52 The number of plankton in a liter of lake water has a mean value of 7. What is the probability that the number of plankton in a given liter is within one standard deviation of the mean?

   Ans. .6575

HYPERGEOMETRIC RANDOM VARIABLE

5.53 Since 1977 twenty-four states have executed at least one death row inmate. In a study concerning capital punishment, ten of the fifty states are randomly selected. Let X represent the number of states in the ten that have executed at least one death row inmate. Identify success and failure. What are the values of N, k, N – k, and n.

   Ans. Success is that at least one death row inmate has been executed since 1977. Failure is that no execution has occurred in the state since 1977. N = 50, k = 24, N – k = 26, and n = 10.

5.54 If success and failure are interchanged in problem 5.53, how is X changed and what are the values of N, k, N – k, and n for X?

   Ans. X is the number of states in the ten that have executed no one since 1977. N = 50, k = 26, N – k = 24, and n = 10.

HYPERGEOMETRIC PROBABILITY FORMULA

5.55 Thirty diabetics have volunteered for a medical study. Ten of the diabetics have high blood pressure. Five are selected for a preliminary screening for the study. What is the probability that none of the five selected have high blood pressure if the selection is done randomly?

   Ans. .1088

5.56 A box of manufactured products contains 20 items. Three of the items are defective. Let X represent the number of defectives in three randomly selected from the box. Give the probability distribution for X and find the mean value for X using the probability distribution. Show that the mean is also given by $\mu = \dfrac{nk}{N}$.

Ans.

x	0	1	2	3
P(x)	.596491	.357895	.044737	.000877

$\mu = .45$

Chapter 6

Continuous Random Variables and Their Probability Distributions

UNIFORM PROBABILITY DISTRIBUTION

A *continuous random variable* is a random variable capable of assuming all the values in an interval or several intervals of real numbers. Because of the uncountable number of possible values, it is not possible to list all the values and their probabilities for a continuous random variable in a table as is true with a discrete random variable. The probability distribution for a continuous random variable is represented as the area under a curve called the *probability density function,* abbreviated *pdf. A pdf is characterized by the following two basic properties: The graph of the pdf is never below the x axis and the total area under the pdf always equals 1.*

The probability density function shown in Fig. 6-1 is a *uniform probability distribution.* This pdf represents the distribution of flight times between Omaha, Nebraska, and Memphis, Tennessee. The flight time is represented by the letter X. The graph shows that the flight times range from 90 to 100 minutes. The distance from the x axis to the graph remains constant at 0.10 and since the area of a rectangle is given by the length times the width, the area under the pdf is $10 \times 0.10 = 1$. Note that this pdf has the two basic properties given above. The graph of the pdf is never below the x axis and the total area under the pdf is equal to 1.

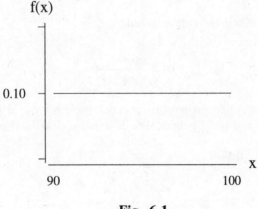

Fig. 6-1

The representation of Fig. 6-1 by an equation is given as follows.

$$f(x) = \begin{cases} .1 & 90 < x < 100 \\ 0 & \text{elsewhere} \end{cases}$$

In general, if a random variable X is uniformly distributed over the interval from a to b, then the pdf is given by formula (*6.1*).

$$f(x) = \begin{cases} \dfrac{1}{(b-a)} & a < x < b \\ 0 & \text{elsewhere} \end{cases} \qquad (6.1)$$

EXAMPLE 6.1 The probability that a flight takes between 92 and 97 minutes is represented as $P(92 < X < 97)$ and is equal to the shaded area shown in Fig. 6-2. The rectangular shaded area has a width equal to 5 and a length equal to .1 and the area is equal to $5 \times .1 = .5$. That is, 50 percent of the flights between Omaha and Memphis will take between 92 and 97 minutes.

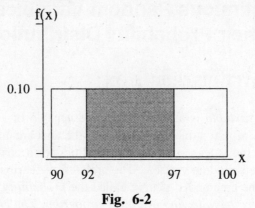

Fig. 6-2

MEAN AND STANDARD DEVIATION FOR THE UNIFORM PROBABILITY DISTRIBUTION

The mean value for a random variable having a uniform probability distribution over the interval from a to b is given by

$$\mu = \frac{a + b}{2} \qquad (6.2)$$

The variance for a uniform random variable is given by

$$\sigma^2 = \frac{(b - a)^2}{12} \qquad (6.3)$$

EXAMPLE 6.2 The weights of 10-pound bags of potatoes packaged by Idaho Farms Inc. are uniformly distributed between 9.75 pounds and 10.75 pounds. The distribution of weights for these bags is shown in Fig. 6-3.

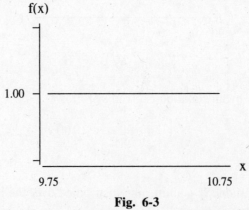

Fig. 6-3

Using formula (6.2), we see that the mean weight per bag is $\mu = \dfrac{a + b}{2} = \dfrac{9.75 + 10.75}{2} = 10.25$ pounds and using

formula (6.3), the standard deviation is $\sigma = \sqrt{\dfrac{(b - a)^2}{12}} = \sqrt{\dfrac{1}{12}} = 0.29$. If X represents the weight per bag, then

P(X > 10.0) corresponds to the proportion of bags that weigh more than 10 pounds. This probability is shown in Fig. 6-4. The shaded rectangle has dimensions 1 and 0.75 and the area is $1 \times 0.75 = 0.75$. Seventy-five percent of the bags weigh more than 10 pounds. To help ensure consumer satisfaction, Idaho Farms instructs their employees to try and never underfill if possible. This is reflected in the average weight of 10.25 pounds per bag.

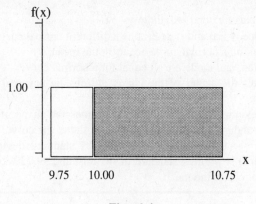

Fig. 6-4

The probability associated with a single value for a continuous random variable is always equal to zero, since there is no area associated with a single point. That is, the probability that X = a is given by

$$P(X = a) = 0 \qquad\qquad (6.4)$$

NORMAL PROBABILITY DISTRIBUTION

The most important and widely used of all continuous distributions is the *normal probability distribution*. Figure 6-5 shows the pdf for a normal distribution having mean μ and standard deviation σ.

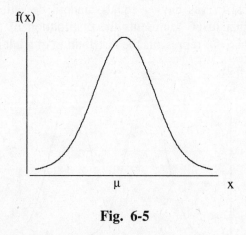

Fig. 6-5

Table 6.1 gives some of the main properties of the normal curve shown in Fig. 6-5. Figure 6-6 shows two different normal curves with the same mean but different standard deviations. The larger the standard deviation, the more disperse are the values about the mean.

Table 6.1

Properties of the Normal Probability Distribution
1. The total area under the normal curve is equal to one.
2. The curve is symmetric about μ and the area under the curve on each side of the mean equals 0.5.
3. The tails of the curve extend indefinitely.
4. Each pair of values for μ and σ determine a different normal curve.
5. The highest point on a normal curve occurs at the mean.
6. The mean, median, and mode are all equal for a normal curve.
7. The mean locates the center of the curve and can be any real number, negative, positive, or zero.
8. The standard deviation is positive, and determines the shape of the normal curve. The larger the standard deviation, the wider and flatter the curve.
9. 68.26% of the area under the curve is within 1 standard deviation of the mean, 95.44% of the area is within 2 standard deviations, and 99.72% of the area is within 3 standard deviations of the mean.

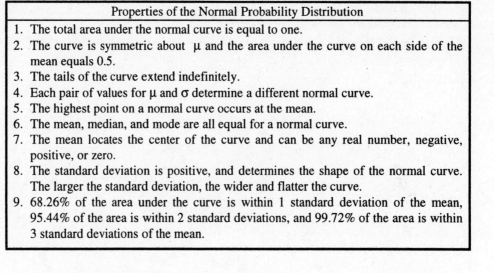

Fig. 6-6

Figure 6-7 shows two normal curves having equal standard deviations but different means. The normal curve with mean equal to 66 represents the distribution of adult female heights and the normal curve with mean equal to 70 represents the distribution of adult male heights.

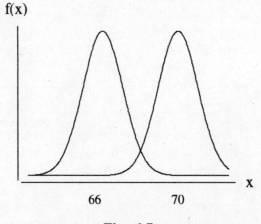

Fig. 6-7

The equation of the pdf of a normal curve having mean μ and standard deviation σ is given in formula (6.5).

$$f(x) = \frac{1}{\sqrt{2\pi}\sigma} e^{-(x-\mu)^2/2\sigma^2} \qquad (6.5)$$

The two constants in the formula, e and π, are important constants in mathematics. The constant e is equal to 2.718281828 . . . and the constant π is equal to 3.141592654 . . . . The equation of the pdf for the normal curve is not used a great deal in practice and is given here for the sake of completeness.

STANDARD NORMAL DISTRIBUTION

The *standard normal distribution* is the normal distribution having mean equal to 0 and standard deviation equal to 1. The letter Z is used to represent the standard normal random variable. The standard normal curve is shown in Fig. 6-8. The curve is centered at the mean, 0, and the z-axis is labeled in standard deviations above and below the mean.

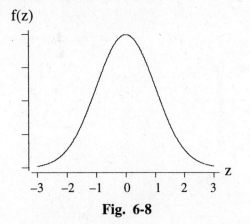

Fig. 6-8

Appendix 2 contains the *standard normal distribution table*. This table gives areas under the standard normal curve for the variable Z ranging from 0 to a positive number z. Some examples will now be given to illustrate how to use this table.

EXAMPLE 6.3 Table 6.2 illustrates how to use the standard normal distribution table to find the area under the standard normal curve between z = 0 and z = 1.65. Figure 6-9 shows the corresponding area as the shaded region under the curve. The value 1.65 may be written as 1.6 + .05, and by locating 1.6 under the column labeled z and then moving to the right of 1.6 until you come under the .05 column you find the area .4505. This is the area shown in Fig. 6-9. We express this area as P(0 < Z < 1.65) = .4505.

Table 6.2

z	.00	.01	...	.05	...	.09
0.0	.0000	.0040	...	.0199	...	.0359
0.1	.0398	.0438	...	.0596	...	.0753
0.2	.0793	.0832	...	.0987	...	.1141
.	.	.	...		...	.
.	.	.	...		...	.
.	.	.	...		...	.
1.6	.4452	.4463	...	**.4505**	...	.4545
.	.	.	...	...	...	.
.	.	.	...	...	...	.
.	.	.	...	...	...	.
3.0	.4987	.4987	...	.4989	...	.4990

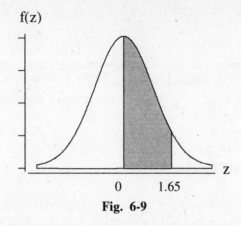

Fig. 6-9

EXAMPLE 6.4 The area under the standard normal curve between $z = -1.65$ and $z = 0$ is represented as $P(-1.65 < Z < 0)$ and is shown in Fig. 6-10. By symmetry, the following probabilities are equal.

$$P(-1.65 < Z < 0) = P(0 < Z < 1.65)$$

From Example 6.3, we know that $P(0 < Z < 1.65) = .4505$ and therefore, $P(-1.65 < Z < 0) = .4505$.

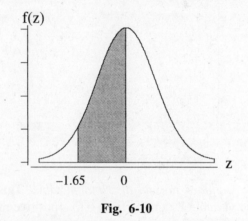

Fig. 6-10

EXAMPLE 6.5 The area under the standard normal curve between $z = -1.65$ and $z = 1.65$ is represented by $P(-1.65 < Z < 1.65)$ and is shown in Fig. 6-11. The probability $P(-1.65 < Z < 1.65)$ is expressible as

$$P(-1.65 < Z < 1.65) = P(-1.65 < Z < 0) + P(0 < Z < 1.65)$$

The probabilities on the right side of the above equation are given in Examples 6.3 and 6.4, and their sum is equal to 0.9010. Therefore, $P(-1.65 < Z < 1.65) = .9010$.

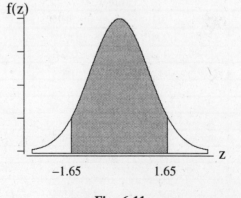

Fig. 6-11

EXAMPLE 6.6 The probability of the event Z < 1.96 is represented by P(Z < 1.96) and is shown in Fig. 6-12.

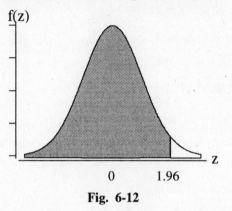

Fig. 6-12

The area shown in Fig. 6-12 is partitioned into two parts as shown in Fig. 6-13. The darker of the two areas is equal to P(Z < 0) = .5, since it is one-half of the total area. The lighter of the two areas is found in the standard normal distribution table to be .4750. The sum of the two areas is .5 + .4750 = .9750. Summarizing,

$$P(Z < 1.96) = P(Z < 0) + P(0 < Z < 1.96) = .5 + .4750 = .9750$$

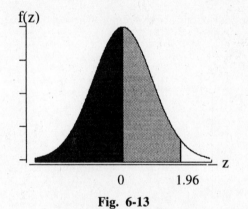

Fig. 6-13

The probability in Example 6.6 can also be found by using CDF of Minitab as follows:

MTB > cdf 1.96;
SUBC > normal 0 1.
Cumulative Distribution Function
Normal with mean = 0 and standard deviation = 1.00000

```
    x      P(X <= x)
1.9600    0.9750
```

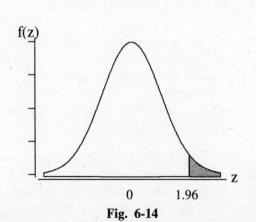

Fig. 6-14

In order to find the probability P(Z > 1.96), shown in Fig. 6-14, we use the complement of the event Z > 1.96 and the result for P(Z < 1.96) given above as follows.

$$P(Z > 1.96) = 1 - P(Z < 1.96) = 1 - .9750 = .0250$$

STANDARDIZING A NORMAL DISTRIBUTION

In order to find areas under a normal distribution having mean μ and standard deviation σ, the normal distribution must be *standardized*. A normal random variable X having mean μ and standard deviation σ is converted or transformed to a standard normal random variable by the formula given in (6.6).

$$Z = \frac{X - \mu}{\sigma} \tag{6.6}$$

EXAMPLE 6.7 Figure 6-15 shows the distribution of adult male weights for a particular age group. The weights, X, are normally distributed with mean 170 pounds and standard deviation equal to 15 pounds. The standardized value for the weight X = 215 is $z = \frac{x - \mu}{\sigma} = \frac{215 - 170}{15} = 3$. For this age group, an individual who weighs 215 pounds is 3 standard deviations above average. The standardized value for 170 pounds is zero, and the standardized value for 125 pounds is –3.

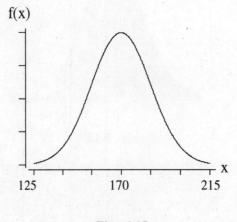

Fig. 6-15

APPLICATIONS OF THE NORMAL DISTRIBUTION

The fact that many real-world phenomena are normally distributed leads to numerous applications of the normal distribution. Applications of the normal distribution usually involve finding areas under a normal curve. *To find the area between two values of x for a normal distribution, first convert both values of x to their respective z values. Then find the area under the standard normal curve between those two z values. The area between the two z values gives the area between the corresponding x values.*

EXAMPLE 6.8 In a study involving stress-induced blood pressure, volunteers played a computer game called the color-word interference task. The game was set so that everyone made errors about 17% of the time. The average increase in systolic blood pressure was 10 points of systolic pressure, and the standard deviation was 3 points. The percent experiencing an increase of 16 points or more is found by evaluating P(X > 16) and multiplying by 100. The probability is shown as the shaded area in Fig. 6-16. The z value corresponding to x = 16 is $z = \dfrac{16-10}{3} = 2$. The area shown in Fig. 6-16 is the same as the area shown in Fig. 6-17.

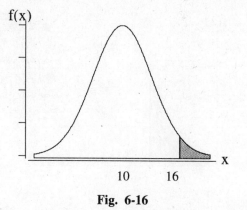

Fig. 6-16

The equality of the areas in Figures 6-16 and 6-17 is expressible in terms of probability as follows:

$$P(X > 16) = P(Z > 2)$$

Using the standard normal distribution table, we find that $P(Z > 2) = .5 - .4772 = .0228$. The percent experiencing an increase of 16 systolic points or more is 2.28%.

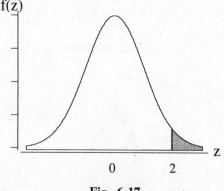

Fig. 6-17

EXAMPLE 6.9 The time between release from prison and conviction for another crime for individuals under 40 is normally distributed with a mean equal to 30 months and a standard deviation equal to 6 months. The percentage of these individuals convicted for another crime within two years of their release from prison is represented as $P(X < 24)$ times 100. The probability is shown as the shaded area in Fig. 6-18. The event $X < 24$ is equivalent to $Z = \dfrac{X-30}{6} < \dfrac{24-30}{6} = -1$. The probability $P(Z < -1)$ is shown as the shaded area in Fig. 6-19.

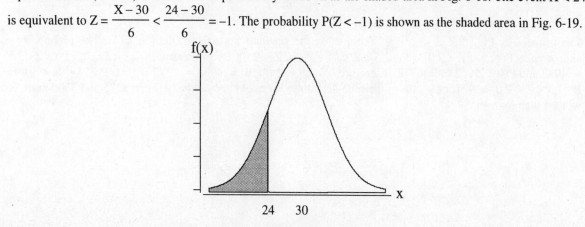

Fig. 6-18

The shaded area shown in Fig. 6-19 is $P(Z < -1)$, and by symmetry $P(Z < -1) = P(Z > 1)$. The probability $P(Z > 1)$ is found using the standard normal distribution table as $.5 - .3413 = .1587$. That is $P(X < 24) = P(Z < -1) = .1587$ or 15.87% commit a crime within two years of their release.

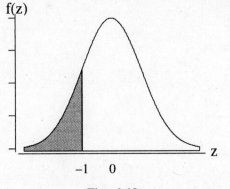

Fig. 6-19

EXAMPLE 6.10 A study determined that the difference between the price quoted to women and men for used cars is normally distributed with mean \$400 and standard deviation \$50. To clarify, let X be the amount quoted to a woman minus the amount quoted to a man for a given used car. Then, the population distribution for X is normal with $\mu = \$400$ and $\sigma = \$50$. The percent of the time that quotes for used cars are \$275 to \$500 more for women than men is given by $P(275 < X < 500)$ times 100. The probability $P(275 < X < 500)$ is shown as the shaded area in Fig. 6-20. The Z value corresponding to X = 275 is $z = \dfrac{275 - 400}{50} = -2.5$ and the Z value corresponding to X = 500 is $z = \dfrac{500 - 400}{50} = 2.0$. The probability $P(-2.5 < Z < 2.0)$ is shown in Fig. 6-21.

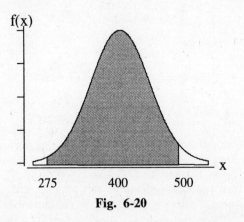

Fig. 6-20

The probability $P(-2.5 < Z < 2.0)$ is found using the standard normal distribution table. The probability is expressed as $P(-2.5 < Z < 2.0) = P(-2.5 < Z < 0) + P(0 < Z < 2.0)$. By symmetry, $P(-2.5 < Z < 0) = P(0 < Z < 2.5) = .4938$ and $P(0 < Z < 2.0) = .4772$. Therefore, $P(-2.5 < Z < 2.0) = .4938 + .4772 = .971$. $P(275 < X < 500) = P(-2.5 < Z < 2.0) = .971$, or 97.1% of the time the quotes for used cars will be between \$275 and \$500 more for women than for men.

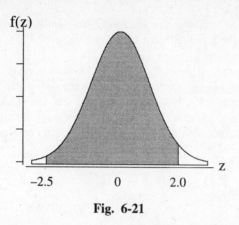

Fig. 6-21

DETERMINING THE Z AND X VALUES WHEN AN AREA UNDER THE NORMAL CURVE IS KNOWN

In many applications, we are concerned with finding the z value or the x value when the area under a normal curve is known. The following examples illustrate the techniques for solving these type problems.

EXAMPLE 6.11 Find the positive value z such that the area under the standard normal curve between 0 and z is .4951. Figure 6-22 shows the area and the location of z on the horizontal axis. Table 6.3 gives the portion of the standard normal distribution table needed to find the z value.

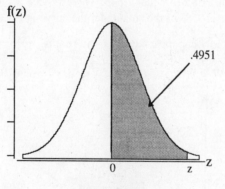

Fig. 6-22

We search the interior of the table until we find .4951, the area we are given. This area is shown in bold print in Table 6.3. By going to the beginning of the row and top of the column in which .4951 resides, we see that the value for z is 2.58.

Table 6.3

z	.00	.01	...	.05	...	.08
0.0	.0000	.0040	....	.0199	....	.0319
0.1	.0398	.0438	...	.0596	....	.0714
0.2	.0793	.0832	...	.0987	....	.1103
.	.	.	...		...	.
.	.	.	...		...	.
.	.	.	...		...	.
2.5	.4938	.4940	...	.4946	...	**.4951**
.	.	.	...	...	...	.
.	.	.	...	...	...	.
.	.	.	...	...	...	.
3.0	.4987	.4987	...	.4989	...	.4990

To find the value of negative z such that the area under the standard normal curve between 0 and z equals .4951, we find the value for positive z as above and then take z to be –2.58.

EXAMPLE 6.12 The mean number of passengers that fly per day is equal to 1.75 million and the standard deviation is 0.25 million per day. If the number of passengers flying per day is normally distributed, the distribution and ninety-fifth percentile, P_{95}, is as shown in Fig. 6-23. The shaded area is equal to 0.95. We have a known area and need to find P_{95}, a value of X. Since we must always use the standard normal tables to solve problems involving any normal curve, we first draw a standard normal curve corresponding to Fig. 6-23. This is shown in Fig. 6-24.

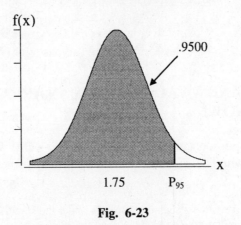

Fig. 6-23

Using the technique shown in Example 6.11, we find the area .4500 in the interior of the standard normal distribution table and find the value of z to equal 1.645. There is 95% of the area under the standard normal curve to the left of 1.645 and there is 95% of the area under the curve in Fig. 6-23 to the left of P_{95}. Therefore if P_{95} is standardized, the standardized value must equal 1.645. That is, $\dfrac{P_{95} - 1.75}{.25} = 1.645$ and solving for P_{95} we find $P_{95} = 1.75 + .25 \times 1.645 = 2.16$ million.

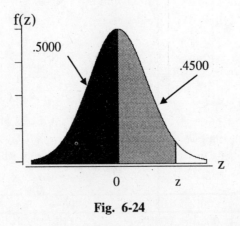

Fig. 6-24

The value of z in Fig. 6-24 can also be found by using INVCDF of Minitab as follows.

```
MTB > invcdf .95;
SUBC > normal 0 1.
Inverse Cumulative Distribution Function
Normal with mean = 0 and standard deviation = 1.00000

  P(X <= x)        x
   0.9500       1.6449
```

NORMAL APPROXIMATION TO THE BINOMIAL DISTRIBUTION

Figure 6-25 shows the binomial distribution for X, the number of heads to occur in 10 tosses of a coin. This distribution is shown as the shaded area under the histogram-shaped figure. Superimposed upon this binomial distribution is a normal curve. The mean of the binomial distribution is $\mu = np = 10 \times .5 = 5$ and the standard deviation of the binomial distribution is $\sigma = \sqrt{npq} = \sqrt{10 \times .5 \times .5} = \sqrt{2.5} = 1.58$. The normal curve, which is shown, has the same mean and standard deviation as the binomial distribution. When a normal curve is fit to a binomial distribution in this manner, this is called the *normal approximation to the binomial distribution*. The shaded area under the binomial distribution is equal to one and so is the total area under the normal curve.

The normal approximation to the binomial distribution is appropriate whenever $np \geq 5$ and $nq \geq 5$.

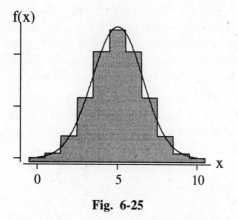

Fig. 6-25

EXAMPLE 6.13 Using the table of binomial probabilities, the probability of 4 to 6 heads inclusive is as follows: $P(4 \leq X \leq 6) = .2051 + .2461 + .2051 = .6563$. This is the shaded area shown in Fig. 6-26.

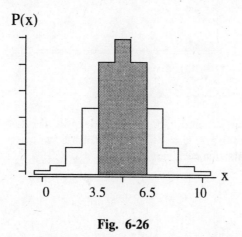

Fig. 6-26

The normal approximation to this area is shown in Fig. 6-27. To account for the area of all three rectangles, note that X must go from 3.5 to 6.5. The area under the normal curve for X between 3.5 and 6.5 is found by determining the area under the standard normal curve for Z between $z = \dfrac{3.5 - 5}{1.58} = -.95$ and $z = \dfrac{6.5 - 5}{1.58} = .95$. This area is $2 \times .3289 = .6578$. Note that the approximation, .6578, is extremely close to the exact answer, .6563.

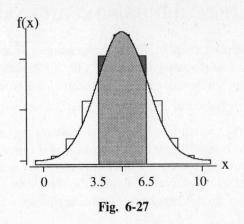

Fig. 6-27

EXAMPLE 6.14 To aid in the early detection of breast cancer women are urged to perform a self-exam monthly. Thirty-eight percent of American women perform this exam monthly. In a sample of 315 women, the probability that 100 or fewer perform the exam monthly is $P(X \leq 100)$. Even Minitab has difficulty performing this extremely difficult computation involving binomial probabilities. However, this probability is quite easy to approximate using the normal approximation. The mean value for X is $\mu = np = 315 \times .38 = 119.7$ and the standard deviation is $\sigma = \sqrt{npq} = \sqrt{315 \times .38 \times .62} = 8.61$. A normal curve is constructed with mean 119.7 and standard deviation 8.61. In order to cover all the area associated with the rectangle at $X = 100$ and all those less than 100, the normal curve area associated with x less than 100.5 is found. Figure 6-28 shows the normal curve area we need to find.

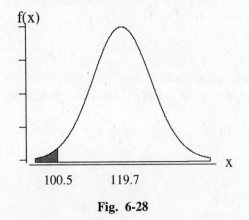

Fig. 6-28

The corresponding area under the standard normal curve is shown in Fig. 6-29. The area under the standard normal curve for $Z < -2.23$ is $.5000 - .4871 = .0129$. The probability is extremely small that 100 or fewer in the 315 will perform the breast examination each month.

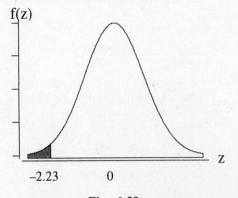

Fig. 6-29

EXPONENTIAL PROBABILITY DISTRIBUTION

The *exponential probability distribution* is a continuous probability distribution that is useful in describing the time it takes to complete some task. The pdf for an exponential probability distribution is given by formula (*6.7*), where μ is the mean of the probability distribution and e = 2.71828 to five decimal places.

$$f(x) = \frac{1}{\mu} e^{-x/\mu} \quad \text{for } x \geq 0 \tag{6.7}$$

The graph for the pdf of a typical exponential distribution is shown in Fig. 6-30.

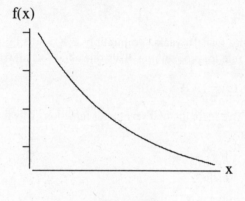

Fig. 6-30

EXAMPLE 6.15 The exponential random variable can be used to describe the following characteristics: the time between logins on the internet, the time between arrests for convicted felons, the lifetimes of electronic devices, and the shelf life of fat free chips.

PROBABILITIES FOR THE EXPONENTIAL PROBABILITY DISTRIBUTION

A standardized table of probabilities does not exist for the exponential distribution and to find areas under the exponential distribution curve requires the use of calculus. However, formula (*6.8*) is useful in solving many problems involving the exponential distribution.

$$P(X \leq a) = 1 - e^{-a/\mu} \tag{6.8}$$

The area corresponding to the probability given in formula (*6.8*) is shown as the shaded area in Fig. 6-31.

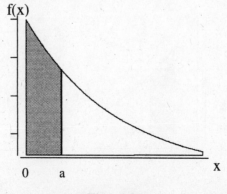

Fig. 6-31

EXAMPLE 6.16 Suppose the time till death after infection with HIV, the AIDS virus, is exponentially distributed with mean equal to 8 years. If X represents the time till death after infection with HIV, then the percent who die within five years after infection with HIV is found by multiplying P(X ≤ 5) by 100. The probability is found as follows: $P(X \leq 5) = 1 - e^{-.625} = 1 - .535 = .465$. Using CDF of Minitab, we have the following as an alternative solution.

MTB > cdf 5;
SUBC > exponential 8.
Cumulative Distribution Function
Exponential with mean = 8.00000

        x   P(X <= x)
    5.0000   0.4647

To find the percent who live more than 10 years, we multiply P(X > 10) by 100. In order to utilize formula (6.8), we use the complementary rule for probabilities. This rule allows us to write P(X > 10) as follows:

$$P(X > 10) = 1 - P(X \leq 10) = 1 - (1 - e^{-1.25}) = e^{-1.25} = .287$$

That is, 28.7% of the individuals live more than 10 years after infection. This probability is shown as the shaded area in Fig. 6-32.

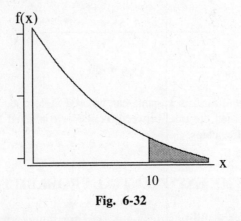

Fig. 6-32

To find the percent who live between 2 and 4 years after infection, we multiply P(2 < X < 4) by 100. To use formula (6.8) to find this probability, we express P(2 < X < 4) as follows:

$$P(2 < X < 4) = P(X < 4) - P(X < 2) = (1 - e^{-.5}) - (1 - e^{-.25}) = e^{-.25} - e^{-.5} = .172$$

That is, 17.2% live between 2 and 4 years after infection. This probability is shown as the shaded area in Fig. 6-33.

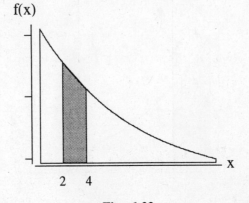

Fig. 6-33

Solved Problems

UNIFORM PROBABILITY DISTRIBUTION

6.1   The price for a gallon of whole milk is uniformly distributed between $2.25 and $2.75 during July in the U.S. Give the equation and graph the pdf for X, the price per gallon of whole milk during July. Also determine the percent of stores that charge more than $2.70 per gallon.

Ans.   The equation of the pdf is $f(x) = \begin{cases} 2 & 2.25 < x < 2.75 \\ 0 & \text{elsewhere} \end{cases}$. The graph is shown in Fig. 6-34.

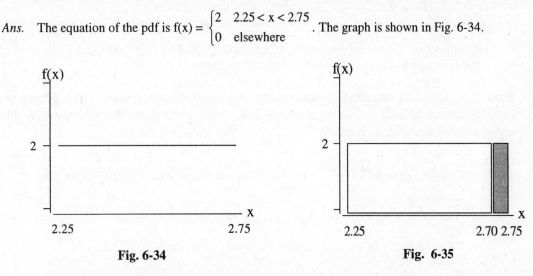

Fig. 6-34   Fig. 6-35

The percent of stores charging a higher price than $2.70 is P(X > 2.70) times 100. The probability P(X > 2.70) is the shaded area in Fig. 6-35. This area is $2 \times .05 = .10$. Ten percent of all milk outlets sell a gallon of milk for more than $2.70.

6.2   The time between release from prison and the commission of another crime is uniformly distributed between 0 and 5 years for a high-risk group. Give the equation and graph the pdf for X, the time between release and the commission of another crime for this group. What percent of this group will commit another crime within two years of their release from prison?

Ans.   The equation of the pdf is $f(x) = \begin{cases} .2 & 0 < x < 5 \\ 0 & \text{elsewhere} \end{cases}$. The graph of the pdf is shown in Fig. 6-36. The

percent who commit another crime within two years is given by P(X < 2) times 100. This probability is shown as the shaded area in Fig. 6-37, and is equal to $2 \times .2 = .4$. Forty percent will commit another crime within two years.

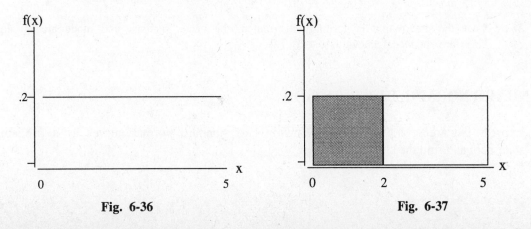

Fig. 6-36   Fig. 6-37

MEAN AND STANDARD DEVIATION FOR THE UNIFORM PROBABILITY DISTRIBUTION

6.3   Find the mean and standard deviation of the milk prices in problem 6.1. What percent of the prices are within one standard deviation of the mean?

Ans.   The mean is given by $\mu = \dfrac{a+b}{2} = \dfrac{2.25+2.75}{2} = 2.50$ and the standard deviation is given by

$\sqrt{\dfrac{(b-a)^2}{12}} = \sqrt{\dfrac{(2.75-2.25)^2}{12}} = .144$. The one standard deviation interval about the mean goes

from 2.36 to 2.64 and the probability of the interval is $(2.64 - 2.36) \times 2 = .56$. Fifty-six percent of the prices are within one standard deviation of the mean.

6.4   Find the mean and standard deviation of the times between release from prison and the commission of another crime in problem 6.2. What percent of the times are within two standard deviations of the mean?

Ans.   The mean is given by $\mu = \dfrac{a+b}{2} = \dfrac{0+5}{2} = 2.50$ and the standard deviation is given by $\sqrt{\dfrac{(b-a)^2}{12}}$

$= \sqrt{\dfrac{(5-0)^2}{12}} = 1.44$. A 2 standard deviation interval about the mean goes from $-.38$ to 5.38 and

100% of the times are within 2 standard deviations of the mean.

NORMAL PROBABILITY DISTRIBUTION

6.5   The mean net worth of all Hispanic individuals aged 51–61 in the U.S. is $80,000, and the standard deviation of the net worths of such individuals is $20,000. If the net worths are normally distributed, what percent have net worths between: (*a*) $60,000 and $100,000; (*b*) $40,000 and $120,000; (*c*) $20,000 and $140,000?

Ans.   (*a*)   68.26% have net worths between $60,000 and $100,000.
          (*b*)   95.44% have net worths between $40,000 and $120,000.
          (*c*)   99.72% have net worths between $20,000 and $140,000.

6.6   If the median amount of money that parents in the age group 51–61 gave a child in the last year is $1,725 and the amount that parents in this age group give a child is normally distributed, what is the modal amount that parents in this age group give a child?

Ans.   Since the distribution is normally distributed, the mean, median, and mode are all equal. Therefore, the modal amount is also $1,725.

STANDARD NORMAL DISTRIBUTION

6.7   Express the areas shown in the following two standard normal curves as a probability statement and find the area of each one.

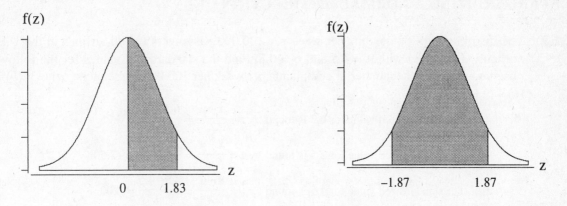

Ans. The area under the curve on the left is represented as P(0 < Z < 1.83) and from the standard normal distribution table is equal to .4664. The area under the curve on the right is represented as P(–1.87 < Z < 1.87) and from the standard normal distribution table is 2 × .4693 = .9386.

6.8 Represent the following probabilities as shaded areas under the standard normal curve and explain in words how you find the areas: (*a*) P(Z < –1.75); (*b*) P(Z < 2.15).

Ans. The probability in part (*a*) is shown as the shaded area in Fig. 6-38 and the probability in part (*b*) is shown in Fig 6-39.

To find the shaded area in Fig. 6-38, we note that P(Z < –1.75) = P(Z > 1.75) because of the symmetry of the normal curve. In addition, P(Z > 1.75) = .5 – P(0 < Z < 1.75) since the total area to the right of 0 is .5. From the standard normal distribution table, P(0 < Z < 1.75) is equal to .4599. Therefore, P(Z < –1.75) = P(Z > 1.75) = .5 – .4599 = .0401.

To find the shaded area in Fig. 6-39, we note that P(Z < 2.15) = P(Z < 0) + P(0 < Z < 2.15). The probability P(Z < 0) = .5 because of the symmetry of the normal curve. From the standard normal distribution table, P(0 < Z < 2.15) = .4842. Therefore, P(Z < 2.15) = .5 + .4842 = .9842. The solution to part (*b*) using Minitab is as follows:

MTB > cdf 2.15;
SUBC > normal 0 1.
Cumulative Distribution Function
Normal with mean = 0 and standard deviation = 1.00000

    x   P(X <= x)
2.1500   0.9842

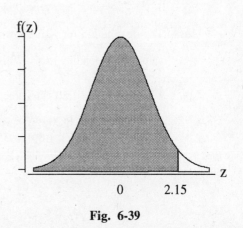

Fig. 6-38   **Fig. 6-39**

STANDARDIZING A NORMAL DISTRIBUTION

6.9 The distribution of complaints per week per 100,000 passengers for all airlines in the U.S. is normally distributed with $\mu = 4.5$ and $\sigma = 0.8$. Find the standardized values for the following observed values of the number of complaints per week per 100,000 passengers: (a) 6.3; (b) 2.5; (c) 4.5; (d) 8.0.

Ans. (a) The standardized value for 6.3 is found by $z = \dfrac{x - \mu}{\sigma} = \dfrac{6.3 - 4.5}{.8} = 2.25$.

(b) The standardized value for 2.5 is found by $z = \dfrac{x - \mu}{\sigma} = \dfrac{2.5 - 4.5}{.8} = -2.50$.

(c) The standardized value for 4.5 is found by $z = \dfrac{x - \mu}{\sigma} = \dfrac{4.5 - 4.5}{.8} = 0.00$.

(d) The standardized value for 8.0 is found by $z = \dfrac{x - \mu}{\sigma} = \dfrac{8.0 - 4.5}{.8} = 4.38$.

6.10 Personal injury awards are normally distributed with a mean equal to \$62,000 and a standard deviation equal to \$13,500. Find the amount of the award corresponding to the following standardized values: (a) 2.0; (b) –3.0; (c) 0.0; (d) 4.5.

Ans. (a) The amount corresponding to standardized value 2.0 is $x = \mu + z\sigma = 62,000 + 2.0 \times 13,500 = 89,000$.

(b) The amount corresponding to standardized value –3.0 is $x = \mu + z\sigma = 62,000 - 3.0 \times 13,500 = 21,500$.

(c) The amount corresponding to standardized value 0.0 is $x = \mu + z\sigma = 62,000 + 0.0 \times 13,500 = 62,000$.

(d) The amount corresponding to standardized value 4.5 is $x = \mu + z\sigma = 62,000 + 4.5 \times 13,500 = 122,750$.

APPLICATIONS OF THE NORMAL DISTRIBUTION

6.11 In a sociological study concerning family life, it is found that the age at first marriage for men is normally distributed with a mean equal to 23.7 years and a standard deviation equal to 3.5 years. Determine the percent of men for whom the age at first marriage is between 20 and 30 years of age. If X represents the age at first marriage for men, draw a normal curve for X and show the shaded area for $P(20 < X < 30)$ as well as the corresponding area under the standard normal curve.

Ans. The distribution for X is shown in Fig. 6-40. The shaded area represents $P(20 < X < 30)$. The z value corresponding to x = 20 is $z = \dfrac{20 - 23.7}{3.5} = -1.06$ and the z value corresponding to x = 30 is $z = \dfrac{30 - 23.7}{3.5} = 1.80$. The area shown in Fig. 6-41 is equal to the area under the curve shown in Fig. 6-40, that is, $P(20 < X < 30) = P(-1.06 < Z < 1.80)$. Utilizing the standard normal distribution table, $P(-1.06 < Z < 1.80) = .3554 + .4641 = .8195$. That is, about 82% of the first marriages for men occur for men between 20 and 30 years of age.

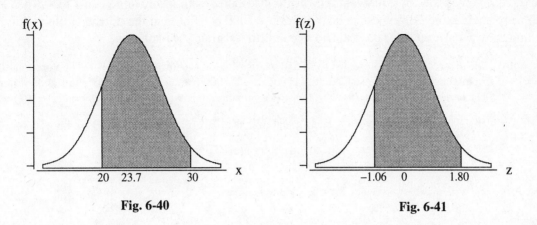

Fig. 6-40   **Fig. 6-41**

6.12 The net worth of senior citizens is normally distributed with mean equal to $225,000 and standard deviation equal to $35,000. What percent of senior citizens have a net worth less than $300,000?

Ans. Let X represent the net worth of senior citizens in thousands of dollars. The percent of senior citizens with a net worth less than $300,000 is found by multiplying P(X < 300) times 100. The probability P(X < 300) is shown in Fig. 6-42. The event X < 300 is equivalent to the event $Z < \dfrac{300 - 225}{35} = 2.14$. The probability that Z < 2.14 is represented as the shaded area in Fig. 6-43. The probability that Z is less than 2.14 is found by adding P(0 < Z < 2.14) to .5, which equals .5 + .4838 = .9838. We can conclude that 98.38% of the senior citizens have net worths less than $300,000.

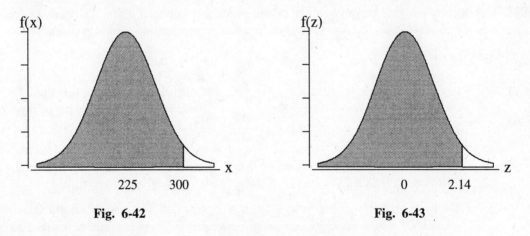

Fig. 6-42   **Fig. 6-43**

DETERMINING THE Z AND X VALUES WHEN AN AREA UNDER THE NORMAL CURVE IS KNOWN

6.13 Find the value for positive number a such that P(–a < Z < a) = .95.

Ans. The symmetry of the normal curve implies that P(0 < Z < a) = .4750. This area is found in the interior of the standard normal distribution table and the z value corresponding to this area is 1.96. The area under the standard normal curve corresponding to –1.96 < Z < 1.96 is .9500.

6.14 U.S. divorce rates, by county, are normally distributed with mean value equal to 4.5 per 1000 people and standard deviation equal to 1.3 per 1000 people. Find the third quartile for divorce rates by county per 1000 people. Draw graphs to illustrate your solution.

Ans.   The divorce rate is represented by X and its distribution is shown in Fig. 6-44. The third quartile is represented by Q_3. The shaded area is equal to .7500. The third quartile for the standard normal distribution is .67 and is shown in Fig. 6-45. Summarizing, we have $P(X < Q_3) = P(Z < .67) = .75$.

The standardized value of Q_3 must equal .67. That is, $\dfrac{Q_3 - 4.5}{1.3} = .67$ or $Q_3 = 4.5 + 1.3 \times .67 = 5.37$. Seventy-five percent of the divorce rates are 5.37 or below.

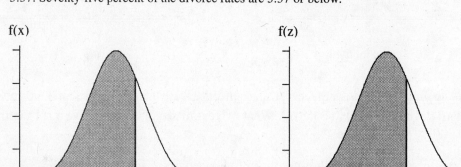

| Fig. 6-44 | Fig. 6-45 |

NORMAL APPROXIMATION TO THE BINOMIAL DISTRIBUTION

6.15 Use Minitab to find the probability of the event that between 8 and 11 heads inclusive occur when a coin is flipped 20 times. Find the normal approximation to this probability.

Ans.

```
MTB > cdf 11;                          MTB > cdf 7;
SUBC > binomial 20 .5.                 SUBC > binomial 20 .5.
Cumulative Distribution Function       Cumulative Distribution Function
Binomial with n = 20 and p = 0.500     Binomial with n = 20 and p = 0.500
    x      P(X <= x)                       x      P(X <= x)
  11.00    0.7483                         7.00    0.1316
```

To find the probability of the event $8 \le X \le 11$ note that $P(8 \le X \le 11) = P(X \le 11) - P(X \le 7)$. Therefore, $P(8 \le X \le 11) = .7483 - .1316 = .6167$.

The mean of the binomial distribution is $\mu = np = 20 \times .5 = 10$, and the standard deviation is $\sigma = \sqrt{npq} = \sqrt{5} = 2.24$. A normal curve with mean equal to 10 and standard deviation 2.24 is fit to the binomial distribution and the area under this normal curve is found for X ranging between 7.5 and 11.5. This area is shown in Fig. 6-46. The standardized value for x = 7.5 is z = −1.12 and the standardized value for x = 11.5 is z = .67. The area between z = −1.12 and z = .67 is shown in Fig. 6-47. The area is now found using Minitab.

```
MTB > cdf 11.5;                        MTB > cdf 7.5;
SUBC > normal 10 2.24.                 SUBC > normal 10 2.24.
Cumulative Distribution Function       Cumulative Distribution Function
Normal with mean = 10.00 and           Normal with mean = 10.00 and
standard deviation = 2.240             standard deviation = 2.240
    x      P(X <= x)                       x      P(X <= x)
  11.50    0.7485                         7.50    0.1322
```

The area between 7.5 and 11.5 is the difference .7485 − .1322 = .6163.

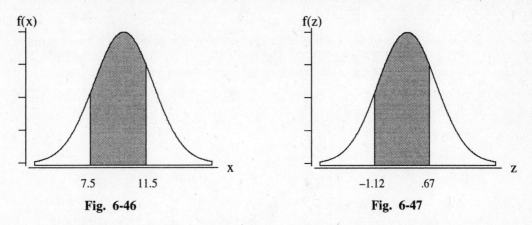

Fig. 6-46   Fig. 6-47

6.16 In the United States, 80% of TV-owning homes also own a VCR. Use the normal approximation to the binomial distribution to find the probability that 425 or more in a sample of 500 TV-owning homes also own a VCR.

> *Ans.* If we let X represent the number of TV-owning homes which also own a VCR, then we are to find $P(X \geq 425)$. A normal curve having mean = np = 400 and standard deviation = $\sqrt{npq}$ = 8.94 is fit to the binomial and we need to find the area under the normal curve to the right of 424.5. The solution using Minitab is as follows.
>
> MTB > cdf 424.5;
> SUBC > normal 400 8.94.
> Cumulative Distribution Function
> Normal with mean = 400.00 and standard deviation = 8.94
>   x   P(X <= x)
> 424.50   0.9969
>
> Since .9969 is the probability that X is less than 424.5, the probability that X exceeds 424.5 is 1 − .9969 = .0031.

EXPONENTIAL PROBABILITY DISTRIBUTION

6.17 Which of the following statements best describes the exponential probability distribution?
  (a) The exponential probability distribution is skewed to the right.
  (b) The exponential probability distribution is skewed to the left.
  (c) The exponential probability distribution is mound shaped.

> *Ans.* (a) The exponential distribution is skewed to the right.

6.18 What is the equation of the exponential pdf having mean equal to 5?

> *Ans.* The equation is $f(x) = .2e^{-.2x}$ for $x \geq 0$.

PROBABILITIES FOR THE EXPONENTIAL PROBABILITY DISTRIBUTION

6.19 The lifetimes in years for a particular brand of cathode ray tube are exponentially distributed with a mean of 5 years. What percent of the tubes have lifetimes between 5 and 8 years? Draw a graph of the pdf and shade the area which represents the probability of the event $5 < X < 8$, where X represents the lifetimes.

Ans.   The shaded area in Fig. 6-48 represents $P(5 < X < 8)$. The probability is found as follows:
$P(5 < X < 8) = P(X < 8) - P(X < 5) = (1 - e^{-1.6}) - (1 - e^{-1}) = e^{-1} - e^{-1.6} = .3679 - .2019 = .166.$
Approximately 16.6% of the tubes have lifetimes between 5 and 8 years.

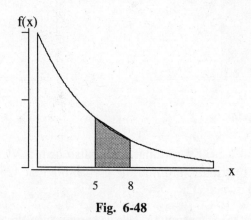

Fig. 6-48

6.20   For the cathode ray tubes in problem 6.19, use Minitab to determine the percent of tubes that have lifetimes of less than 10 years.

Ans.   MTB > cdf 10;
          SUBC > exponential 5.
          Cumulative Distribution Function
          Exponential with mean = 5.00000
              x   P(X <= x)
          10.00   0.8647

The percentage of tubes having lifetimes less than 10 years is 86.5%.

Supplementary Problems

UNIFORM PROBABILITY DISTRIBUTION

6.21   The RND function is a computer language function which uniformly generates random numbers between 0 and 1.
(a)   What percent of the random numbers generated by RND are less than .35?
(b)   What percent of the random numbers generated by RND are between .20 and .55?
(c)   What percent of the random numbers generated by RND are either less than .14 or greater than .81?

Ans.   (a) 35%   (b) 35%   (c) 33%

6.22   In a psychological study involving personality types and career selections, it is found that the time required to complete a task is uniformly distributed over the interval from 5.0 to 7.5 minutes.
(a)   What is the probability that the task is completed in less than 4 minutes?
(b)   What is the probability that the task is completed in 7.0 or more minutes?
(c)   What is the probability that it requires more than 10.0 minutes to complete the task?

Ans.   (a) 0   (b) .2   (c) 0

MEAN AND STANDARD DEVIATION FOR THE UNIFORM PROBABILITY DISTRIBUTION

6.23 What is the mean and standard deviation of the random numbers generated by RND in problem 6.21?

Ans. $\mu = .50$ $\sigma = .29$

6.24 What percent of the distribution in problem 6.23 is included within one standard deviation of the mean?

Ans. 57.6%

NORMAL PROBABILITY DISTRIBUTION

6.25 The heights of adult males are normally distributed with a mean equal to 70 inches and a standard deviation equal to 3 inches. Give the pdf for the normal curve describing this distribution.

Ans. $f(x) = \dfrac{1}{3\sqrt{2\pi}} e^{-(x-70)^2/18}$ for all values of x

6.26 A normal distribution has mean equal to a and standard deviation equal to b and c is a positive number. If it is known that $P(X > a + c) = d$, find the following probabilities:

(a) $P(X < a + c)$ (b) $P(X < a - c)$ (c) $P(a - c < X < a + c)$ (d) $P(0 < X < a + c)$

Ans. (a) $1 - d$ (b) d   (c) $1 - 2d$ (d) $.5 - d$

STANDARD NORMAL DISTRIBUTION

6.27 Find the following probabilities concerning the standard normal random variable Z.

(a) $P(0 < Z < 2.13)$ (b) $P(-1.45 < Z < 2.10)$ (c) $P(Z > 2.88)$ (d) $P(Z \geq -2.01)$

Ans. (a) .4834   (b) .9086   (c) .0020   (d) .9778

6.28 Find the probabilities of the following events involving the standard normal random variable Z.

(a) $Z > 4.50$ (b) $-4.00 < Z < 5.50$ (c) $Z < -1.25$ or $Z > 2.35$
(d) $Z < 1.15$ and $Z > -1.15$ (e) $Z < -2.45$ and $Z > 1.11$ (f) the complement of $Z < 1.44$

Ans. (a) approx 0   (b) approx 1   (c) .1150   (d) .7498   (e) 0   (f) .0749

STANDARDIZING A NORMAL DISTRIBUTION

6.29 The marriage rate per 1,000 population per county has a normal distribution with mean 8.9 and standard deviation 1.7. Find the standardized values for the following marriage rates per 1,000 per county.

(a) 6.5   (b) 8.8   (c) 12.5   (d) 13.5

Ans. (a) −1.41   (b) −0.06   (c) 2.12   (d) 2.71

6.30 The hospital cost for individuals involved in accidents who do not wear seat belts is normally distributed with mean $7,500 and standard deviation $1,200.

(a) Find the cost for an individual whose standardized value is 2.5.
(b) Find the cost for an individual whose bill is 3 standard deviations below the average.

Ans. (a) $10,500   (b) $3,900

APPLICATIONS OF THE NORMAL DISTRIBUTION

6.31 The average TV-viewing time per week for children ages 2 to 11 is 22.5 hours and the standard deviation is 5.5 hours. Assuming the viewing times are normally distributed, find the following.
(a) What percent of the children have viewing times less than 10 hours per week?
(b) What percent of the children have viewing times between 15 and 25 hours per week?
(c) What percent of the children have viewing times greater than 40 hours per week?

Ans.   (a) 1.16%   (b) 58.67%   (c) less than 0.1%

6.32 The amount that airlines spend on food per passenger is normally distributed with mean $8.00 and standard deviation $2.00.
(a) What percent spend less than $5.00 per passenger?
(b) What percent spend between $6.00 and $10.00?
(c) What percent spend more than $12.50?

Ans.   (a) 6.68%   (b) 68.26%   (c) 1.22%

DETERMINING THE Z AND X VALUES WHEN AN AREA UNDER THE NORMAL CURVE IS KNOWN

6.33 Find the value of a in each of the following probability statements involving the standard normal variable Z.
(a) $P(0 < Z < a) = .4616$ (b) $P(Z < a) = .8980$ (c) $P(-a < Z < a) = .8612$
(d) $P(Z < a) = .1894$ (e) $P(Z > a) = .1894$ (f) $P(Z = a) = .5000$

Ans.   (a) 1.77   (b) 1.27   (c) 1.48   (d) –0.88   (e) 0.88   (f) no solutions

6.34 The GMAT test is required for admission to most graduate programs in business. In a recent year, the GMAT test scores were normally distributed with mean value 550 and standard deviation 100.
(a) Find the first quartile for the distribution of GMAT test scores.
(b) Find the median for the distribution of GMAT test scores.
(c) Find the ninety-fifth percentile for the distribution of GMAT test scores.

Ans.   (a) 483   (b) 550   (c) 715

NORMAL APPROXIMATION TO THE BINOMIAL DISTRIBUTION

6.35 For which of the following binomial distributions is the normal approximation to the binomial distribution appropriate?
(a) n = 15, p = .2   (b) n = 40, p = .1   (c) n = 500, p = .05   (d) n = 50, p = .3

Ans.   (c) and (d)

6.36 Thirteen percent of students took a college remedial course in 1992–1993. Assuming this is still true, what is the probability that in 350 randomly selected students:
(a) Less than 40 take a remedial course
(b) Between 40 and 50, inclusive, take a remedial course
(c) More than 55 take a remedial course

Ans.   (a) .1711   (b) .6141   (c) .0559

EXPONENTIAL PROBABILITY DISTRIBUTION

6.37 The random variable X has an exponential distribution with pdf $f(x) = .1e^{-.1x}$, for $x > 0$. What is the mean for this random variable?

Ans.   10

6.38 What is the largest value an exponential random variable may assume?

Ans.   There is no upper limit for an exponential distribution.

PROBABILITIES FOR THE EXPONENTIAL PROBABILITY DISTRIBUTION

6.39 The time that a family lives in a home between purchase and resale is exponentially distributed with a mean equal to 5 years. Let X represent the time between purchase and resale. (*a*) Find $P(X < 3)$. (*b*) Find $P(X > 5)$.

Ans.   (*a*) .4512   (*b*) .3679

6.40 The time between orders received at a mail order company are exponentially distributed with a mean equal to 0.5 hour. What is the probability that the time between orders is between 1 and 2 hours?

Ans.   .1170

Chapter 7

Sampling Distributions

SIMPLE RANDOM SAMPLING

In order to obtain information about some population, either a *census* of the whole population is taken or a *sample* is chosen from the population and the information is inferred from the sample. The second approach is usually taken, since it is much cheaper to obtain a sample than to conduct a census. In choosing a sample, it is desirable to obtain one that is representative of the population. The average weight of the football players at a college would not be a representative estimate of the average weight of students attending the college, for example. A *simple random sample* of size n from a population of size N is one selected in such a way that every sample of size n has the same chance of occurring. In *simple random sampling with replacement,* a member of the population can be selected more than once. In *simple random sampling without replacement,* a member of the population can be selected at most once. Simple random sampling without replacement is the most common type of simple random sampling.

EXAMPLE 7.1 Consider the population consisting of the world's five busiest airports. This population consists of the following: A: Chicago O'Hare, B: Atlanta Hartsfield, C: London Heathrow, D: Dallas–Fort Worth, and E: Los Angeles Intl. The number of possible samples of size 2 from this population of size 5 is given by the combination of 5 items selected two at a time, that is, $\binom{5}{2} = \dfrac{5!}{2!3!} = 10$. In simple random sampling, each possible pair would have probability 0.1 of being the pair selected. That is, Chicago O'Hare and Atlanta Hartsfield would have probability 0.1 of being chosen, Chicago O'Hare and London Heathrow would have probability 0.1 of being chosen, etc. One way of ensuring that each pair would have an equal chance of being selected would be to write the names of the five airports on separate sheets of paper and select two of the sheets randomly from a box.

USING RANDOM NUMBER TABLES

The technique of writing names on slips of paper and selecting them from a box is not practical for most real world situations. Tables of random numbers are available in a variety of sources. The digits 0 through 9 occur randomly throughout a random number table with each digit having an equal chance of occurring. Table 7.1 is an example of a random number table. This particular table has 50 columns and 20 rows. To use a random number table, first randomly select a starting position and then move in any direction to select the numbers.

EXAMPLE 7.2 The money section of *USA Today* gives the 1,900 most active New York Stock Exchange issues. The random numbers in Table 7.1 can be used to randomly select 10 of these issues. Imagine that the issues are numbered from 0001 to 1900. Suppose we randomly decide to start in row 1 and columns 21 through 24. The four-digit number located here is 0345. Reading down these four columns and discarding any number exceeding 1900, we obtain the following eight random numbers between 0001 and 1900: 0345, 1304, 0990, 1580, 1461, 1064, 0676, and 0347. To obtain our other two numbers, we proceed to row 1 and columns 26 through 29. Reading down this column, we find 1149 and 1074. To obtain the 10 stock issues, we read down the columns and select the ones located in positions 345, 347, 676, 990, 1064, 1074, 1149, 1304, 1461, and 1580.

Table 7.1

Random Numbers (25 rows, 50 columns)									
01–05	06–10	11–15	16–20	21–25	26–30	31–35	36–40	41–45	46–50
87032	26561	44020	06061	03453	22484	55858	61768	12676	29353
98340	94192	81975	69931	13047	28533	34529	02625	11020	65106
70363	95651	64089	31921	09900	81554	53640	92109	00459	09599
09749	91862	12659	63079	91937	58272	90766	09950	27996	29679
84223	87730	91759	93041	48757	89477	84221	38566	96274	29195
27190	92922	86046	09124	42493	26551	76639	15763	18068	38998
47087	68993	24807	70755	36834	19522	59510	83888	51540	39119
47335	69753	44311	33070	15800	92668	78460	39356	91692	34824
31031	52240	10346	02133	42534	84923	44548	75222	89959	92119
60072	37318	07550	04411	43925	11499	93024	72791	60190	29692
90202	45248	84967	67293	14612	99573	69573	98695	51303	44925
91887	83092	39204	23539	98551	48427	25425	43864	10714	08308
08264	04860	05919	28393	21460	28370	43026	78296	58382	08276
46655	67610	35334	44369	10649	10744	50515	01372	55081	31121
30428	33957	53553	22925	06766	37433	45349	46565	47011	46762
55238	40718	83328	97613	23119	77718	16016	58590	03726	03091
64993	84882	03067	19953	21077	27665	10583	62587	36875	00638
90420	80152	10418	26576	40361	82421	61952	62713	04890	01032
44621	76402	04778	58739	03474	00570	28368	60340	95227	39059
15988	94013	71898	05785	72883	17772	57471	75775	95202	06545

USING THE COMPUTER TO OBTAIN A SIMPLE RANDOM SAMPLE

Most computer statistical software packages can be used to select random numbers and to some extent have replaced random number tables. As the capability and availability of computers continue to increase, many of the statistical tables are becoming obsolete.

EXAMPLE 7.3 Minitab can be used to select the random sample of stock issues in Example 7.2. The commands are as follows.

```
MTB > set c1
DATA > 1:1900
DATA > end
MTB > sample 10 c1 c2
MTB > print c2

Data Display
C2
 1227  969 1834 1441  423  897  824  664  414   77
```

The first three lines of command put the numbers 1 through 1900 into column 1. The command on the fourth line asks for a sample of size 10 from the numbers in column c1 and asks that the selected numbers be placed into column c2. The print command causes the random numbers to be printed. These are the numbers of the 10 stocks to be selected.

SYSTEMATIC RANDOM SAMPLING

Systematic random sampling consists of choosing a sample by randomly selecting the first element and then selecting every kth element thereafter. The systematic method of selecting a sample often saves time in selecting the sample units.

EXAMPLE 7.4 In order to obtain a systematic sample of 50 of the nation's 3,143 counties, divide 3,143 by 50 to obtain 62.86. Round 62.86 down to obtain 62. From a list of the 3,143 counties, select one of the first 62 counties at random. Suppose county number 35 is selected. To obtain the other 49 counties, add 62 to 35 to obtain 97, add 2×62 to 35 to obtain 159, and continue in this fashion until the number $49 \times 62 + 35 = 3,073$ is obtained. The counties numbered 35, 97, 159, . . . , 3073 would represent a systematic sample from the nations counties.

CLUSTER SAMPLING

In *cluster sampling*, the population is divided into clusters and then a random sample of the clusters are selected. The selected clusters may be completely sampled or a random sample may be obtained from the selected clusters.

EXAMPLE 7.5 A large company has 30 plants located throughout the United States. In order to access a new total quality plan, the 30 plants are considered to be clusters and five of the plants are randomly selected. All of the quality control personnel at the five selected plants are asked to evaluate the total quality plan.

STRATIFIED SAMPLING

In *stratified sampling*, the population is divided into *strata* and then a random sample is selected from each strata. The strata may be determined by income levels, different stores in a supermarket chain, different age groups, different governmental law enforcement agencies, and so forth.

EXAMPLE 7.6 Super Value Discount has 10 stores. To assess job satisfaction, one percent of the employees at each of the 10 stores are administered a job satisfaction questionnaire. The 10 stores are the strata into which the population of all employees at Super Value Discount are divided. The results at the 10 stores are combined to evaluate the job satisfaction of the employees.

SAMPLING DISTRIBUTION OF THE SAMPLING MEAN

The mean of a population, μ, is a parameter that is often of interest but usually the value of μ is unknown. In order to obtain information about the population mean, a sample is taken and the sample mean, $\overline{x}$, is calculated. The value of the sample mean is determined by the sample actually selected. The sample mean can assume several different values, whereas the population mean is constant. The set of all possible values of the sample mean along with the probabilities of occurrence of the possible values is called the *sampling distribution of the sampling mean*. The following example will help illustrate the sampling distribution of the sample mean.

EXAMPLE 7.7 Suppose the five cities with the most African-American-owned businesses measured in thousands is given in Table 7.2.

Table 7.2

City	Number of African-American-owned businesses, in thousands
A: New York	42
B: Washington, D.C.	39
C: Los Angeles	36
D: Chicago	33
E: Atlanta	30

If X represents the number of African-American-owned businesses in thousands for this population consisting of five cities, then the probability distribution for X is shown in Table 7.3.

Table 7.3

x	30	33	36	39	42
P(x)	.2	.2	.2	.2	.2

The population mean is $\mu = \Sigma\, xP(x) = 30 \times .2 + 33 \times .2 + 36 \times .2 + 39 \times .2 + 42 \times .2 = 36$, and the variance is given by $\sigma^2 = \Sigma x^2\, P(x) - \mu^2 = 900 \times .2 + 1089 \times .2 + 1296 \times .2 + 1521 \times .2 + 1764 \times .2 - 1296 = 1314 - 1296 = 18$. The population standard deviation is the square root of 18, or 4.24.

The number of samples of size 3 possible from this population is equal to the number of combinations possible when selecting three cities from five. The number of possible samples is $C_3^5 = \dfrac{5!}{3!2!} = 10$. Using the letters A, B, C, D, and E rather than the name of the cities, Table 7.4 gives all the possible samples of three cities, the sample values, and the means of the samples.

Table 7.4

Samples	Number of businesses in the samples	Sample mean $\overline{x}$
A, B, C	42, 39, 36	39
A, B, D	42, 39, 33	38
A, B, E	42, 39, 30	37
A, C, D	42, 36, 33	37
A, C, E	42, 36, 30	36
A, D, E	42, 33, 30	35
B, C, D	39, 36, 33	36
B, C, E	39, 36, 30	35
B, D, E	39, 33, 30	34
C, D, E	36, 33, 30	33

The sampling distribution of the mean is obtained from Table 7.4. For random sampling, each of the samples in Table 7.4 is equally likely to be selected. The probability of selecting a sample with mean 39 is .1 since only one of 10 samples has a mean of 39. The probability of selecting a sample with mean 36 is .2, since two of the samples have a mean equal to 36. Table 7.5 gives the sampling distribution of the sample mean.

Table 7.5

$\overline{x}$	33	34	35	36	37	38	39
$P(\overline{x})$	.1	.1	.2	.2	.2	.1	.1

SAMPLING ERROR

When the sample mean is used to estimate the population mean, an error is usually made. This error is called the *sampling error*, and is defined to be the absolute difference between the sample mean and the population mean. The sampling error is defined by

$$\text{sampling error} = |\,\overline{x} - \mu\,| \qquad\qquad (7.1)$$

EXAMPLE 7.8 In Example 7.7, the population of the five cities with the most African-American-owned businesses is given. The mean of this population is 36. Table 7.5 gives the possible sample means for all samples of size 3 selected from this population. Table 7.6 gives the sampling errors and probabilities associated with all the different sample means.

Table 7.6

Sample mean	Sampling error	Probability
33	3	.1
34	2	.1
35	1	.2
36	0	.2
37	1	.2
38	2	.1
39	3	.1

From Table 7.6, it is seen that the probability of no sampling error in this scenario is .20. There is a 60% chance that the sampling error is 1 or less.

MEAN AND STANDARD DEVIATION OF THE SAMPLE MEAN

Since the sample mean has a distribution, it is a random variable and has a mean and a standard deviation. The mean of the sample mean is represented by the symbol $\mu_{\overline{x}}$ and the standard deviation of the sample mean is represented by $\sigma_{\overline{x}}$. The standard deviation of the sample mean is referred to as the *standard error of the mean*. Example 7.9 illustrates how to find the mean of the sample mean and the standard error of the mean.

EXAMPLE 7.9 In Example 7.7, the sampling distribution of the mean shown in Table 7.7 was obtained.

Table 7.7

$\overline{x}$	33	34	35	36	37	38	39
$P(\overline{x})$	.1	.1	.2	.2	.2	.1	.1

The mean of the sample mean is found as follows:

$$\mu_{\overline{x}} = \Sigma\,\overline{x}\,P(\overline{x}) = 33 \times .1 + 34 \times .1 + 35 \times .2 + 36 \times .2 + 37 \times .2 + 38 \times .1 + 39 \times .1 = 36$$

The variance of the sample mean is found as follows:

$$\sigma_{\overline{x}}^2 = \Sigma\,\overline{x}^2\,P(\overline{x}) - \mu_{\overline{x}}^2$$

$$\sigma_{\overline{x}}^2 = 1089 \times .1 + 1156 \times .1 + 1225 \times .2 + 1296 \times .2 + 1369 \times .2 + 1444 \times .1 + 1521 \times .1 - 1296 = 3$$

The standard error of the mean, $\sigma_{\overline{x}}$, is equal to $\sqrt{3} = 1.73$.

The relationship between the mean of the sample mean and the population mean is expressed by

$$\mu_{\bar{x}} = \mu \qquad (7.2)$$

The relationship between the variance of the sample mean and the population variance is expressed by formula (7.3), where N is the population size and n is the sample size.

$$\sigma_{\bar{x}}^2 = \frac{\sigma^2}{n} \times \frac{N-n}{N-1} \qquad (7.3)$$

EXAMPLE 7.10 In Example 7.7, the population consisting of the five cities with the most African-American-owned businesses was introduced. The population mean, μ, is equal to 36 and the variance, σ^2, is equal to 18. In Example 7.9, the mean of the sample mean, $\mu_{\bar{x}}$, was shown to equal 36 and the variance of the sample mean, $\sigma_{\bar{x}}^2$, was shown to equal 3. It is seen that $\mu = \mu_{\bar{x}} = 36$, illustrating formula (7.2). To illustrate formula (7.3), note that

$$\frac{\sigma^2}{n} \times \frac{N-n}{N-1} = \frac{18}{3} \times \frac{5-3}{5-1} = 3 = \sigma_{\bar{x}}^2$$

The standard error of the mean is found taking the square root of both sides of formula (7.3), and is given by

$$\sigma_{\bar{x}} = \frac{\sigma}{\sqrt{n}} \times \sqrt{\frac{N-n}{N-1}} \qquad (7.4)$$

The term $\sqrt{\dfrac{N-n}{N-1}}$ is called the *finite population correction factor*. If the sample size n is less than 5% of the population size, i.e., n < .05N, the finite population correction factor is very near one and is omitted in formula (7.4). If n < .05N, the standard error of the mean is given by

$$\sigma_{\bar{x}} = \frac{\sigma}{\sqrt{n}} \qquad (7.5)$$

EXAMPLE 7.11 The mean cost per county in the United States to maintain county roads is $785 thousand per year and the standard deviation is $55 thousand. Approximately 4% of the counties are randomly selected and the mean cost for the sample is computed. The number of counties is 3,143 and the sample size is 125. The standard error of the mean using formula (7.4) is:

$$\sigma_{\bar{x}} = \frac{55}{\sqrt{125}} \times \sqrt{\frac{3,143-125}{3,143-1}} = 4.91935 \times .98 = 4.82 \text{ thousand}$$

The standard error of the mean using formula (7.5) is:

$$\sigma_{\bar{x}} = \frac{55}{\sqrt{125}} = 4.92 \text{ thousand}$$

Ignoring the finite population correction factor in this case changes the standard error by a small amount.

SHAPE OF THE SAMPLING DISTRIBUTION OF THE SAMPLE MEAN AND THE CENTRAL LIMIT THEOREM

If samples are selected from a population which is normally distributed with mean μ and standard deviation σ, then the distribution of sample means is normally distributed and the mean of this distribution is $\mu_{\bar{x}} = \mu$, and the standard deviation of this distribution is $\sigma_{\bar{x}} = \dfrac{\sigma}{\sqrt{n}}$. The shape of the distribution of the sample means is normal or bell-shaped regardless of the sample size.

The *central limit theorem* states that when sampling from a large population of any distribution shape, the sample means have a normal distribution whenever the sample size is 30 or more. Furthermore, the mean of the distribution of sample means is $\mu_{\bar{x}} = \mu$, and the standard deviation of this distribution is $\sigma_{\bar{x}} = \dfrac{\sigma}{\sqrt{n}}$. It is important to note that when sampling from a nonnormal distribution, $\bar{x}$ has a normal distribution only if the sample size is 30 or more. The central limit theorem is illustrated graphically in Figure 7-1.

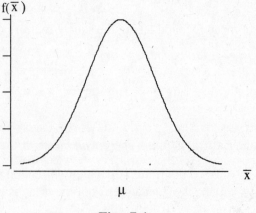

Fig. 7-1

Figure 7-1 illustrates that for samples greater than or equal to 30, $\bar{x}$ has a distribution that is bell-shaped and centers at μ. The spread of the curve is determined by $\sigma_{\bar{x}}$.

EXAMPLE 7.12 If a large number of samples each of size n, where n is 30 or more, are selected and the means of the samples are calculated, then a histogram of the means will be bell-shaped regardless of the shape of the population distribution from which the samples are selected. However, if the sample size is less than 30, the histogram of the sample means may not be bell-shaped unless the samples are selected from a bell-shaped distribution.

APPLICATIONS OF THE SAMPLING DISTRIBUTION OF THE SAMPLE MEAN

The distribution properties of the sample mean are used to evaluate the results of sampling, and form the underpinnings of many of the statistical inference techniques found in the remaining chapters of this text. The examples in this section illustrate the usefulness of the central limit theorem.

EXAMPLE 7.13 A government report states that the mean amount spent per capita for police protection for cities exceeding 150,000 in population is $500 and the standard deviation is $75. A criminal justice research study found that for 40 such randomly selected cities, the average amount spent per capita for this sample for police protection is \$465. If the government report is correct, the probability of finding a sample mean that is \$35 or more below the national average is given by $P(\overline{X} < 465)$. The central limit theorem assures us that because the sample size exceeds 30, the sample mean has a normal distribution. Furthermore, if the government report is correct, the mean of $\overline{X}$ is \$500 and the standard error is $\dfrac{75}{\sqrt{40}} = \$11.86$. The area to the left of \$465 under the normal curve for $\overline{X}$ is the same as the area to the left of $Z = \dfrac{465-500}{11.86} = -2.95$. We have the following equality. $P(\overline{X} < 465) = P(Z < -2.95) = .0016$. This result suggests that either we have a highly unusual sample or the government claim is incorrect. Figures 7-2 and 7-3 illustrate the solution graphically.

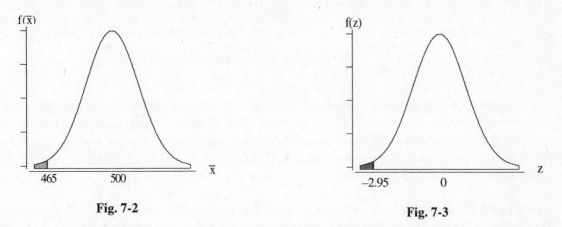

Fig. 7-2   **Fig. 7-3**

In Example 7.13, the value 465 was transformed to a z value by subtracting the population mean 500 from 465, and then dividing by the standard error 11.86. The equation for transforming a sample mean to a z value is shown in formula (7.6):

$$z = \frac{\overline{x} - \mu}{\sigma_{\overline{x}}} \qquad (7.6)$$

EXAMPLE 7.14 A machine fills containers of coffee labeled as 113 grams. Because of machine variability, the amount per container is normally distributed with $\mu = 113$ and $\sigma = 1$. Each day, 4 of the containers are selected randomly and the mean amount in the 4 containers is determined. If the mean amount in the four containers is either less than 112 grams or greater than 114 grams, the machine is stopped and adjusted. Since the distribution of fills is normally distributed, the sample mean is normally distributed even for a sample as small as four. The mean of the sample mean is $\mu_{\overline{x}} = 113$ and the standard error is $\sigma_{\overline{x}} = .5$. The machine is adjusted if $\overline{x} < 112$ or if $\overline{x} > 114$. The probability the machine is adjusted is equal to the sum $P(\overline{x} < 112) + P(\overline{x} > 114)$ since we add probabilities of events connected by the word *or*. To evaluate these probabilities, we use formula (7.6) to express the events involving $\overline{x}$ in terms of z as follows:

$$P(\overline{x} < 112) = P\left(\frac{\overline{x}-113}{.5} < \frac{112-113}{.5}\right) = P(z < -2.00) = .0228$$

$$P(\overline{x} > 114) = P\left(\frac{\overline{x}-113}{.5} > \frac{114-113}{.5}\right) = P(z > 2.00) = .0228$$

The probability that the machine is adjusted is $2 \times .0228 = .0456$. It is seen that if this sampling technique is used to monitor this process, there is a 4.56% chance that the machine will be adjusted even though it is maintaining an average fill equal to 113 grams.

SAMPLING DISTRIBUTION OF THE SAMPLE PROPORTION

A *population proportion* is the proportion or percentage of a population that has a specified characteristic or attribute. The population proportion is represented by the letter p. The *sample proportion* is the proportion or percentage in the sample that has a specified characteristic or attribute. The sample proportion is represented by either the symbol $\hat{p}$ or $\overline{p}$. We shall use the symbol $\overline{p}$ to represent the sample proportion in this text.

EXAMPLE 7.15 The nation's work force at a given time is 133,018,000 and the number of unemployed is 7,355,000. The proportion unemployed is $p = \dfrac{7,355,000}{133,018,000} = .055$ and the *jobless rate* is 5.5%. A sample of size 65,000 is selected from the nation's work force and 3,900 are unemployed in the sample. The sample proportion unemployed is $\overline{p} = \dfrac{3,900}{65,000} = .06$ and the *sample jobless rate* is 6%.

The population proportion p is a *parameter* measured on the complete population and is constant over some time interval. The sample proportion $\overline{p}$ is a *statistic* measured on a sample and is considered to be a random variable whose value is dependent on the sample chosen. The set of all possible values of a sample proportion along with the probabilities corresponding to those values is called the *sampling distribution of the sample proportion*.

EXAMPLE 7.16 According to recent data, the nation's five most popular theme parks are shown in Table 7.8. The table gives the name of the theme park and indicates whether or not the attendance exceeds 10 million per year.

Table 7.8

Theme park	Attendance exceeds 10 million
A: Disneyland (Anaheim)	yes
B: Magic Kingdom (Disney World)	yes
C: Epcot (Disney World)	yes
D: Disney/MGM Studios (Disney World)	no
E: Universal Studios Florida (Orlando)	no

For this population of size N = 5, the proportion of theme parks with attendance exceeding 10 million is $p = \dfrac{3}{5} = $ .60 or 60%. There are 5 samples of size 4 possible when selected from this population. These samples, the theme parks exceeding 10 million (yes or no), the sample proportion, and the probability associated with the sample proportion, are given in Table 7.9.

Table 7.9

Sample	Exceeds 10 million	Sample proportion	Probability
A, B, C, D	y, y, y, n	.75	.2
A, B, C, E	y, y, y, n	.75	.2
A, B, D, E	y, y, n, n	.50	.2
A, C, D, E	y, y, n, n	.50	.2
B, C, D, E	y, y, n, n	.50	.2

For each sample, the sampling error, $|p - \overline{p}|$, is either .10 or .15. From Table 7.9, it is seen that the probability associated with sample proportion .75 is .2 + .2 = .4 and the probability associated with sample proportion .50 is .2 + .2 + .2 = .6. The sampling distribution for $\overline{p}$ is given in Table 7.10.

Table 7.10

$\overline{p}$	.50	.75
$P(\overline{p})$	.6	.4

For larger populations and samples, the sampling distribution of the sample proportion is more difficult to construct, but the technique is the same.

MEAN AND STANDARD DEVIATION OF THE SAMPLE PROPORTION

Since the sample proportion has a distribution, it is a random variable and has a mean and a standard deviation. The mean of the sample proportion is represented by the symbol $\mu_{\overline{p}}$ and the standard deviation of the sample proportion is represented by $\sigma_{\overline{p}}$. The standard deviation of the sample proportion is called the *standard error of the proportion*. Example 7.17 illustrates how to find the mean of the sample proportion and the standard error of the proportion.

EXAMPLE 7.17 For the sampling distribution of the sample proportion shown in Table 7.10, the mean is

$$\mu_{\overline{p}} = \Sigma \ \overline{p}P(\overline{p}) = .5 \times .6 + .75 \times .4 = 0.6$$

The variance of the sample proportion is

$$\sigma_{\overline{p}}^2 = \Sigma \ \overline{p}^2 P(\overline{p}) - \mu_{\overline{p}}^2 = .25 \times .6 + .5625 \times .4 - .36 = 0.015$$

The standard error of the proportion, $\sigma_{\overline{p}}$, is equal to $\sqrt{0.015} = 0.122$.

The relationship between the mean of the sample proportion and the population proportion is expressed by

$$\mu_{\overline{p}} = p \qquad\qquad (7.7)$$

The standard error of the sample proportion is related to the population proportion, the population size, and the sample size by

$$\sigma_{\overline{p}} = \sqrt{\frac{p(1-p)}{n}} \times \sqrt{\frac{N-n}{N-1}} \qquad\qquad (7.8)$$

The term $\sqrt{\dfrac{N-n}{N-1}}$ is called the *finite population correction factor*. If the sample size n is less than 5% of the population size, i.e., $n < .05N$, the finite population correction factor is very near one and is omitted in formula (7.8).

If $n < .05N$, the *standard error of the proportion* is given by formula (7.9), where $q = 1 - p$.

$$\sigma_{\overline{p}} = \sqrt{\frac{pq}{n}} \qquad\qquad (7.9)$$

EXAMPLE 7.18 In Example 7.16, dealing with the five most popular theme parks, it was shown that $p = 0.6$ and in Example 7.17, it was shown that $\mu_{\overline{p}} = 0.6$ illustrating formula (7.7). In Example 7.17 it was also shown that $\sigma_{\overline{p}} = \sqrt{0.015} = 0.122$. To illustrate formula (7.8), note that

$$\sqrt{\frac{p(1-p)}{n}} \times \sqrt{\frac{N-n}{N-1}} = \sqrt{\frac{.6 \times .4}{4}} \times \sqrt{\frac{5-4}{5-1}} = \sqrt{0.015} = 0.122 = \sigma_{\bar{p}}$$

EXAMPLE 7.19 Suppose 80% of all companies use e-mail. In a survey of 100 companies, the standard error of the sample proportion using e-mail is $\sigma_{\bar{p}} = \sqrt{\frac{pq}{n}} = \sqrt{\frac{.8 \times .2}{100}} = .04$. The finite population correction factor is not needed since there is a very large number of companies, and it is reasonable to assume that $n < .05N$.

SHAPE OF THE SAMPLING DISTRIBUTION OF THE SAMPLE PROPORTION AND THE CENTRAL LIMIT THEOREM

When the sample size satisfies the inequalities $np > 5$ and $nq > 5$, the sampling distribution of the sample proportion is normally distributed with mean equal to p and standard error $\sigma_{\bar{p}}$. This result is sometimes referred to as the *central limit theorem for the sample proportion*. This result is illustrated in Fig. 7-4.

EXAMPLE 7.20 Approximately 20% of the adults 25 and older have a bachelor's degree in the United States. If a large number of samples of adults 25 and older, each of size 100, were taken across the country, then the sample proportions having a bachelor's degree would vary from sample to sample. The distribution of sample proportions would be normally distributed with a mean equal to 20% and a standard error equal to $\sqrt{\frac{20 \times 80}{100}} =$ 4%. According to the empirical rule, approximately 68% of the sample proportions would fall between 16% and 24%, approximately 95% of the sample proportions would fall between 12% and 28%, and approximately 99.7% would fall between 8% and 32%. The sample proportion distribution may be assumed to be normally distributed since $np = 20$ and $nq = 80$ are both greater than 5.

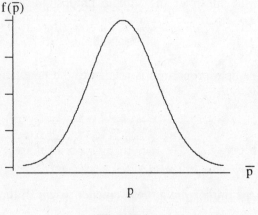

Fig. 7-4

APPLICATIONS OF THE SAMPLING DISTRIBUTION OF THE SAMPLE PROPORTION

The theory underlying the sample proportion is utilized in numerous statistical applications. The *margin of error, control chart limits,* and many other useful statistical techniques make use of the sampling distribution theory connected with the sample proportion.

EXAMPLE 7.21 It is estimated that 42% of women and 36% of men ages 45 to 54 are overweight, that is, 20% over their desirable weight. The probability that one-half or more in a sample of 50 women ages 45 to 54 are overweight is expressed as $P(\overline{p} \geq .5)$. The distribution of the sample proportion is normal since $np = 50 \times$ .42 = 21 > 5 and $nq = 50 \times .58 = 29 > 5$. The mean of the distribution is .42 and the standard error is $\sigma_{\overline{p}} = \sqrt{\dfrac{pq}{n}}$ $= \sqrt{\dfrac{.42 \times .58}{50}} = .07$. The probability, $P(\overline{p} \geq .5)$, is shown as the shaded area in Fig. 7-5.

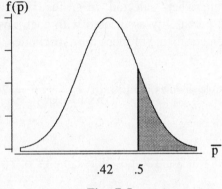

Fig. 7-5

To find the area under the normal curve shown in Fig. 7-5, it is necessary to transform the value .5 to a standard normal value. The value for z is $z = \dfrac{.5 - .42}{.07} = 1.14$. The area to the right of $\overline{p} = .5$ is equal to the area to the right of $z = 1.14$ and is given as follows.

$$P(\overline{p} \geq .5) = P(z > 1.14) = .5 - .3729 = .1272$$

In Example 7.21, the $\overline{p}$ value equal to .5 was transformed to a z value by subtracting the mean value of $\overline{p}$ from .5 and then dividing the result by the standard error of the proportion. The equation for transforming a sample proportion value to a z value is given by

$$z = \frac{\overline{p} - \mu_{\overline{p}}}{\sigma_{\overline{p}}} \tag{7.10}$$

EXAMPLE 7.22 It is estimated that 1 out of 5 individuals over 65 have Parkinsonism, that is, signs of Parkinson's disease. The probability that 15% or less in a sample of 100 such individuals have Parkinsonism is represented as $P(\overline{p} \leq 15\%)$. Since $np = 100 \times .20 = 20 > 5$ and $nq = 100 \times .80 = 80 > 5$, it may be assumed that $\overline{p}$ has a normal distribution. The mean of $\overline{p}$ is 20% and the standard error is $\sigma_{\overline{p}} = \sqrt{\dfrac{pq}{n}} = \sqrt{\dfrac{20 \times 80}{100}} = 4\%$. Formula (7.10) is used to find an event involving z which is equivalent to the event $\overline{p} \leq 15$. The event $\overline{p} \leq 15$ is equivalent to the event $z = \dfrac{\overline{p} - 20}{4} < \dfrac{15 - 20}{4} = -1.25$, and therefore we have $P(\overline{p} \leq 15\%) = P(z < -1.25) = .5 - .3944 = .1056$.

Solved Problems

SIMPLE RANDOM SAMPLING

7.1   The top nine medical doctor specialties in terms of median income are as follows: radiology, surgery, anesthesiology, obstetrics/gynecology, pathology, internal medicine, psychiatry, family practice, and pediatrics. How many simple random samples of size 4, chosen without replacement, are possible when selected from the population consisting of these nine specialties? What is the probability associated with each possible simple random sample of size 4 from the population consisting of these nine specialties?

Ans: The number of possible samples is equal to $\binom{9}{4} = \dfrac{9!}{4!5!} = 126$. Each possible sample has probability $\dfrac{1}{126} = .00794$ of being selected.

7.2   In problem 7.1, how many of the 126 possible samples of size 4 would include the specialty *pediatrics*?

Ans: The number of samples of size 4 which include the specialty *pediatrics* would equal the number of ways the other three specialties in the sample could be selected from the other eight specialties in the population. The number of ways to select 3 from 8 is $\binom{8}{3} = \dfrac{8!}{3!5!} = 56$. One such sample is the sample consisting of the following: (radiology, surgery, pathology, *pediatrics*). There are 55 other such samples containing the specialty *pediatrics*.

USING RANDOM NUMBER TABLES

Table 7.11

1. Alabama	18. Kentucky	35. North Dakota
2. Alaska	19. Louisiana	36. Ohio
3. Arizona	20. Maine	37. Oklahoma
4. Arkansas	21. Maryland	38. Oregon
5. California	22. Massachusetts	39. Pennsylvania
6. Colorado	23. Michigan	40. Rhode Island
7. Connecticut	24. Minnesota	41. South Carolina
8. Delaware	25. Mississippi	42. South Dakota
9. District of Columbia	26. Missouri	43. Tennessee
10. Florida	27. Montana	44. Texas
11. Georgia	28. Nebraska	45. Utah
12. Hawaii	29. Nevada	46. Vermont
13. Idaho	30. New Hampshire	47. Virginia
14. Illinois	31. New Jersey	48. Washington
15. Indiana	32. New Mexico	49. West Virginia
16. Iowa	33. New York	50. Wisconsin
17. Kansas	34. North Carolina	51. Wyoming

7.3   Use Table 7.1 of this chapter to obtain a sample of size 5 from the population consisting of the 50 states and the District of Columbia. Assume that the 51 members of the population are listed in alphabetical as shown in Table 7.11. Start with the two digits in columns 11 and 12, and line 1 of Table 7.1. Read down these two columns until five unique states have been selected. Give the numbers and the selected states.

Ans: The random numbers are 44, 12, 24, 10, and 07. From an alphabetical listing of the 50 states and the District of Columbia in Table 7.11, the following sample is obtained: (Connecticut, Florida, Hawaii, Minnesota, and Texas)

7.4 The random numbers in Table 7.1 are used to select 15 faculty members from an alphabetical listing of the 425 faculty members at a university. The total faculty listing goes from 001: Lawrence Allison to 425: Carol Ziebarth. A random start position is determined to be columns 31 through 33 and line 1. Read down these columns until you reach the end. Then go to columns 36 through 38, line 1 and read down these columns until you reach the end. If necessary, go to columns 41 through 43 and line 1 and read down these columns until you obtain 15 unique numbers. Give the 15 numbers which determine the 15 selected faculty members.

Ans: The numbers, in the order obtained from Table 7.1 are: 345, 254, 160, 105, 283, 026, 099, 385, 157, 393, 013, 126, 110, 004, and 279.

USING THE COMPUTER TO OBTAIN A SIMPLE RANDOM SAMPLE

7.5 In reference to problem 7.3, use Minitab to obtain a sample of 5 of the 50 states plus the District of Columbia.

Ans: MTB > set c1
DATA > 1:51
DATA > end
MTB > sample 5 c1 c2
MTB > print c2

Data Display
C2
3   21   38   44   36

From the alphabetical listing of the states shown in Table 7.11, the following sample is obtained:

(Arizona, Maryland, Ohio, Oregon, and Texas)

Texas is the only state which is common to the samples obtained in problems 7.3 and 7.5.

7.6 In reference to problem 7.4, use Minitab to obtain a sample of 15 of the 425 faculty members.

Ans: MTB > set c1
DATA > 1:425
DATA > end
MTB > sample 15 c1 c2
MTB > sort c2 put into c3
MTB > print c3

Data Display
C3
58   73   111   112   187   228   239   283   285   319   322   325   364   384   394
The faculty member corresponding to number 283 is the only one that is common to the samples obtained in problems 7.4 and 7.6.

SYSTEMATIC RANDOM SAMPLING

7.7 In reference to problem 7.3, choose a systematic sample of 5 of the 50 states plus the District of Columbia.

   Ans: Divide 51 by 5 to obtain 10.2. Round 10.2 down to 10 and select a number randomly between 1 and 10. Suppose the number 7 is obtained. The other 4 members of the sample are 17, 27, 37, and 47. From the alphabetical listing of the states given in Table 7.11, the following sample is selected:

(Connecticut, Kansas, Montana, Oklahoma, Virginia)

CLUSTER SAMPLING

7.8 A large school district wishes to obtain a measure of the mathematical competency of the junior high students in the district. The district consists of 40 different junior high schools. Describe how you could use cluster sampling to obtain a measure of the mathematical competency of the junior high school students in the district.

   Ans: Randomly select a small number, say 5, of the 40 junior high schools. Then administer a test of mathematical competency to all students in the 5 selected schools. The test results from the 5 schools constitute a cluster sample.

STRATIFIED SAMPLING

7.9 Refer to problem 7.8. Explain how you could use stratified sampling to determine the mathematical competency of the junior high students in the district.

   Ans: Consider each of the 40 junior high schools as a stratum. Randomly select a sample from each junior high proportional to the number of students in that junior high school and administer the mathematical competency test to the selected students. Note that even though stratified sampling could be used, cluster sampling as described in problem 7.8 would be easier to administer.

SAMPLING DISTRIBUTION OF THE SAMPLING MEAN

7.10 The five cities with the most African-American-owned businesses was given in Table 7.2 and is reproduced below.

Table 7.2

City	Number of African-American-owned businesses, in thousands
A: New York	42
B: Washington, D.C.	39
C: Los Angeles	36
D: Chicago	33
E: Atlanta	30

List all samples of size 4 and find the mean of each sample. Also, construct the sampling distribution of the sample mean.

Ans: Table 7.12 lists the 5 samples of size 4, along with the sample values for each sample, and the mean of each sample, and Table 7.13 gives the sampling distribution of the sample mean.

Table 7.12

Samples	Number of businesses in the samples	Sample mean
A, B, C, D	42, 39, 36, 33	37.50
A, B, C, E	42, 39, 36, 30	36.75
A, B, D, E	42, 39, 33, 30	36.00
A, C, D, E	42, 36, 33, 30	35.25
B, C, D, E	39, 36, 33, 30	34.50

Table 7.13

$\overline{x}$	34.50	35.25	36.00	36.75	37.50
$P(\overline{x})$	.2	.2	.2	.2	.2

7.11 Consider sampling from an infinite population. Suppose the number of firearms of any type per household in this population is distributed uniformly with 25% having no firearms, 25% having exactly one, 25% having exactly two, and 25% having exactly three. If X represents the number of firearms per household, then X has the distribution shown in Table 7.14.

Table 7.14

x	0	1	2	3
P(x)	.25	.25	.25	.25

Give the sampling distribution of the sample mean for all samples of size two from this population.

Table 7.15

Possible sample values x_1 and x_2	Sample mean	Probability
0, 0	0	.0625
0, 1	0.5	.0625
0, 2	1.0	.0625
0, 3	1.5	.0625
1, 0	0.5	.0625
1, 1	1.0	.0625
1, 2	1.5	.0625
1, 3	2.0	.0625
2, 0	1.0	.0625
2, 1	1.5	.0625
2, 2	2.0	.0625
2, 3	2.5	.0625
3, 0	1.5	.0625
3, 1	2.0	.0625
3, 2	2.5	.0625
3, 3	3.0	.0625

Ans: Suppose x_1 and x_2 represents the two possible values for the two households sampled. Table 7.15 gives the possible values for x_1 and x_2, the mean of the sample values, and the probability associated with each pair of sample values. Since there are 16 different possible sample pairs each one has probability $\frac{1}{16} = .0625$. The possible values for the sample mean are 0, 0.5, 1.0, 1.5, 2.0,

2.5, and 3.0. The probability associated with each sample mean value can be determined from Table 7.15. For example, the sample mean equals 1.5 for four different pairs and the probability for the value 1.5 is obtained by adding the probabilities associated with (0, 3), (1, 2), (2, 1), and (3, 0). That is, $P(\overline{x} = 1.5) = .0625 + .0625 + .0625 + .0625 = .25$. The sampling distribution is shown in Table 7.16.

Table 7.16

$\overline{x}$	0	0.5	1.0	1.5	2.0	2.5	3.0
$P(\overline{x})$	.0625	.1250	.1875	.2500	.1875	.1250	.0625

SAMPLING ERROR

7.12 Give the sampling error associated with each of the samples in problem 7.10.

Ans: The mean number of African-American-owned businesses for the five cities is 36,000. There are five different possible samples as shown in Table 7.17. Each of the sample means shown in Table 7.17 is an estimate of the population mean. The sampling error is $|\overline{x} - 36|$, and is shown for each sample in the last column in Table 7.17.

Table 7.17

Samples	Sample mean	Sampling error
A, B, C, D	37.50	1.50
A, B, C, E	36.75	0.75
A, B, D, E	36.00	0
A, C, D, E	35.25	0.75
B, C, D, E	34.50	1.50

7.13 Give the minimum and the maximum sampling error encountered in the sampling described in problem 7.11 and the probability associated with the minimum and maximum sampling error.

Ans: The population described in problem 7.11 has a uniform distribution and the mean is as follows: $\mu = 0 \times .25 + 1 \times .25 + 2 \times .25 + 3 \times .25 = 1.5$. The minimum error is 0 and occurs when $\overline{x} = 1.5$. The probability of a sampling error equal to 0 is equal to the probability that the sample mean is equal to 1.5 which is .25 as shown in Table 7.16. The maximum sampling error is 1.5 and occurs when $\overline{x} = 0$ or $\overline{x} = 3.0$. The probability associated with the maximum sampling error is .0625 + .0625 = .125.

MEAN AND STANDARD DEVIATION OF THE SAMPLE MEAN

7.14 Find the mean and variance of the sampling distribution of the sample mean derived in problem 7.10 and given in Table 7.13. Verify that formulas (7.2) and (7.3) hold for this problem.

Ans: Table 7.13 is reprinted below for ease of reference. The mean of the sample mean is

$$\mu_{\overline{x}} = \Sigma \ \overline{x} P(\overline{x}) = 34.50 \times .2 + 35.25 \times .2 + 36.00 \times .2 + 36.75 \times .2 + 37.50 \times .2 = 36$$

The variance of the sample mean is $\sigma_{\overline{x}}^2 = \Sigma \ \overline{x}^2 P(\overline{x}) - \mu_{\overline{x}}^2$.

$$\Sigma \ \overline{x}^2 P(\overline{x}) = 1190.25 \times .2 + 1242.5625 \times .2 + 1296 \times .2 + 1350.5625 \times .2 + 1406.25 \times .2$$
$$= 1297.125 \text{ and } \mu_{\overline{x}}^2 = 1296, \text{ and therefore } \sigma_{\overline{x}}^2 = 1297.125 - 1296 = 1.125.$$

Table 7.13

$\overline{x}$	34.50	35.25	36.00	36.75	37.50
$P(\overline{x})$	.2	.2	.2	.2	.2

In Example 7.7 it is shown that $\mu = 36$ and $\sigma^2 = 18$. Formula (7.2) is verified since $\mu = \mu_{\overline{x}} = 36$. To verify formula (7.3), note that $\dfrac{\sigma^2}{n} \times \dfrac{N-n}{N-1} = \dfrac{18}{4} \times \dfrac{5-4}{5-1} = 1.125 = \sigma_{\overline{x}}^2$.

7.15 Find the mean and variance of the sampling distribution of the sample mean derived in problem 7.11 and given in Table 7.16. Verify that formulas (7.2) and (7.3) hold for this problem.

Ans: Table 7.16 is reprinted below for ease of reference. The mean of the sample mean is
$$\mu_{\overline{x}} = \Sigma \overline{x}\ P(\overline{x})$$
$$= 0 \times .0625 + 0.5 \times .125 + 1 \times .1875 + 1.5 \times .25 + 2 \times .1875 + 2.5 \times .125 + 3 \times .0625$$
$= 1.5$. The variance of the sample mean is $\sigma_{\overline{x}}^2 = \Sigma\ \overline{x}^2 P(\overline{x}) - \mu_{\overline{x}}^2$.

$$\Sigma\ \overline{x}^2 P(\overline{x}) = 0 \times .0625 + .25 \times .125 + 1 \times .1875 + 2.25 \times .25 + 4 \times .1875 + 6.25 \times .125 + 9 \times .0625$$
$= 2.875$ and $\mu_{\overline{x}}^2 = 2.25$, and therefore $\sigma_{\overline{x}}^2 = 2.875 - 2.25 = 0.625$.

Table 7.16

$\overline{x}$	0	0.5	1.0	1.5	2.0	2.5	3.0
$P(\overline{x})$	.0625	.1250	.1875	.2500	.1875	.1250	.0625

In order to verify formulas (7.2) and (7.3), we need to find the values for μ and σ^2 in problem 7.11. The population distribution was given in table 7.14 and is shown below.

Table 7.14

x	0	1	2	3
P(x)	.25	.25	.25	.25

The population mean is $\mu = \Sigma\ xP(x) = 0 \times .25 + 1 \times .25 + 2 \times .25 + 3 \times .25 = 1.5$, and the variance is given by $\sigma^2 = \Sigma\ x^2 P(x) - \mu^2 = 0 \times .25 + 1 \times .25 + 4 \times .25 + 9 \times .25 - 2.25 = 1.25$.

Formula (7.2) is verified, since $\mu = \mu_{\overline{x}} = 1.5$. To verify formula (7.3), note that since the population is infinite, the finite population correction factor is not needed and $\dfrac{\sigma^2}{n} = \dfrac{1.25}{2} = 0.625$ $= \sigma_{\overline{x}}^2$. Anytime the population is infinite, the finite population correction factor is omitted and formula (7.3) simplifies to $\sigma_{\overline{x}}^2 = \dfrac{\sigma^2}{n}$.

SHAPE OF THE SAMPLING DISTRIBUTION OF THE MEAN AND THE CENTRAL LIMIT THEOREM

7.16 The portfolios of wealthy people over the age of 50 produce yearly retirement incomes which are normally distributed with mean equal to $125,000 and standard deviation equal to $25,000. Describe the distribution of the means of samples of size 16 from this population.

Ans: Since the population of yearly retirement incomes are normally distributed, the distribution of sample means will be normally distributed for any sample size. The distribution of sample means has mean \$125,000 and standard error $\sigma_{\bar{x}} = \dfrac{\sigma}{\sqrt{n}} = \dfrac{25,000}{\sqrt{16}} = \$6,250$.

7.17 The mean age of nonresidential buildings is 30 years and the standard deviation of the ages of nonresidential buildings is 5 years. The distribution of the ages is not normally distributed. Describe the distribution of the means of samples from the distribution of ages of nonresidential buildings for sample sizes $n = 10$ and $n = 50$.

Ans: The distribution type for the sample mean is unknown for small samples, i.e., $n < 30$. Therefore we cannot say anything about the distribution of $\bar{x}$ when $n = 10$. For large samples, i.e., ≥ 30, the central limit theorem assures us that $\bar{x}$ has a normal distribution. Therefore, for $n = 50$, the sample mean has a normal distribution with mean = 30 years and standard error $\sigma_{\bar{x}} = \dfrac{\sigma}{\sqrt{n}} = \dfrac{5}{\sqrt{50}} = 0.707$ years.

APPLICATIONS OF THE SAMPLING DISTRIBUTION OF THE SAMPLE MEAN

7.18 In problem 7.16, find the probability of selecting a sample of 16 wealthy individuals whose portfolios produce a mean retirement income exceeding \$135,000.

Ans: We are asked to determine the probability that $\bar{x}$ exceeds \$135,000. From problem 7.16, we know that $\bar{x}$ has a normal distribution with mean \$125,000 and standard error \$6,250. The event $\bar{x} > 135,000$ is equivalent to the event $z = \dfrac{\bar{x} - 125,000}{6,250} > \dfrac{135,000 - 125,000}{6,250} = 1.60$. Since the event $\bar{x} > 135,000$ is equivalent to the event $z > 1.60$, the two events have equal probabilities. That is, $P(\bar{x} > 135,000) = P(Z > 1.60)$. Using the standard normal distribution table, the probability $P(z > 1.60)$ is equal to $.5 - .4452 = .0548$. That is, $P(\bar{x} > 135,000) = P(Z > 1.60) = .0548$.

7.19 In problem 7.17, determine the probability that a random sample of 50 nonresidential buildings will have a mean age of 27.5 years or less.

Ans: We are asked to determine $P(\bar{x} < 27.5)$. From problem 7.17, we know that $\bar{x}$ has a normal distribution with mean 30 years and standard error 0.707 years. Since the event $\bar{x} < 27.5$ is equivalent to the event $z = \dfrac{\bar{x} - 30}{.707} < \dfrac{27.5 - 30}{.707} = -3.54$, we have $P(\bar{x} < 27.5) = P(z < -3.54)$. From the standard normal distribution table, the probability is less than .001.

SAMPLING DISTRIBUTION OF THE SAMPLE PROPORTION

7.20 Approximately 8.8 million of the 12.8 million individuals receiving Aid to Families with Dependent children are 18 or younger. What proportion of the individuals receiving such aid are 18 or younger?

Ans: The population consists of all individuals receiving Aid to Families with Dependent Children. The proportion having the characteristic that an individual receiving such aid is 18 or younger is represented by p and is equal to $\dfrac{8.8}{12.8} = 0.69$.

7.21 Table 7.18 describes a population consisting of five states and indicates whether or not there is at least one woman on death row for that state.

Table 7.18

State	At least one woman on death row
A: Alabama	yes
B: California	yes
C: Colorado	no
D: Kentucky	no
E: Nebraska	no

For this population 40% of the states have at least one woman on death row, that is, p = 0.4. List all samples of size 2 from this population and find the proportion having at least one woman on death row for each sample. Use this listing to derive the sampling distribution for the sample proportion.

Ans: Table 7.19 lists all possible samples of size 2, indicates whether or not the states in the sample have at least one woman on death row, gives the sample proportion for each sample, and gives the probability for each sample.

Table 7.19

Sample	At least one woman on death row	Sample proportion	Probability
A, B	y, y	1	.1
A, C	y, n	.5	.1
A, D	y, n	.5	.1
A, E	y, n	.5	.1
B, C	y, n	.5	.1
B, D	y, n	.5	.1
B, E	y, n	.5	.1
C, D	n, n	0	.1
C, E	n, n	0	.1
D, E	n, n	0	.1

From Table 7.19, it is seen that $\overline{p}$ takes on the values 0, .5, and 1 with probabilities .3, .6, and .1, respectively. The probability that $\overline{p} = 0$ is obtained by adding the probabilities for the samples (C, D), (C, E), and (D, E) since $\overline{p} = 0$ if any one of these samples is selected. The other two probabilities are obtained similarly. Table 7.20 gives the sampling distribution for $\overline{p}$.

Table 7.20

$\overline{p}$	0	.5	1
$P(\overline{p})$	.3	.6	.1

MEAN AND STANDARD DEVIATION OF THE SAMPLE PROPORTION

7.22 Find the mean and variance of the sample proportion in problem 7.21 and verify that formula (7.7) and formula (7.8) are satisfied.

Ans: The mean of the sample proportion is $\mu_{\overline{p}} = \sum \overline{p} P(\overline{p}) = 0 \times .3 + .5 \times .6 + 1 \times .1 = .4$, and the

variance of the sample proportion is $\sigma_{\overline{p}}^2 = \sum \overline{p}^2 P(\overline{p}) - \mu_{\overline{p}}^2 = 0 \times .3 + .25 \times .6 + 1 \times .1 - (.4)^2$

$= .09$. Formula (7.7) is satisfied since $\mu_{\overline{p}} = p = .4$. To verify formula (7.8), we need to show that

$\sigma_{\overline{p}} = \sqrt{\dfrac{p(1-p)}{n}} \times \sqrt{\dfrac{N-n}{N-1}}$. Since the variance of the sample proportion is .09, the standard

deviation is the square root of .09 or .3. Therefore, to verify formula (7.8), all we need do is show that the right-hand side of the equation equals .3.

$$\sqrt{\frac{p(1-p)}{n}} \times \sqrt{\frac{N-n}{N-1}} = \sqrt{\frac{.4 \times .6}{2}} \times \sqrt{\frac{5-2}{5-1}} = .3 = \sigma_{\overline{p}}$$

7.23 Suppose 5% of all adults in America have 10 or more credit cards. Find the standard error of the sample proportion in a sample of 1,000 American adults who have 10 or more credit cards.

Ans: Since the sample size is less than 5% of the population size, the standard error is $\sigma_{\overline{p}} = \sqrt{\dfrac{pq}{n}} =$

$\sqrt{\dfrac{.05 \times .95}{1,000}} = 0.0069.$

SHAPE OF THE SAMPLING DISTRIBUTION OF THE SAMPLE PROPORTION AND THE CENTRAL LIMIT THEOREM

7.24 A sample of size n is selected from a large population. The proportion in the population with a specified characteristic is p. The proportion in the sample with the specified characteristic is $\overline{p}$. In which of the following does $\overline{p}$ have a normal distribution?
(a) n = 20, p = .9   (b) n = 15, p = .4   (c) n = 100, p = .97   (d) n = 1000, p = .01

Ans: (a) np = 18, and nq = 2. Since both np and nq do not exceed 5, we are not sure that the distribution of $\overline{p}$ is normal.
(b) np = 6, and nq = 9. Since both np and nq exceed 5, the distribution of $\overline{p}$ is normal.
(c) np = 97, and nq = 3. Since both np and nq do not exceed 5, we are not sure that the distribution of $\overline{p}$ is normal.
(d) np = 10, and nq = 990. Since both np and nq exceed 5, the distribution of $\overline{p}$ is normal.

APPLICATIONS OF THE SAMPLING DISTRIBUTION OF THE SAMPLE PROPORTION

7.25 Approximately 15% of the population is left-handed. What is the probability that in a sample of 50 randomly chosen individuals, 30% or more in the sample will be left-handed? That is, what is the probability of finding 15 or more left-handers in the 50?

Ans: The sample proportion, $\overline{p}$, has a normal distribution since np = 50 × .15 = 7.5 > 5 and nq = 50 ×

.85 = 42.5 > 5. The mean of the sample proportion is 15% and the standard error is $\sigma_{\overline{p}} = \sqrt{\dfrac{pq}{n}} =$

$\sqrt{\dfrac{15 \times 85}{50}} = 5.05\%$. To find the probability that $\overline{p} \geq 30\%$, we first transform $\overline{p}$ to a standard normal by the transormation $z = \dfrac{\overline{p} - \mu_{\overline{p}}}{\sigma_{\overline{p}}}$ as follows. The inequality $\overline{p} \geq 30$ is equivalent to $z =$

$\dfrac{\overline{p} - 15}{5.05} \geq \dfrac{30 - 15}{5.05} = 2.97$. Because $\overline{p} \geq 30$ is equivalent to $z \geq 2.97$ we have $P(\overline{p} \geq 30) =$ $P(z \geq 2.97) = .5 - .4985 = .0015$. That is, 15 or more left-handers will be found in a sample of 50 individuals only about 0.2% of the time.

7.26 Suppose 70% of the population support the ban on assault weapons. What is the probability that between 65% and 75% in a poll of 100 individuals will support the ban on assault weapons?

> *Ans:* We are asked to find $P(65 < \overline{p} < 75)$. The sample proportion $\overline{p}$ has a normal distribution with mean 70% and standard error $\sigma_{\overline{p}} = \sqrt{\dfrac{pq}{n}} = \sqrt{\dfrac{70 \times 30}{100}} = 4.58\%$. The event $65 < \overline{p} < 75$ is equivalent to the event $\dfrac{65 - 70}{4.58} < \dfrac{\overline{p} - 70}{4.58} < \dfrac{75 - 70}{4.58}$ or $-1.09 < z < 1.09$ and since these events are equivalent, they have equal probabilities. That is, $P(65 < \overline{p} < 75) = P(-1.09 < z < 1.09) = 2 \times .3621 = .7242$.

Supplementary Problems

SIMPLE RANDOM SAMPLING

7.27 Rather than have all 25 students in her Statistics class complete a teacher evaluation form, Mrs. Jones decides to randomly select three students and have the department chairman interview the three concerning her teaching after the course grade has been given. How many different samples of size 3 can be selected?

> *Ans.* 2,300

7.28 *USA Today* lists the 1900 most active New York stock exchange issues. How many samples of size three are possible when selected from these 1900 stock issues?

> *Ans.* 1,141,362,300

USING RANDOM NUMBER TABLES

7.29 In a table of random numbers such as Table 7.1 what relative frequency would you expect for each of the digits 0, 1, 2, 3, 4, 5, 6, 7, 8, and 9?

> *Ans.* 0.1

7.30 The 100 U.S. Senators are listed in alphabetical order and then numbered as 00, 01, . . . , 99. Use Table 7.1 to select 10 of the senators. Start with the two digits in columns 31 and 32 and row 6. The first selected senator is numbered 76. Reading down the two columns from the number 76, what are the other 9 two-digit numbers of the other selected senators?

> *Ans.* 59, 78, 44, 93, 69, 25, 43, 50, and 45

USING THE COMPUTER TO OBTAIN A SIMPLE RANDOM SAMPLE

7.31　Use Minitab to select 25 of the 1900 stock issues discussed in problem 7.28. Give the numbers in ascending order of the 25 selected stock issues. Assume the stock issues are numbered from 1 to 1900.

Ans.　MTB > set c1
DATA > 1:1900
DATA > end
MTB > sample 25 from c1 put into c2
MTB > sort c2 put into c3
MTB > print c3

Data Display
C3

26	219	238	288	328	423	766	785	943	1006	1187
1197	1234	1238	1257	1313	1532	1562	1613	1639	1728	1784
1788	1798	1870								

7.32　Use Minitab to select 50 of the 3143 counties in the United States. Give the numbers in ascending order of the 50 selected counties. Assume the counties are in alphabetical order and are numbered from 1 to 3143.

Ans.　MTB > set c1
DATA > 1:3143
DATA > end
MTB > sample 50 from c1 put into c2
MTB > sort c2 put into c3
MTB > print c3

Data Display
C3

33	47	53	139	265	312	321	343	444	519	599
652	658	764	789	829	949	964	1063	1134	1209	1300
1463	1567	1630	1694	1706	1818	1848	2021	2048	2138	2143
2270	2285	2306	2311	2408	2414	2463	2487	2513	2658	2701
2718	2763	2862	2892	2917	2975					

SYSTEMATIC RANDOM SAMPLING

7.33　Describe how the state patrol might obtain a systematic random sample of the speeds of vehicles along a stretch of interstate 80.

Ans.　Use a radar unit to measure the speeds of systematically chosen vehicles along the stretch of interstate 80. For example, measure the speed of every tenth vehicle.

CLUSTER SAMPLING

7.34　A particular city is composed of 850 blocks and each block contains approximately 20 homes. Fifteen of the 850 blocks are randomly selected and each household on the selected block is administered a survey concerning issues of interest to the city council. How large is the population? How large is the sample? What type of sampling is being used?

Ans.　The population consists of 17,000 households. The sample consists of 300 households. Cluster sampling is being used.

STRATIFIED SAMPLING

7.35 A drug store chain has stores located in 5 cities as follows: 20 stores in Los Angeles, 40 stores in New York, 20 stores in Seattle, 10 stores in Omaha, and 30 stores in Chicago. In order to estimate pharmacy sales, the following number of stores are randomly selected from the 5 cities: 4 from Los Angeles, 8 from New York, 4 from Seattle, 2 from Omaha, and 6 from Chicago. What type of sampling is being used?

Ans.   stratified sampling

SAMPLING DISTRIBUTION OF THE SAMPLING MEAN

7.36 The export value in billions of dollars for four American cities for a recent year are as follows: A: Detroit, 28; B: New York, 24; C: Los Angeles, 22; and D: Seattle, 20. If all possible samples of size 2 are selected from this population of four cities and the sample mean value of exports computed for each sample, give the sampling distribution of the sample mean.

Ans. $P(\overline{x}) = \frac{1}{6}$, where $\overline{x} = 21, 22, 23, 24, 25,$ or 26.

7.37 Consider a large population of households. Ten percent of the households have no home computer, 60 percent of the households have exactly one home computer, and 30 percent of the households have exactly two home computers. Construct the sampling distribution for the mean of all possible samples of size 2.

Ans.

$\overline{x}$	0	.5	1	1.5	2
$P(\overline{x})$	.01	.12	.42	.36	.09

SAMPLING ERROR

7.38 In reference to problem 7.36, if the mean of a sample of two cities is used to estimate the mean export value of the four cities, what are the minimum and maximum values for the sampling error?

Ans.   minimum sampling error = .5   maximum sampling error = 2.5

7.39 In reference to problem 7.37, what are the possible sampling error values associated with samples of two households used to estimate the mean number of home computers per household for the population? What is the most likely value for the sampling error?

Ans.   2, .3, .7, .8, and 1.2 The most likely value is .2. The probability that the sampling error equals .2 is .42.

MEAN AND STANDARD DEVIATION OF THE SAMPLE MEAN

7.40 Find the mean and variance of the sampling distribution in problem 7.36. Verify that formulas (*7.2*) and (*7.3*) hold for this sampling distribution.

Ans. $\mu = 23.5$ $\sigma^2 = 8.75$ $\mu_{\overline{x}} = 23.5$ $\sigma^2_{\overline{x}} = 2.917$

$$\mu_{\overline{x}} = \mu \qquad \frac{\sigma^2}{n} \times \frac{N-n}{N-1} = \frac{8.75}{2} \times \frac{4-2}{4-1} = 2.917 = \sigma^2_{\overline{x}}$$

7.41 Find the mean and variance of the sampling distribution in problem 7.37. Verify that formulas (*7.2*) and (*7.3*) hold for this sampling distribution.

Ans. $\mu = 1.2$ $\sigma^2 = 0.36$ $\mu_{\bar{x}} = 1.2$ $\sigma_{\bar{x}}^2 = 0.18$

$\mu_{\bar{x}} = \mu$ Since the population is infinite, formula (*7.3*) becomes $\sigma_{\bar{x}}^2 = \dfrac{\sigma^2}{n}$.

$\dfrac{\sigma^2}{n} = \dfrac{.36}{.2} = 0.18 = \sigma_{\bar{x}}^2$.

SHAPE OF THE SAMPLING DISTRIBUTION OF THE MEAN
AND THE CENTRAL LIMIT THEOREM

7.42   For cities of 100,000 or more, the number of violent crimes per 1,000 residents is normally distributed with mean equal to 30 and standard deviation equal to 7. Describe the sampling distribution of means of samples of size 4 from such cities.

Ans.   The means of samples of size 4 will be normally distributed with mean equal to 30 and standard error equal to 3.5.

7.43   For cities of 100,000 or more, the mean total crime rate per 1,000 residents is 95 and the standard deviation of the total crime rate per 1,000 is 15. The distribution of the total crime rate per 1,000 residents for cities of 100,000 or more is not normally distributed. The distribution is skewed to the right. Describe the sampling distribution of the means of samples of sizes 10 and 50 from such cities.

Ans.   For samples of size 50, the central limit theorem assures us that the distribution of sample means is normally distributed with mean equal to 95 and standard error equal to 2.12. For samples of size 10 from a nonnormal distribution, the distribution form of the sample means is unknown.

APPLICATIONS OF THE SAMPLING DISTRIBUTION OF THE SAMPLE MEAN

7.44   In problem 7.42, find the probability of selecting four cities of population 100,000 or more whose mean number of violent crimes per 1,000 residents exceeds 40 violent crimes per 1,000 residents.

Ans.   0.0021

7.45   The mean number of bumped passengers per 10,000 passengers per day is 1.35 and the standard deviation is 0.25. For a random selection of 40 days, what is the probability that the mean number of bumped passengers per 10,000 passengers for the 40 days will be between 1.25 and 1.50 ?

Ans.   0.9938

SAMPLING DISTRIBUTION OF THE SAMPLE PROPORTION

7.46   The world output of bicycles in 1995 was 114 million. China produced 41 million bicycles in 1995. What proportion of the world's new bicycles in 1995 were produced by China?

Ans. $p = 0.36$

7.47   A company has 38,000 employees and the proportion of the employees who have a college degree equals 0.25. In a sample of 400 of the employees, 30 percent have a college degree. How many of the company employees have a college degree? How many in the sample have a college degree?

Ans.   9,500 of the company employees have a college degree and 120 in the sample have a degree.

MEAN AND STANDARD DEVIATION OF THE SAMPLE PROPORTION

7.48 One percent of the vitamin C tablets produced by an industrial process are broken. The process fills containers with 1,000 tablets each. At regular intervals, a container is selected and 100 of the 1,000 tablets are inspected for broken tablets. What is the mean value and standard deviation of $\overline{p}$, where $\overline{p}$ represents the proportion of broken tablets in the samples of 100 selected tablets?

Ans. The mean value of $\overline{p}$ is .01 and the standard deviation of $\overline{p}$ is $\sqrt{\dfrac{.01 \times .99}{100}} \times \sqrt{\dfrac{1,000 - 100}{1,000 - 1}} =$ .0094.

7.49 A survey reported that 20% of pregnant women smoke, 19% drink , and 13% use crack cocaine or other drugs. Assuming the survey results are correct, what is the mean and standard deviation of $\overline{p}$, where $\overline{p}$ is the proportion of smokers in samples of 300 pregnant women?

Ans. The mean value of $\overline{p}$ is 20% and the standard deviation of $\overline{p}$ is $\sqrt{\dfrac{20 \times 80}{300}} = 2.31\%$.

SHAPE OF THE SAMPLING DISTRIBUTION OF THE SAMPLE PROPORTION AND THE CENTRAL LIMIT THEOREM

7.51 For a sample of size 50, give the range of population proportion values for which $\overline{p}$ has an approximate normal distribution.

Ans. $.1 < p < .9$

APPLICATIONS OF THE SAMPLING DISTRIBUTION OF THE SAMPLE PROPORTION

7.52 Thirty-five percent of the athletes who participated in the 1996 summer Olympics in Atlanta are female. What is the probability of randomly selecting a sample of 50 of these athletes in which over half of those selected are female?

Ans. $P(\overline{p} > .50) = P(z > 2.22) = .0132$

7.53 It is estimated that 12% of Native Americans have diabetes. What is the probability of randomly selecting 100 Native Americans and finding 5 or fewer in the hundred whom are diabetic?

Ans. $P(\overline{p} \leq .05) = P(z < -2.15) = .0158$

7.54 A pair of dice is tossed 180 times. What is the probability that the sum on the faces is equal to 7 on 20% or more of the tosses?

Ans. $P(\overline{p} \geq .20) = .1170$

Chapter 8

Estimation and Sample Size Determination: One Population

POINT ESTIMATE

Estimation is the assignment of a numerical value to a population parameter or the construction of an interval of numerical values likely to contain a population parameter. The value of a sample statistic assigned to an unknown population parameter is called a *point estimate* of the parameter.

EXAMPLE 8.1 The mean starting salary for 10 randomly selected new graduates with a Masters of Business Administration (MBA) at Fortune 500 companies is found to be $56,000. Fifty-six thousand dollars is a point estimate of the mean starting salary for all new MBA degree graduates at Fortune 500 companies. The median cost for 350 homes selected from across the United States is found to equal $115,000. The value of the sample median, $115,000, is a point estimate of the median cost of a home in the United States. A survey of 950 households finds that 35% have a home computer. Thirty-five percent is a point estimate of the percentage of homes that have a home computer.

INTERVAL ESTIMATE

In addition to a point estimate, it is desirable to have some idea of the size of the sampling error, that is the difference between the population parameter and the point estimate. By utilizing the standard error of the sample statistic and its sampling distribution, an *interval estimate* for the population parameter may be developed. A *confidence interval* is an interval estimate that consists of an interval of numbers obtained from the point estimate of the parameter along with a percentage that specifies how confident we are that the value of the parameter lies in the interval. The confidence percentage is called the *confidence level*. This chapter is concerned with the techniques for finding confidence intervals for population means and population proportions.

CONFIDENCE INTERVAL FOR THE POPULATION MEAN: LARGE SAMPLES

According to the central limit theorem, the sample mean, $\overline{x}$, has a normal distribution provided the sample size is 30 or more. Furthermore, the mean of the sample mean equals the mean of the population and the standard error of the mean equals the population standard deviation divided by the square root of the sample size. In Chapter 7, the variable given in formula (7.6) (and reproduced below) was shown to have a standard normal distribution provided n ≥ 30.

$$z = \frac{\overline{x} - \mu}{\sigma_{\overline{x}}} \tag{7.6}$$

Since 95% of the area under the standard normal curve is between z = −1.96 and z = 1.96, and since the variable in formula (7.6) has a standard normal distribution, we have the result shown in formula (8.1)

$$P(-1.96 < \frac{\overline{x} - \mu}{\sigma_{\overline{x}}} < 1.96) = .95 \qquad (8.1)$$

The inequality, $-1.96 < \dfrac{\overline{x} - \mu}{\sigma_{\overline{x}}} < 1.96$, is solved for μ and the result is given in formula (8.2).

$$\overline{x} - 1.96\sigma_{\overline{x}} < \mu < \overline{x} + 1.96\sigma_{\overline{x}} \qquad (8.2)$$

The interval given in formula (8.2) is called a 95% confidence interval for the population mean, μ. The general form for the interval is shown in formula (8.3), where z represents the proper value from the standard normal distribution table as determined by the desired confidence level.

$$\overline{x} - z\sigma_{\overline{x}} < \mu < \overline{x} + z\sigma_{\overline{x}} \qquad (8.3)$$

EXAMPLE 8.2   The mean age of policyholders at Mutual Insurance Company is estimated by sampling the records of 75 policyholders. The standard deviation of ages is known to equal 5.5 years and has not changed over the years. However, it is unknown if the mean age has remained constant. The mean age for the sample of 75 policyholders is 30.5 years. The standard error of the ages is $\sigma_{\overline{x}} = \dfrac{\sigma}{\sqrt{n}} = \dfrac{5.5}{\sqrt{75}} = .635$ years. In order to find a 90% confidence interval for μ, it is necessary to find the value of z in formula (8.3) for confidence level equal to 90%. If we let c be the correct value for z, then we are looking for that value of c which satisfies the equation $P(-c < z < c) = .90$. Or, because of the symmetry of the z curve, we are looking for that value of c which satisfies the equation $P(0 < z < c) = .45$. From the standard normal distribution table, we find $P(0 < z < 1.64) = .4495$ and $P(0 < z < 1.65) = .4505$. The interpolated value for c is 1.645, which we round to 1.65. Figure 8-1 illustrates the confidence level and the corresponding values of z. Now the 90% confidence interval is computed as follows. The lower limit of the interval is $\overline{x} - 1.65\sigma_{\overline{x}} = 30.5 - 1.65 \times .635 = 29.5$ years, and the upper limit is $\overline{x} + 1.65\sigma_{\overline{x}} = 31.5$. We are 90% confident that the mean age of all 250,000 policyholders is between 29.5 and 31.5 years. It is important to note that μ either is or is not between 29.5 and 31.5 years. To say we are 90% confident that the mean age of all policyholders is between 29.5 and 31.5 years means that if this study were conducted a large number of times and a confidence interval were computed each time, then 90% of all the possible confidence intervals would contain the true value of μ.

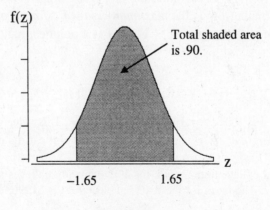

Fig. 8-1

Since it is time consuming to determine the correct value of z in formula (8.3), the values for the most often used confidence levels are given in Table 8.1. They are found in the same manner as illustrated in Example 8.2.

Table 8.1

Confidence level	Z value
80	1.28
90	1.65
95	1.96
99	2.58

EXAMPLE 8.3 The 911 response time for terrorists bomb threats was investigated. No historical data existed concerning the standard deviation or mean for the response times. When σ is unknown and the sample size is 30 or more, the standard deviation of the sample itself is used in place of σ when constructing a confidence interval for μ. A sample of 35 response times was obtained and it was found that the sample mean was 8.5 minutes and the standard deviation was 4.5 minutes. The estimated standard error of the mean is represented by $s_{\overline{x}}$ and is equal to $s_{\overline{x}} = \dfrac{s}{\sqrt{n}} = \dfrac{4.5}{\sqrt{35}} = .76$. From Table 8.1, the value for z is 2.58. The lower limit for the 99% confidence interval is $\overline{x} - 2.58 \times s_{\overline{x}} = 8.5 - 2.58 \times .76 = 6.54$ and the upper limit is $8.5 + 2.58 \times .76 = 10.46$. We are 99% confident that the mean response time is between 6.54 minutes and 10.46 minutes. The true value of μ may or may not be between these limits. However, 99% of all such intervals contain μ. It is this fact that gives us 99% confidence in the interval from 6.54 to 10.46.

For large samples, no assumption is made concerning the shape of the population distribution. If the population standard deviation is known, it is used in formula (8.3). If the population standard deviation is unknown, it is estimated by using the sample standard deviation. The value for z in formula (8.3) is found in the standard normal distribution table or, if applicable, by using Table 8.1.

MAXIMUM ERROR OF ESTIMATE FOR THE POPULATION MEAN

The inequality in formula (8.3) may be expressed as shown:

$$| \overline{x} - \mu | < z\sigma_{\overline{x}} \qquad (8.4)$$

The left-hand side of formula (8.4) is the *sampling error* when $\overline{x}$ is used as a point estimate of μ. The right-hand side of formula (8.4) is the *maximum error of estimate or margin of error* when $\overline{x}$ is used as a point estimate of μ. That is, when $\overline{x}$ is used as a point estimate of μ, the maximum error of estimate or margin of error, E, is

$$E = z\sigma_{\overline{x}} \qquad (8.5)$$

When the confidence level is 95%, $z = 1.96$ and $E = 1.96\sigma_{\overline{x}}$. This value of E, $1.96\sigma_{\overline{x}}$, is called the *95% margin of error* or simply *margin of error* when $\overline{x}$ is used as a point estimate of μ.

EXAMPLE 8.4 The annual college tuition costs for 40 community colleges selected from across the United States are given in Table 8.2. The mean for these 40 sample values is $1396, the standard deviation of the 40 values is $655, and the estimated standard error is $s_{\overline{x}} = \dfrac{s}{\sqrt{n}} = \dfrac{655}{\sqrt{40}} = \$104$. The mean tuition cost for all community colleges in the United States is represented by μ. A point estimate of μ is given by $1,396. The margin of error associated with this estimate is $1.96 \times 104 = \$204$. The 95% confidence interval for μ, based upon these data goes from $1396 - 204 = \$1,192$ to $1396 + 204 = \$1,600$. It is worth noting that the margin of error is actually ± $204, since the error may occur in either direction. Some publications give the margin of error as E and some give it as ± E. We shall omit the ± sign when giving the margin of error.

Table 8.2

1200	850	1750	930
850	3000	1650	1640
1700	2100	900	1320
1500	500	2050	1750
700	500	1780	2500
1200	1950	675	2310
1500	1000	1080	2900
2000	950	680	1875
1950	560	900	1450
750	500	1500	950

To compute the confidence interval for μ using Minitab, set the data given in Table 8.2 into column c1 and use the command **standard deviation c1** to compute the sample standard deviation. The command **zinterval 95% confidence, sigma = 655.44, data in c1** uses the data in c1 to compute a 95% confidence interval for μ. The output is shown below. The confidence interval extends from 1193 to 1599. The interval computed in Example 8.4 extends from 1192 to 1600. The difference in the answers is due to rounding.

```
MTB > set c1
DATA >  1200    850   1700   1500    700   1200   1500   2000   1950    750    850
DATA >  3000   2100    500    500   1950   1000    950    560    500   1750   1650
DATA >   900   2050   1780    675   1080    680    900   1500    930   1640   1320
DATA >  1750   2500   2310   2900   1875   1450    950
DATA > end
MTB > standard deviation c1
```

Column Standard Deviation

```
Standard deviation of C1 = 655.44
MTB > name c1 'cost'
MTB > zinterval 95% confidence, sigma = 655.44, data in c1
```

Confidence Intervals
The assumed sigma = 655

Variable	N	Mean	St Dev	SE Mean	95.0 % C.I.
Cost	40	1396	655	104	(1193, 1599)

The *width of a confidence interval* is equal to the upper limit of the interval minus the lower limit of the interval. In Example 8.4, the width of the 95% confidence interval is $1599 - 1193 = \$406$.

THE t DISTRIBUTION

When the sample size is less than 30 and the estimated standard error, $s_{\bar{x}} = \dfrac{s}{\sqrt{n}}$, is used in place of $\sigma_{\bar{x}}$ in formula (*8.3*), the width of the confidence interval for μ will generally be incorrect. The *t distribution*, also known as the *Student-t distribution*, is used rather than the standard normal distribution to find confidence intervals for μ when the sample size is less than 30. In this section we will discuss the properties of the t distribution, and in the next section, we will discuss the use of the t distribution for confidence intervals when the sample size is small. The t distribution is used in many different statistical applications.

The t distribution is actually a family of probability distributions. A parameter called the *degrees of freedom*, and represented by df, determines each separate t distribution. The t distribution curves, like the standard normal distribution curve, are centered at zero. However, the standard deviation of each t distribution exceeds one and is dependent upon the value of df, whereas the standard deviation of the standard normal distribution equals one. Figure 8-2 compares the standard normal curve with the t distribution having 10 degrees of freedom. Generally speaking, the t distribution curves have a lower maximum height and thicker tails than the standard normal curve as shown in Fig. 8-2. All t distribution curves are symmetrical about zero.

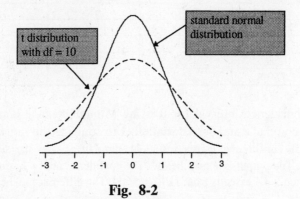

Fig. 8-2

Appendix 3 gives right-hand tail areas under t distribution curves for degrees of freedom varying from 1 to 40. This table is referred to as the *t distribution table*. Table 8.3 contains the rows corresponding to degrees of freedom equal to 5, 10, 15, 20, and 25 selected from the t distribution table in Appendix 3.

Table 8.3

df	Area in the right tail under the t distribution curve					
	.10	**.05**	.025	.01	.005	.001
5	1.476	2.015	2.571	3.365	4.032	5.893
10	1.372	**1.812**	2.228	2.764	3.169	4.144
15	1.341	1.753	2.131	2.602	2.947	3.733
20	1.325	1.725	2.086	2.528	2.845	3.552
25	1.316	1.708	2.060	2.485	2.787	3.450

EXAMPLE 8.5 The t distribution having 10 degrees of freedom is shown in Fig. 8-3. To find the shaded area in the right-hand tail to the right of t = 1.812, locate the degrees of freedom, 10, under the df column in Table 8.3. The t value, 1.812, is located under the column labeled .05 in Table 8.3. The shaded area is equal to .05. Since the total area under the curve is equal to 1, the area under this curve to the left of t = 1.812 is 1 − .05 = .95. The area under this curve between t = −1.812 and t = 1.812 is 1 − .05 −.05 = .90, since there is .05 to the right of t = 1.812 and .05 to the left of t = −1.812.

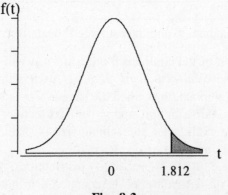

Fig. 8-3

EXAMPLE 8.6 A t distribution curve with df = 20 is shown in Fig. 8-4. Suppose we wish to find the shaded area under the curve between t = −2.528 and t = 2.086. From Table 8.3, the area to the right of t = 2.086 is .025, and the area to the right of t = 2.528 is .01. By symmetry, the area to the left of t = −2.528 is also .01. Since the total area under the curve is 1, the shaded area is 1 − .025 − .01 = .965.

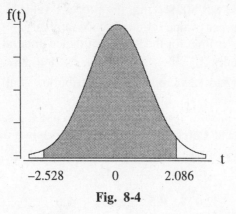

Fig. 8-4

CONFIDENCE INTERVAL FOR THE POPULATION MEAN: SMALL SAMPLES

When a small sample (n < 30) is taken from a normally distributed population and the population standard deviation, σ, is unknown, a confidence interval for the population mean, μ, is given by formula (8.6):

$$\bar{x} - ts_{\bar{x}} < \mu < \bar{x} + ts_{\bar{x}} \qquad (8.6)$$

In formula (8.6), $\bar{x}$ is a point estimate of the population mean, $s_{\bar{x}}$ is the estimated standard error, and t is determined by the confidence level and the degrees of freedom. The degrees of freedom is given by formula (8.7), where n is the sample size.

$$df = n - 1 \qquad (8.7)$$

EXAMPLE 8.7 The distance traveled, in hundreds of miles by automobile, was determined for 20 individuals returning from vacation. The results are given in Table 8.4.

Table 8.4

12.5	15.0	10.0	15.5
8.0	5.0	15.0	9.5
16.0	20.0	9.0	17.5
12.0	13.0	16.0	7.5
9.5	12.0	12.0	12.5

For the data in Table 8.4, $\bar{x}$ = 12.375, s = 3.741, and $s_{\bar{x}}$ = 0.837. To find a 90% confidence interval for the mean vacation travel distance of all such individuals, we need to find the value of t in formula (8.6). The degrees of freedom is df = n − 1 = 20 − 1 = 19. Using df = 19 and the t distribution table, we must find the t value for which the area under the curve between −t and t is .90. This means the area to the left of −t is .05 and the area to the right of t is also .05. Table 8.5, which is selected from the t distribution table, indicates that the area to the right of 1.729 is .05. This is the proper value of t for the 90% confidence level when df = 19.

Table 8.5

df	\multicolumn{6}{c}{Area in the right tail under the t distribution curve}

df	.10	**.05**	.025	.01	.005	.001
19	1.328	**1.729**	2.093	2.539	2.861	3.579

The following technique is also often recommended for finding the t value for a confidence interval: *To find the t value for a given confidence interval, subtract the confidence level from 1 and divide the answer by 2 to find the correct area in the right-hand tail.* In this Example, the confidence level is 90% or .90. Subtracting .90 from 1, we get .10. Now, dividing .10 by 2, we get .05 as the right-hand tail area. From Table 8.5, we see that the proper t value is 1.729. Many students prefer this technique for finding the proper t value for confidence intervals.

Using formula (8.6), the lower limit for the 90% confidence interval is $\bar{x} - ts_{\bar{x}} = 12.375 - 1.729 \times .837 = 10.928$ and the upper limit is $\bar{x} + ts_{\bar{x}} = 12.375 + 1.729 \times .837 = 13.822$. A 90% confidence interval for μ extends from 1,093 to 1,382 miles. The distribution of all vacation travel distances are assumed to be normally distributed.

To find a 90% confidence interval for μ using Minitab, use the set command to enter the data given in Table 8.4. The command **tinterval 90 percent confidence data in c1** is used to find the confidence interval. The output is shown below. The confidence interval is (10.928, 13.822).

```
MTB > set c1
DATA > 12.5    8.0    16.0    12.0    9.5    15.0    5.0    20.0    13.0    12.0    10.0
DATA > 15.0    9.0    16.0    12.0    15.5    9.5    17.5    7.5    12.5
DATA > end
MTB > name c1 'distance'
MTB > tinterval 90 percent confidence data in c1
```

Confidence Intervals

Variable	N	Mean	St Dev	SE Mean	90.0 % C.I.
Distance	20	12.375	3.741	0.837	(10.928, 13.822)

EXAMPLE 8.8 The health-care costs, in thousands of dollars for 20 males aged 75 or over, are shown in Table 8.6. Formula (8.6) may not be used to set a confidence interval on the mean health-care cost for such individuals, since the sample data indicate that the distribution of such health-care costs is not normally distributed. The $515,000 and the $950,000 costs are far removed from the remaining health-care costs. This indicates that the distribution of health-care costs for males aged 75 or over is skewed to the right and therefore is not normally distributed. Formula (8.6) is applicable only if it is reasonable to assume that the population characteristic is normally distributed.

Table 8.6

8.5	515.0	950.0	5.5
8.0	5.0	15.0	19.5
16.0	15.0	9.0	7.5
12.0	13.0	6.0	37.5
2.5	2.0	12.0	12.5

EXAMPLE 8.9 The number of square feet per mall devoted to children's apparel was determined for 15 malls selected across the United States. The summary statistics for these 15 malls are as follows: $\bar{x} = 2,700$, s = 450, and $s_{\bar{x}} = 116.19$. To determine a 99% confidence interval for μ, we need to find the value t where the area under the t distribution curve between $-t$ and t is .99 and the degrees of freedom $= n - 1 = 15 - 1 = 14$. Since the area between $-t$ and t is .99, the area to the left of $-t$ is .005 and the area to the right of t is .005. Table 8.7 is taken from the t distribution table and shows that the area to the right of 2.977 is .005. The lower limit for the 99% confidence interval is $\bar{x} - ts_{\bar{x}} = 2,700 - 2.977 \times 116.19 = 2,354$ ft^2 and the upper limit is $\bar{x} + ts_{\bar{x}} = 2,700 + 2.977 \times 116.19 = 3,046$ ft^2. The 99% confidence interval for μ is (2,354 to 3,046).

Table 8.7

df	Area in the right tail under the t distribution curve					
	.10	.05	.025	.01	**.005**	.001
14	1.345	1.761	2.145	2.624	**2.977**	3.787

CONFIDENCE INTERVAL FOR THE POPULATION PROPORTION: LARGE SAMPLES

In chapter 7, the variable shown in formula (7.10) and reproduced below was shown to have a standard normal distribution provided np > 5 and nq > 5.

$$z = \frac{\overline{p} - \mu_{\overline{p}}}{\sigma_{\overline{p}}} \qquad (7.10)$$

Since 95% of the area under the standard normal curve is between –1.96 and 1.96, and since the variable in formula (7.10) has a standard normal distribution, we have the result shown below:

$$P\left(-1.96 < \frac{\overline{p} - \mu_{\overline{p}}}{\sigma_{\overline{p}}} < 1.96\right) = .95 \qquad (8.8)$$

From Chapter 7, we also know that $\mu_{\overline{p}} = p$ and $\sigma_{\overline{p}} = \sqrt{\dfrac{pq}{n}}$. Substituting for $\mu_{\overline{p}}$ and $\sigma_{\overline{p}}$ in the inequality $-1.96 < \dfrac{\overline{p} - \mu_{\overline{p}}}{\sigma_{\overline{p}}} < 1.96$ given in formula (8.8) and solving the resulting inequality for p, we obtain

$$\overline{p} - 1.96\sqrt{\frac{pq}{n}} < p < \overline{p} + 1.96\sqrt{\frac{pq}{n}} \qquad (8.9)$$

To obtain numerical values for the lower and upper limits of the confidence interval, it is necessary to substitute the sample values $\overline{p}$ and $\overline{q}$ for p and q in the expression for $\sigma_{\overline{p}}$. Making these substitutions, we obtain formula (8.10).

$$\overline{p} - 1.96\sqrt{\frac{\overline{p} \cdot \overline{q}}{n}} < p < \overline{p} + 1.96\sqrt{\frac{\overline{p} \cdot \overline{q}}{n}} \qquad (8.10)$$

The interval given in formula (8.10) is called a 95% confidence interval for p. The general form for the interval is shown in formula (8.11), where z represents the proper value from the standard normal distribution table as determined by the desired confidence level.

$$\overline{p} - z\sqrt{\frac{\overline{p} \cdot \overline{q}}{n}} < p < \overline{p} + z\sqrt{\frac{\overline{p} \cdot \overline{q}}{n}} \qquad (8.11)$$

EXAMPLE 8.10 A study of 75 small-business owners determined that 80% got the money to start their business from personal savings or credit cards. If p represents the proportion of all small-business owners who got the money to start their business from personal savings or credit cards, then a point estimate of p is $\overline{p} = 80\%$. The estimated standard error of the proportion is represented by $s_{\overline{p}}$, and is given by $s_{\overline{p}} = \sqrt{\dfrac{\overline{p} \cdot \overline{q}}{n}} =$

$\sqrt{\dfrac{80 \times 20}{75}}$ = 4.62%. From Table 8.1, the z value for a 99% confidence interval is 2.58. The lower limit for a

99% confidence interval is $\overline{p} - z\sqrt{\dfrac{p \cdot q}{n}}$ = 80 − 2.58 × 4.62 = 68.08% and the upper limit is $\overline{p} + z\sqrt{\dfrac{p \cdot q}{n}}$ = 80 +

2.58 × 4.62 = 91.92%. Formula (8.11) is considered to be valid if the sample size satisfies the inequalities np > 5 and nq > 5. Since p and q are unknown, we check the sample size requirement by checking to see if $n\overline{p}$ > 5 and $n\overline{q}$ > 5. In this case $n\overline{p}$ = 75 × .8 = 60 and $n\overline{q}$ = 75 × .2 = 15, and since both $n\overline{p}$ and $n\overline{q}$ exceed 5, the sample size is large enough to assure the valid use of formula (8.11).

EXAMPLE 8.11 In a random sample of 40 workplace homicides, it was found that 2 of the 40 were due to a personal dispute. A point estimate of the population proportion of workplace homicides due to a personal dispute is $\overline{p} = \frac{2}{40}$ = .05 or 5%. Formula (8.11) may not be used to set a confidence interval on p, the population proportion of homicides due to a personal dispute, since $n\overline{p}$ = 40 × .05 = 2 is less than 5. Note that since p is unknown, np cannot be computed. When p is unknown, $n\overline{p}$ and $n\overline{q}$ are computed to determine the appropriateness of using formula (8.11).

DETERMINING THE SAMPLE SIZE FOR THE ESTIMATION OF THE POPULATION MEAN

The maximum error of estimate when using $\overline{x}$ as a point estimate of μ is defined by formula (8.5) and is restated below.

$$E = z\sigma_{\overline{x}} \qquad (8.5)$$

If the maximum error of estimate is specified, then the sample size necessary to give this maximum error may be determined from formula (8.5). Replacing $\sigma_{\overline{x}}$ by $\dfrac{\sigma}{\sqrt{n}}$ to obtain E = $z\dfrac{\sigma}{\sqrt{n}}$, and then solving for n we obtain formula (8.12). The value obtained for n is always rounded up to the next whole number. This is a conservative approach and sometimes results in a sample size larger than actually needed.

$$n = \dfrac{z^2\sigma^2}{E^2} \qquad (8.12)$$

EXAMPLE 8.12 In Example 8.4, the mean of a sample of 40 tuition costs for community colleges was found to equal to $1,396 and the standard deviation was $655. The margin of error associated with using $1,396 as a point estimate of μ is equal to $204. To reduce the margin of error to $100, a larger sample is needed. The required sample size is n = $\dfrac{1.96^2 \times 655^2}{100^2}$ = 164.8. To obtain a conservative estimate, we round the estimate up to 165. Note that we estimated σ by s, because σ is unknown. The 40 tuition costs obtained in the original survey would be supplemented by 165 − 40 = 125 additional community colleges. The new estimate based on the 165 tuition costs would have a margin of error of $100.

To use formula (8.12) to determine the sample size, either σ or an estimate of σ is needed. In Example 8.12, the original study, based on sample size 40, may be regarded as a pilot sample. When a pilot sample is taken, the sample standard deviation is used in place of σ. In other instances, a historical value of σ may exist. In some instances, the maximum and minimum value of the characteristic being studied may be known and an estimate of σ may be obtained by dividing the range by 4.

EXAMPLE 8.13 A sociologist desires to estimate the mean age of teenagers having abortions. She wishes to estimate μ with a 99% confidence interval so that the maximum error of estimate is E = .1 years. The z value is 2.58. The range of ages for teenagers is $19 - 13 = 6$ years. A rough estimate of the standard deviation is obtained by dividing the range by 4 to obtain 1.5 years. This method for estimating σ works best for mound-shaped distributions. The approximate sample size is n = $\dfrac{2.58^2 \times 1.5^2}{.1^2}$ = 1497.7. Rounding up, we obtain n = 1,498.

DETERMINING THE SAMPLE SIZE FOR THE ESTIMATION OF THE POPULATION PROPORTION

When $\bar{p}$ is used as a point estimate of p, the maximum error of estimate is given by

$$E = z\sigma_{\bar{p}} \tag{8.13}$$

If the maximum error of estimate is specified, then the sample size necessary to give this maximum error may be determined from formula (8.13). If $\sigma_{\bar{p}}$ is replaced by $\sqrt{\dfrac{pq}{n}}$ and the resultant equation is solved for n, we obtain

$$n = \frac{z^2 pq}{E^2} \tag{8.14}$$

Since p and q are usually unknown, they must be estimated when formula (8.14) is used to determine a sample size to give a specified maximum error of estimate. If a reasonable estimate of p and q exists, then the estimate is used in the formula. If no reasonable estimate is known, then both p and q are replaced by .5. This gives a conservative estimate for n. That is, replacing p and q by .5 usually gives a larger sample size than is needed, but it covers all cases so to speak.

EXAMPLE 8.14 A study is undertaken to obtain a precise estimate of the proportion of diabetics in the United States. Estimates ranging from 2% to 5% are found in various publications. The sample size necessary to estimate the population percentage to within 0.1% with 95% confidence is n = $\dfrac{1.96^2 \times .05 \times .95}{.001^2}$ = 182,476. When a range of possible values for p exists, as in this problem, use the value closest to .5 as a reasonable estimate for p. In this example, the value in the range from .02 to .05 closest to .5 is .05. If the prior estimate of p were not used, and .5 is used for p and q, then the computed sample size is n = $\dfrac{1.96^2 \times .5 \times .5}{.001^2}$ = 960,400. Notice that using the prior information concerning p makes a tremendous difference in this example.

Solved Problems

POINT ESTIMATE

8.1 The mean annual salary for public school teachers is $32,000. The mean salary for a sample of 750 public school teachers equals $31,895. Identify the population, the population mean, and the sample mean. Identify the parameter and the point estimate of the parameter.

   Ans. The population consists of all public school teachers. μ = $32,000, $\bar{x}$ = $31,895. The parameter is μ. A point estimate of the parameter μ is $31,895.

8.2 Ninety-eight percent of U.S. homes have a TV. A survey of 2000 homes finds that 1,925 have a TV. Identify the population, the population proportion, and the sample proportion. Identify the parameter and a point estimate of the parameter.

   Ans. The population consists of all U.S. homes. $p = 98\%$, $\bar{p} = 96.25\%$. The parameter is p. A point estimate of p is 96.25%.

INTERVAL ESTIMATE

8.3 When a multiple of the standard error of a point estimate is subtracted from the point estimate to obtain a lower limit and added to the point estimate to obtain an upper limit, what term is used to describe the numbers between the lower limit and the upper limit?

   Ans. interval estimate

CONFIDENCE INTERVAL FOR THE POPULATION MEAN: LARGE SAMPLES

Table 8.8

1515	959	1744	589	762
1432	723	764	1901	364
120	1212	868	168	1396
1270	202	1236	1769	668
312	1118	1631	232	1781
1904	1726	681	171	313
1662	1562	1243	271	1067
1857	1518	392	1062	1367
903	1671	1623	1023	283
1313	395	169	800	1973

8.4 A sample of 50 taxpayers receiving tax refunds is shown in Table 8.8. Find an 80% confidence interval for μ, where μ represents the mean refund for all taxpayers receiving a refund.

   Ans. For the data in Table 8.8, $\Sigma\, x = 51{,}685$, $\Sigma\, x^2 = 70{,}158{,}336$, $\bar{x} = 1033.7$, $s = 584.3$, $s_{\bar{x}} = 82.6$. From Table 8.1, the z value for 80% confidence is 1.28. Lower limit $= 1033.7 - 1.28 \times 82.6 = 927.97$, upper limit $= 1033.7 + 1.28 \times 82.6 = 1139.43$. The 80% confidence interval is (927.97, 1139.43).

8.5 Use Minitab to find the 80% confidence interval for μ in problem 8.4.

   Ans.
```
MTB > set c1
DATA > 1515 1432  120 1270  312 1904 1662 1857  903 1313  959
DATA >  723 1212  202 1118 1726 1562 1518 1671  395 1744  764
DATA >  868 1236 1631  681 1243  392 1623  169  589 1901  168
DATA > 1769  232  171  271 1062 1023  800  762  364 1396  668
DATA > 1781  313 1067 1367  283 1973
DATA > end
MTB > name c1 'refund'
MTB > standard deviation c1
```

Standard deviation of refund $= 584.35$
MTB > zinterval 80 percent confidence sigma $= 584.35$ data in c1

Confidence Intervals
The assumed sigma $= 584$

Variable	N	Mean	St Dev	SE Mean	80.0 % C.I.
Refund	50	1033.7	584.3	82.6	(927.8, 1139.6)

MAXIMUM ERROR OF ESTIMATE FOR THE POPULATION MEAN

8.6 A sample of size 50 is selected from a population having a standard deviation equal to 4. Find the maximum error of estimate associated with the confidence intervals having the following confidence levels: (*a*) 80%; (*b*) 90%; (*c*) 95%; (*d*) 99%.

Ans. The standard error of the mean is $\dfrac{4}{\sqrt{50}}$ = .566. Using the z values given in Table 8.1, the following maximum errors of estimate are obtained.

(*a*) E = 1.28 × .566 = .724   (*b*) E = 1.65 × .566 = .934
(*c*) E = 1.96 × .566 = 1.109   (*d*) E = 2.58 × .566 = 1.460

8.7 Describe the effect of the following on the maximum error of estimate when determining a confidence interval for the mean: (*a*) sample size; (*b*) confidence level; (*c*) variability of the characteristic being measured

Ans. (*a*) The maximum error of estimate decreases when the sample size is increased.
(*b*) The maximum error of estimate increases if the confidence level is increased.
(*c*) The larger the variability of the characteristic being measured, the larger the maximum error of estimate.

THE t DISTRIBUTION

8.8 Table 8.9 contains the row corresponding to 7 degrees of freedom taken from the t distribution table in Appendix 3. For a t distribution curve with df = 7, find the following:
(*a*) Area under the curve to the right of t = 2.998
(*b*) Area under the curve to the left of t = –2.998
(*c*) Area under the curve between t = –2.998 and t = 2.998

Table 8.9

df	\multicolumn{6}{c}{Area in the right tail under the t distribution curve}
	.10	.05	.025	.01	.005	.001
7	1.415	1.895	2.365	2.998	3.499	4.785

Ans. (*a*) .01
(*b*) Since the curve is symmetrical about 0, the answer is the same as in part (*a*), .01
(*c*) Since the total area under the curve is 1, the answer is 1 – .01 – .01 = .98

8.9 Refer to problem 8.8 and Table 8.9 to find the following.
(*a*) Find the t value for which the area under the curve to the right of t is .05.
(*b*) Find the t value for which the area under the curve to the left of t is .05.
(*c*) By considering the answers to parts (*a*) and (*b*), find that positive t value for which the area between –t and t is equal to .90.

Ans.   (a) 1.895 (b) −1.895 (c) 1.895

CONFIDENCE INTERVAL FOR THE POPULATION MEAN: SMALL SAMPLES

8.10 In a transportation study, 20 cities were randomly selected from all cities having a population of 50,000 or more. The number of cars per 1,000 people was determined for each selected city, and the results are shown in Table 8.10. Find a 95% confidence interval for μ, where μ is the mean number of cars per 1,000 people for all cities having a population of 50,000 or more. What assumption is necessary for the confidence interval to be valid?

Table 8.10

409	487	480	535
663	676	494	332
304	565	670	434
535	554	515	665
628	308	319	519

Ans.   For the data in Table 8.10, $\Sigma x = 10{,}092$, $\Sigma x^2 = 5{,}381{,}738$, $\overline{x} = 504.6$, $s = 123.4$, $s_{\overline{x}} = 27.6$. The t distribution is used because the sample is small. The degrees of freedom is df $= 20 - 1 = 19$. The t value in the confidence interval is determined from Table 8.11, which is taken from the t distribution table in Appendix 3. From Table 8.11, we see that the area to the right of 2.093 is .025 and the area to the left of −2.093 is .025, and therefore the area between −2.093 and 2.093 is .95. The lower limit of the 95% confidence interval is $\overline{x} - ts_{\overline{x}} = 504.6 - 2.093 \times 27.6 = 446.8$ and the upper limit is $\overline{x} + ts_{\overline{x}} = 504.6 + 2.093 \times 27.6 = 562.4$. The 95% confidence interval is (446.8, 562.4). It is assumed that the distribution of the number of cars per 1,000 people in cities of 50,000 or over is normally distributed.

Table 8.11

df	Area in the right tail under the t distribution curve					
	.10	.05	**.025**	.01	.005	.001
19	1.328	1.729	**2.093**	2.539	2.861	3.579

8.11 Find the Minitab solution to problem 8.10.

Ans.   MTB > Set c1
DATA >   409   663   304   535   628   487   676   565   554   308   480
DATA >   494   670   515   319   535   332   434   665   519
DATA > end
MTB > name c1 'cars'
MTB > tinterval 95 percent confidence data in c1
Confidence Intervals
Variable   N   Mean   St Dev   SE Mean   95.0 % C.I.
Cars   20   504.6   123.4   27.6   (446.8, 562.4)

CONFIDENCE INTERVAL FOR THE POPULATION PROPORTION: LARGE SAMPLES

8.12 A national survey of 1200 adults found that 450 of those surveyed were pro-choice on the abortion issue. Find a 95% confidence interval for p, the proportion of all adults who are pro-choice on the abortion issue.

Ans. The sample proportion who are pro-choice is $\bar{p} = \frac{450}{1200} = .375$ and the proportion who are not pro-choice or undecided is $\bar{q} = \frac{750}{1200} = .625$. The estimated standard error of the proportion is $\sqrt{\dfrac{\bar{p} \times \bar{q}}{n}} = \sqrt{\dfrac{.375 \times .625}{1200}} = .014$. The z value is 1.96. The lower limit of the 95% confidence interval is $\bar{p} - z\sqrt{\dfrac{\bar{p} \times \bar{q}}{n}} = .375 - 1.96 \times .014 = .348$ and the upper limit is $\bar{p} + z\sqrt{\dfrac{\bar{p} \times \bar{q}}{n}} = .375 + 1.96 \times .014 = .402$. The 95% confidence interval is (.348, .402).

8.13 A national survey of 500 African-Americans found that 62% in the survey favored affirmative action programs. Find a 90% confidence interval for p, the proportion of all African-Americans who favor affirmative action programs.

Ans. The sample percent who favor affirmative action programs is 62%. The sample percent who do not favor affirmative action programs or are undecided is 38%. The estimated standard error of proportion, expressed as a percentage, is $\sqrt{\dfrac{\bar{p} \times \bar{q}}{n}} = \sqrt{\dfrac{62 \times 38}{500}} = 2.17\%$. The z value is 1.65. The lower limit of the 90% confidence interval is $\bar{p} - z\sqrt{\dfrac{\bar{p} \times \bar{q}}{n}} = 62 - 1.65 \times 2.17 = 58.4\%$ and the upper limit is $\bar{p} + z\sqrt{\dfrac{\bar{p} \times \bar{q}}{n}} = 62 + 1.65 \times 2.17 = 65.6\%$. The 90% confidence interval is (58.4%, 65.6%).

DETERMINING THE SAMPLE SIZE FOR THE ESTIMATION OF THE POPULATION MEAN

8.14 A machine fills containers with corn meal. The machine is set to put 680 grams in each container on the average. The standard deviation is equal to 0.5 gram. The average fill is known to shift from time to time. However, the variability remains constant. That is, σ remains constant at 0.5 gram. In order to estimate μ, how many containers should be selected from a large production run so that the maximum error of estimate equal 0.2 gram with probability 0.95?

Ans. The sample size is determined by use of the formula $n = \dfrac{z^2 \sigma^2}{E^2}$. The value for σ is known to equal 0.5, E is specified to be 0.2 and for probability equal to 0.95, the z value is 1.96. Therefore, the sample size is $n = \dfrac{z^2 \sigma^2}{E^2} = \dfrac{1.96^2 \times 5^2}{.2^2} = 24.01$. The required sample size is obtained by rounding the answer up to 25.

8.15 A pilot study of 250 individuals found that the mean annual health-care cost per person was $2550 and the standard deviation was $1350. How large a sample is needed to estimate the true annual health-care cost with a maximum error of estimate equal to $100 with probability equal to 0.99?

Ans. The sample size is determined by use of the formula $n = \dfrac{z^2 \sigma^2}{E^2}$. The value for σ is estimated from the pilot study to be $1350. E is specified to be $100 and for probability equal to 0.99, the z value is 2.58. Therefore, the sample size is $n = \dfrac{z^2 \sigma^2}{E^2} = \dfrac{2.58^2 \times 1350^2}{100^2} = 1213.13$. The required sample size is obtained by rounding the answer up to 1214. If the results from the pilot study are used, then an additional $1214 - 250 = 964$ individuals need to be surveyed.

DETERMINING THE SAMPLE SIZE FOR THE ESTIMATION OF THE POPULATION PROPORTION

8.16 A large study is undertaken to estimate the percentage of students in grades 9 through 12 who use cigarettes. Other studies have indicated that the percent ranges between 30% and 35%. How large a sample is needed in order to estimate the true percentage for all such students with a maximum error of estimate equal to 0.5% with a probability of 0.90?

Ans. The sample size is given by $n = \dfrac{z^2 pq}{E^2}$. The specified value for E is 0.5%. The z value for probability equal to 0.90 is 1.65. The previous studies indicate that p is between 30% and 35%. When a range of likely values for p exist, the value closest to 50% is used. This value gives a conservative estimate of the sample size. The sample size is $n = \dfrac{1.65^2 \times 35 \times 65}{.5^2} = 24{,}774.75$. The sample size is 24,775.

8.17 Find the sample size in problem 8.16 if no prior estimate of the population proportion exists.

Ans. The sample size is given by $n = \dfrac{z^2 pq}{E^2}$. The specified value for E is 0.5%. The z value for probability equal to 0.90 is 1.65. When no prior estimate for p exists, p and q are set equal to 50% in the sample size formula. Therefore, $\dfrac{1.65^2 \times 50 \times 50}{.5^2} = 27{,}225$. Note that 2,450 fewer subjects are needed when the prior estimate of 35% is used.

Supplementary Problems

POINT ESTIMATE

8.18 Table 8.8 in problem 8.4 contains a sample of 50 tax refunds received by taxpayers. Give point estimates for the population mean, population median, and the population standard deviation.

Ans. $1033.70, the mean of the 50 tax refunds, is a point estimate of the population mean.
$1064.50, the median of the 50 tax refunds, is a point estimate of the population median.
$584.30, the standard deviation of the 50 tax refunds, is a point estimate of the population standard deviation.

8.19 Table 8.10 in problem 8.10 contains a sample of the number of cars per 1,000 people for 20 cities having a population of 50,000 or more. Give a point estimate of the proportion of all such cities with 600 or more cars per 1,000 people.

Ans. In this sample, there are 5 cities with 600 or more cars per 1,000. The point estimate of p is .25 or 25%.

INTERVAL ESTIMATE

8.20 What happens to the width of an interval estimate when the sample size is increased?

   Ans.   The width of an interval estimate decreases when the sample size is increased.

CONFIDENCE INTERVAL FOR THE POPULATION MEAN: LARGE SAMPLES

8.21 One hundred subjects in a psychological study had a mean score of 35 on a test instrument designed to measure anger. The standard deviation of the 100 test scores was 10. Find a 99% confidence interval for the mean anger score of the population from which the sample was selected.

   Ans.   The confidence interval is $\bar{x} - z\sigma_{\bar{x}} < \mu < \bar{x} + z\sigma_{\bar{x}}$.
   The lower limit is $\bar{x} - z\sigma_{\bar{x}} = 35 - 2.58 \times 1 = 32.42$ and the upper limit is $\bar{x} - z\sigma_{\bar{x}} = 35 + 2.58 \times 1 = 37.58$. The interval is (32.42, 37.58).

8.22 The width of a large sample confidence interval is equal to $w = (\bar{x} + z\sigma_{\bar{x}}) - (\bar{x} - z\sigma_{\bar{x}}) = 2z\sigma_{\bar{x}}$. Replacing $\sigma_{\bar{x}}$ by $\dfrac{\sigma}{\sqrt{n}}$, w is given as $w = 2z\dfrac{\sigma}{\sqrt{n}}$. Discuss the effect of the confidence level, the standard deviation, and the sample size on the width.

   Ans.   The width varies directly as the confidence level, directly as the variability of the characteristic being measured, and inversely as the square root of the sample size.

MAXIMUM ERROR OF ESTIMATE FOR THE POPULATION MEAN

8.23 The U.S. abortion rate per 1,000 female residents in the age group 15 – 44 was determined for 35 different cities having a population of 25,000 or more across the U.S. The mean for the 35 cities was equal to 24.5 and the standard deviation was equal to 4.5. What is the maximum error of estimate when a 90% confidence interval is constructed for μ?

   Ans.   The maximum error of estimate is $E = z\sigma_{\bar{x}}$. For a 90% confidence interval, z = 1.65. The estimated standard error is 0.76. E = 1.65 × .76 = 1.25.

8.24 A study concerning one-way commuting times to work was conducted in Omaha, Nebraska. The mean commuting time for 300 randomly selected workers was 45 minutes. The margin of error for the study was 7 minutes. What is the 95% confidence interval for μ?

   Ans.   The 95% confidence interval is $\bar{x} - 1.96\sigma_{\bar{x}} < \mu < \bar{x} + 1.96\sigma_{\bar{x}}$. The margin of error is $E = 1.96\sigma_{\bar{x}} = 7$ and therefore, the lower limit of the interval is 38 and the upper limit is 52.

THE T DISTRIBUTION

8.25 Use the t distribution table and the standard normal distribution table to find the values for a, b, c, and d. What distribution does the t distribution approach as the degrees of freedom increases?

Table 8.12

Distribution type	Area under the curve to the right of this value is .025
t distribution, df = 10	a
t distribution, df = 20	b
t distribution, df = 30	c
standard normal	d

Ans.   a = 2.228   b = 2.086   c = 2.042   d = 1.96

The t distribution approaches the standard normal distribution when the degrees of freedom are increased. This is illustrated by the results in Table 8.12.

8.26   The standard deviation of the t distribution is equal to $\sqrt{\dfrac{df}{df-2}}$, for df > 2. Find the standard deviation for the t distributions having df values equal to 5,10, 20, 30, 50, and 100. What value is the standard deviation getting close to when the degrees of freedom get larger?

Ans.   The standard deviations are shown in Table 8.13. The standard deviation is approaching one as the degrees of freedom get larger. The t distribution approaches the standard normal distribution when the degrees of freedom are increased.

Table 8.13

df	Standard deviation of the t distribution with this value for df
5	1.29
10	1.12
20	1.05
30	1.04
50	1.02
100	1.01

CONFIDENCE INTERVAL FOR THE POPULATION MEAN: SMALL SAMPLES

8.27   In order to estimate the lifetime of a new bulb, ten were tested and the mean lifetime was equal to 835 hours with a standard deviation equal to 45 hours. A stem-and-leaf of the lifetimes indicated that it was reasonable to assume that lifetimes were normally distributed. Determine a 99% confidence interval for μ, where μ represents the mean lifetime of the population of lifetimes.

Ans.   The confidence interval is $\overline{x} - ts_{\overline{x}} < \mu < \overline{x} + ts_{\overline{x}}$, where $\overline{x}$ = 835, s = 45, $s_{\overline{x}}$ = 14.23, and t = 3.250. The lower limit is $\overline{x} - ts_{\overline{x}}$ = 835 – 46.25 = 788.75 hours and the upper limit is $\overline{x} + ts_{\overline{x}}$ = 835 + 46.25 = 881.25 hours.

8.28   The heights of 15 randomly selected buildings in Chicago are given in Table 8.14. Find a 90% confidence interval on the mean height of buildings in Chicago.

Table 8.14

525	500	390
300	715	425
1,127	255	500
625	1,454	375
475	814	750

Ans.   A Minitab histogram of the data in Table 8.14 is shown in Fig. 8-5. Since the heights are not normally distributed, the confidence interval using the t distribution is not appropriate.

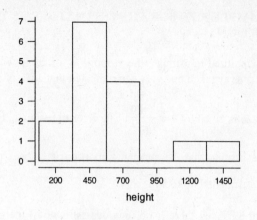

Fig. 8-5

CONFIDENCE INTERVAL FOR THE POPULATION PROPORTION: LARGE SAMPLES

8.29   In a survey of 900 adults, 360 responded "yes" to the question "Have you attended a major league baseball game in the last year?" Determine a 90% confidence interval for p, the proportion of all adults who attended a major league baseball game in the last year.

> *Ans.*   The sample proportion attending a game in the last year is $\overline{p}$ = .40 and the proportion not attending is $\overline{q}$ = .60. The z value for a 90% confidence interval is 1.65. The confidence interval for p is $\overline{p} - z\sqrt{\dfrac{\overline{p} \cdot \overline{q}}{n}} < p < \overline{p} + z\sqrt{\dfrac{\overline{p} \cdot \overline{q}}{n}}$. The maximum error of estimate is $E = z\sqrt{\dfrac{\overline{p} \cdot \overline{q}}{n}} = .027$. The lower limit of the confidence interval is .40 − .027 = .373 and the upper limit is .40 + .027 = .427. The 90% interval is (.373, .427).

8.30   Use the data in Table 8.8, found in problem 8.4, to find a 99% confidence interval for the proportion of all taxpayers receiving a refund who receive a refund of more than $500.

> *Ans.*   Thirty-seven of the fifty refunds in Table 8.8 exceed $500. The sample proportion exceeding $500 is .74. The z value for a 99% confidence interval is 2.58. The confidence interval for p is $\overline{p} - z\sqrt{\dfrac{\overline{p} \cdot \overline{q}}{n}} < p < \overline{p} + z\sqrt{\dfrac{\overline{p} \cdot \overline{q}}{n}}$. The maximum error of estimate is $E = z\sqrt{\dfrac{\overline{p} \cdot \overline{q}}{n}} = .16$ The lower limit is .74 − .16 = .58 and the upper limit is .74 + .16 = .90. The 99% interval is (.58, .90).

DETERMINING THE SAMPLE SIZE FOR THE ESTIMATION OF THE POPULATION MEAN

8.31   The estimated standard deviation of commuting distances for workers in a large city is determined to be 3 miles in a pilot study. How large a sample is needed to estimate the mean commuting distance of all workers in the city to within .5 mile with 95% confidence?

> *Ans.*   The sample size is given by $n = \dfrac{z^2 \sigma^2}{E^2}$. For this problem, E = .5, z = 1.96, and σ is estimated to be 3. The sample size is determined to be 139.

8.32   In problem 8.31, what size sample is required to estimate μ to within .1 mile with 95% confidence?

> *Ans.*   n = 3,458

DETERMINING THE SAMPLE SIZE FOR THE ESTIMATION OF THE POPULATION PROPORTION

8.33 What sample size is required to estimate the proportion of adults who wear a beeper to within 3% with probability 0.95? A survey taken one year ago indicated that 15% of all adults wore a beeper.

Ans. The sample size is given by $n = \dfrac{z^2 pq}{E^2}$. The error of estimate, E, is given to be 3%, and the z value is 1.96. If we assume that the current proportion is "close" to 15%, then $n = \dfrac{1.96^2 \times 15 \times 85}{3^2} = 544.2$, which is rounded up to 545.

8.34 Suppose in problem 8.33 that there is no previous estimate of the population proportion. What sample size would be required to estimate p to within 3% with probability 0.95?

Ans. The sample size is given by $n = \dfrac{z^2 pq}{E^2}$. The error of estimate, E, is given to be 3%, and the z value is 1.96. With no previous estimate for p, use p = q = 50%. $n = \dfrac{1.96^2 \times 50 \times 50}{3^2} = 1,067.1$, which is rounded up to 1,068.

Chapter 9

Tests of Hypotheses: One Population

NULL HYPOTHESIS AND ALTERNATIVE HYPOTHESIS

There are a number of mail order companies in the U.S. Consider such a company called L. C. Stephens Inc. The mean sales per order for L. C. Stephens, determined from a large database last year, is known to equal $155. From the same database, the standard deviation is determined to be $\sigma = \$50$. The company wishes to determine whether the mean sales per order for the current year is different from the historical mean of $155. If μ represents the mean sales per order for the current year, then the company is interested in determining if $\mu \neq \$155$. The *alternative hypothesis,* also often called the *research hypothesis*, is related to the purpose of the research. In this instance, the purpose of the research is to determine whether $\mu \neq \$155$. The research hypothesis is represented symbolically by H_a: $\mu \neq \$155$. The negation of the research hypothesis is called the *null hypothesis*. In this case, the negation of the research hypothesis is that $\mu = \$155$. The null hypothesis is represented symbolically by H_0: $\mu = \$155$. The company wishes to conduct a *test of hypothesis* to determine which of the two hypotheses is true. A test of hypothesis where the research hypothesis is of the form H_a : $\mu \neq$ (some constant) is called a *two-tailed test.* The reason for this term will be clear when the details of the test procedure are discussed.

EXAMPLE 9.1 A tire manufacturer claims that the mean mileage for their premium brand tire is 60,000 miles. A consumer organization doubts the claim and decides to test it. That is, the purpose of the research from the perspective of the consumer organization is to test if $\mu < 60,000$, where μ represents the mean mileage of all such tires. The research hypothesis is stated symbolically as H_a: $\mu < 60,000$ miles. The negation of the research hypothesis is $\mu = 60,000$. The null hypothesis is stated symbolically as H_0: $\mu = 60,000$. A test of hypothesis where the research hypothesis is of the form H_a: $\mu <$ (some constant) is called a *one-tailed test,* a *left-tailed test,* or a *lower-tailed test.*

EXAMPLE 9.2 The National Football League (NFL) claims that the average cost for a family of four to attend an NFL game is $200. This figure includes ticket prices and snacks. A sports magazine feels that this figure is too low and plans to perform a test of hypothesis concerning the claim. The research hypothesis is H_a: $\mu > \$200$, where μ is the mean cost for all such families attending an NFL game. The null hypothesis is H_0: $\mu = \$200$. A test of hypothesis where the research hypothesis is of the form H_a: $\mu >$ (some constant) is called a *one-tailed test,* a *right-tailed test,* or an *upper-tailed test.*

TEST STATISTIC, CRITICAL VALUES, REJECTION, AND NONREJECTION REGIONS

Consider again the test of hypothesis to be performed by the L. C. Stephens mail order company discussed in the previous section. The two hypotheses are restated as follows:

H_0: $\mu = \$155$ (the mean sales per order this year is $155)
H_a: $\mu \neq \$155$ (the mean sales per order this year is not $155)

To test this hypothesis, a sample of 100 orders for the current year are selected and the mean of the sample is used to decide whether to reject the null hypothesis or not. A statistic, which is used to decide whether to reject the null hypothesis or not is called a *test statistic*. The central limit theorem assures us that if the null hypothesis is true, then $\overline{x}$ will equal \$155 on the average and the standard error will equal $\sigma_{\overline{x}} = \dfrac{50}{\sqrt{100}} = 5$, assuming that the standard deviation equals 50 and has remained constant. Furthermore, the sample mean has a normal distribution. If the value of the test statistic, $\overline{x}$, is "close" to \$155, then we would likely not reject the null hypothesis. However, if the computed value of $\overline{x}$ is considerably different from \$155, we would reject the null hypothesis since this outcome supports the truth of the research hypothesis. The critical decision is how different does $\overline{x}$ need to be from \$155 in order to reject the null hypothesis. Suppose we decide to reject the null hypothesis if $\overline{x}$ differs from \$155 by two standard errors or more. Two standard errors is equal to $2 \times \sigma_{\overline{x}} = \$10$. Since the distribution of the sample mean is bell-shaped, the empirical rule assures us that the probability that $\overline{x}$ differs from \$155 by two standard errors or more is approximately equal to 0.05. That is, if $\overline{x} < \$145$ or if $\overline{x} > \$165$, then we will reject the null hypothesis. Otherwise, we will not reject the null hypothesis. The values \$145 and \$165 are called *critical values*. The critical values divide the possible values of the test statistic into two regions called the *rejection region* and the *nonrejection region*. The critical values along with the regions are shown in Fig. 9-1.

rejection region |	nonrejection region	| rejection region	$\overline{x}$
	145	165	

Fig. 9-1

EXAMPLE 9.3 In Example 9.1, the null and research hypothesis are stated as follows:

H_0: $\mu = 60,000$ miles (the tire manufacturer's claim is correct)
H_a: $\mu < 60,000$ miles (the tire manufacturer's claim is false)

Suppose it is known that the standard deviation of tire mileages for this type of tire is 7,000 miles. If a sample of 49 of the tires are road tested for mileage and the manufacturer's claim is correct, then $\overline{x}$ will equal 60,000 miles on the average and the standard error will equal $\sigma_{\overline{x}} = \dfrac{7,000}{\sqrt{49}} = 1,000$ miles. Furthermore, because of the large sample size, the sample mean will have a normal distribution. Suppose the critical value is chosen two standard errors below 60,000; then the rejection and nonrejection regions will be as shown in Fig. 9-2.

rejection region |	nonrejection region	$\overline{x}$
58,000		

Fig. 9-2

When the consumer organization tests the 49 tires, if the mean mileage is less than 58,000 miles for the sample, then the manufacturer's claim will be rejected.

EXAMPLE 9.4 In Example 9.2, the null hypothesis and the research hypothesis are stated as follows:

H_0: $\mu = \$200$ (the NFL claim is correct)
H_a: $\mu > \$200$ (the NFL claim is not correct)

Suppose the standard deviation of the cost for a family of four to attend an NFL game is known to equal \$30. If a sample of 36 costs for families of size 4 are obtained, then $\overline{x}$ will be normally distributed with mean equal to

$200 and standard error equal to $\dfrac{30}{\sqrt{36}}$ = \$5 if the NFL claim is correct. Suppose the critical value is chosen three standard errors above \$200; then the rejection and nonrejection regions are as shown in Fig. 9-3.

nonrejection region | rejection region $\overline{x}$

215

Fig. 9-3

If the mean of the sample of costs for the 36 families of size 4 exceeds \$215, then the NFL claim will be rejected.

TYPE I AND TYPE II ERRORS

Consider again the test of hypothesis to be performed by the L. C. Stephens mail order company discussed in the previous two sections. The two hypotheses are restated as follows:

$$H_0: \mu = \$155 \text{ (the mean sales per order this year is \$155)}$$
$$H_a: \mu \neq \$155 \text{ (the mean sales per order this year is not \$155)}$$

The decision to reject or not reject the null hypothesis is to be based on the sample mean. The critical values and the rejection and nonrejection regions are given in Fig. 9-1. If the current year mean, μ, is \$155 but the sample mean falls in the rejection region, resulting in the null hypothesis being rejected, then a *Type I error* is made. If the current year mean is not equal to \$155, but the sample mean falls in the nonrejection region, resulting in the null hypothesis not being rejected, then a *Type II error* is made. That is, if the null hypothesis is true but the statistical test results in rejection of the null, then a Type I error occurs. If the null hypothesis is false but the statistical test results in not rejecting the null, then a Type II error occurs. The errors as well as the possible correct conclusions are summarized in Table 9.1.

Table 9.1

Conclusion	H_0 True	H_0 False
Do not reject H_0	Correct conclusion	Type II error
Reject H_0	Type I error	Correct conclusion

The first two letters of the Greek alphabet are used in statistics to represent the probabilities of committing the two types of errors. The definitions are as follows.

$$\alpha = \text{probability of making a Type I error}$$
$$\beta = \text{probability of making a Type II error}$$

The calculation of a Type I error will now be illustrated, and the calculation of Type II errors will be illustrated in a later section. The term *level of significance* is also used for α.

The level of significance, α, is defined by formula (*9.1*):

$$\alpha = P(\text{rejecting the null hypothesis when the null hypothesis is true}) \qquad (9.1)$$

In the current discussion of the L. C. Stephens mail order company, the level of significance may be expressed as $\alpha = P(\overline{x} < 145 \text{ or } \overline{x} > 165 \text{ and } \mu = 155)$. In order to evaluate this probability, recall that for large samples, $z = \dfrac{\overline{x} - \mu}{\sigma_{\overline{x}}}$ has a standard normal distribution. As previously discussed, the standard

error is \$5 and assuming that the null hypothesis is true, $z = \dfrac{\overline{x} - 155}{5}$ has a standard normal

distribution. The event $\overline{x} < 145$ is equivalent to the event $\dfrac{\overline{x} - 155}{5} < \dfrac{145 - 155}{5} = -2$ or $z < -2$. The

event $\overline{x} > 165$ is equivalent to the event $\dfrac{\overline{x} - 155}{5} > \dfrac{165 - 155}{5} = 2$ or $z > 2$. Therefore, α may be

expressed as $\alpha = P(\overline{x} < 145) + P(\overline{x} > 165) = P(z < -2) + P(z > 2)$. From the standard normal
distribution table, $P(z < -2) = P(z > 2) = .5 - .4772 = .0228$. And therefore, $\alpha = 2 \times .0228 = .0456$.
When using the mean of a sample of 100 order sales and the rejection and nonrejection regions
shown in Fig. 9-1 to decide whether the overall mean sales per order has changed, 4.56% of the time
the company will conclude that the mean has changed when, in fact, it has not.

EXAMPLE 9.5 In Example 9.3, the null and research hypothesis for the tire manufacturer's claim is as
follows:

$$H_0: \mu = 60,000 \text{ miles (the tire manufacturer's claim is correct)}$$
$$H_a: \mu < 60,000 \text{ miles (the tire manufacturer's claim is false)}$$

The rejection and nonrejection regions were illustrated in Fig. 9-2. The level of significance is determined as
follows:

$$\alpha = P(\text{rejecting the null hypothesis when the null hypothesis is true})$$

Since the null is rejected when the sample mean is less than 58,000 miles and the null is true when $\mu = 60,000$,

$$\alpha = P(\overline{x} < 58,000 \text{ when } \mu = 60,000)$$

Transforming the sample mean to a standard normal, the expression for α now becomes

$$\alpha = P\left(\dfrac{\overline{x} - 60,000}{1,000} < \dfrac{58,000 - 60,000}{1,000}\right)$$

Simplifying, and finding the area to the left of -2 under the standard normal curve, we find

$$\alpha = P(z < -2) = .5 - .4772 = .0228$$

If the tire manufacturer's claim is tested by determining the mileages for 49 tires and rejecting the claim when
the mean mileage for the sample is less than 58,000 miles, then the probability of rejecting the manufacturer's
claim, when it is correct, equals .0228.

EXAMPLE 9.6 In Example 9.4, the null and research hypothesis for the NFL's claim is as follows:

$$H_0: \mu = \$200 \text{ (the NFL claim is correct)}$$
$$H_a: \mu > \$200 \text{ (the NFL claim is not correct)}$$

The rejection and nonrejection regions were illustrated in Fig. 9-3. The level of significance is determined as
follows:

$$\alpha = P(\text{rejecting the null hypothesis when the null hypothesis is true})$$

Since the null is rejected when the sample mean is greater than \$215 and the null is true when $\mu = \$200$,

$$\alpha = P(\overline{x} > 215 \text{ when } \mu = 200)$$

Transforming the sample mean to a standard normal, the expression for α now becomes

$$\alpha = P\left(\frac{\overline{x} - 200}{5} > \frac{215 - 200}{5}\right)$$

Simplifying, and finding the area to the right of 3 under the standard normal curve, we find

$$\alpha = P(z > 3) = .5 - .4986 = .0014$$

The probability of falsely rejecting the NFL's claim when using a sample of 36 families and the rejection and nonrejection regions shown in Fig. 9-3 to test the hypothesis is .0014.

Thus far, we have considered how to find the level of significance when the rejection and nonrejection regions are specified. Often, the level of significance is specified and the rejection and nonrejection regions are required. Suppose in the mail order example that the level of significance is specified to be $\alpha = .01$. Because the hypothesis test is two-tailed, the level of significance is divided into two halves equal to .005 each. We need to find the value a, where $P(z > a) = .005$ and $P(z < -a) = .005$. The probability $P(z > a) = .005$ implies that $P(0 < z < a) = .5 - .005 = .495$. Using the standard normal table, we see that $P(0 < z < 2.57) = .4949$ and $P(0 < z < 2.58) = .4951$. The interpolated value is a = 2.575, which we round to 2.58. The rejection region is therefore as follows: z < -2.58 or z > 2.58. Figure 9-4 shows the rejection regions as well as the area under the standard normal curve associated with the regions.

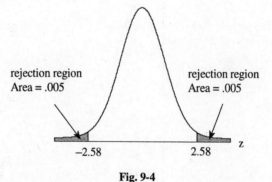

rejection region   rejection region
Area = .005   Area = .005

-2.58   2.58   z

Fig. 9-4

To determine the rejection region in terms of $\overline{x}$, the equation $z = \dfrac{\overline{x} - 155}{5}$ may be used. The inequality z < -2.58 is replaced by $\dfrac{\overline{x} - 155}{5} < -2.58$. Multiplying both sides of the inequality by 5 and then adding 155 to both sides, the inequality $\dfrac{\overline{x} - 155}{5} < -2.58$ is seen to be equivalent to $\overline{x} < 142.1$. The inequality z > 2.58 is replaced by $\dfrac{\overline{x} - 155}{5} > 2.58$ which is equivalent to $\overline{x} > 167.9$. Figure 9-5 shows the rejection regions as well as the area under the normal curve associated with the sample means.

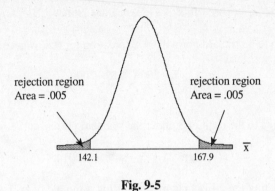

Fig. 9-5

EXAMPLE 9.7 Suppose we wish to test the hypothesis in Example 9.5 at a level of significance equal to .01. Since this is a lower-tailed test, we need to find a standard normal value a such that $P(z < a) = .01$. From the standard normal distribution table, we find that $P(0 < z < 2.33) = .4901$. Therefore $P(z > 2.33) = .01$. Because of the symmetry of the standard normal curve, $P(z < -2.33)$ also equals .01. Hence $z < -2.33$ is a rejection region of size $\alpha = .01$. To find the rejection region in terms of the sample mean, we note from Example 9.5 that if the null hypothesis is true, then $z = \dfrac{\overline{x} - 60,000}{1,000}$ has a standard normal distribution. Substituting for z in the inequality $z < -2.33$, we have $\dfrac{\overline{x} - 60,000}{1,000} < -2.33$ or solving for $\overline{x}$, we find $\overline{x} < 57,670$ miles as the rejection region. Figure 9-6 shows the rejection region as well as the area under the standard normal curve associated with the region.

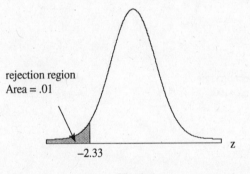

Fig. 9-6

Figure 9-7 shows the rejection region as well as the area under the normal curve associated with the sample means.

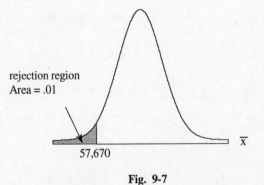

Fig. 9-7

EXAMPLE 9.8 Suppose we wish to test the hypothesis in Example 9.6 at a level of significance equal to .10. Since this is an upper-tailed test, we need to find a standard normal value a such that $P(z > a) = .10$. From the standard normal distribution table, we find that $P(0 < z < 1.28) = .3997$. Therefore $P(z > 1.28) = .10$. Hence $z > 1.28$ is a rejection region of size $\alpha = .10$. To find the rejection region in terms of the sample mean, we note from Example 9.6 that if the null hypothesis is true, then $z = \dfrac{\overline{x} - 200}{5}$ has a standard normal distribution. Substituting for z in the inequality $z > 1.28$, we have $\dfrac{\overline{x} - 200}{5} > 1.28$ or solving for $\overline{x}$, we find $\overline{x} > \$206.40$ as the rejection region. Figure 9-8 shows the rejection region as well as the area under the standard normal curve associated with the region.

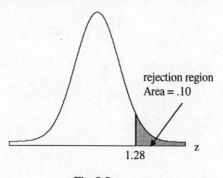

rejection region
Area = .10

1.28

Fig. 9-8

Figure 9-9 shows the rejection region as well as the area under the normal curve associated with the sample means.

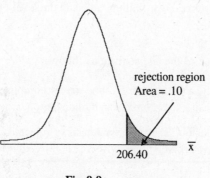

rejection region
Area = .10

206.40

Fig. 9-9

HYPOTHESIS TESTS ABOUT A POPULATION MEAN: LARGE SAMPLES

This section summarizes the material presented in the proceeding sections. The techniques presented are appropriate when the sample size is 30 or more.

EXAMPLE 9.9　The mean age of policyholders at World Life Insurance Company, determined two years ago, was found to equal 32.5 years and the standard deviation was found to equal 5.5 years. It is reasonable to believe that the mean age has increased. However, some of the older policyholders are now deceased and some younger policyholders have been added. The company determines the ages of 50 current policyholders in order to decide whether the mean age has changed. If μ represents the current mean of all policyholders, the null and research hypothesis are stated as follows:

H_0: $\mu = 32.5$ (the mean age has not changed)
H_a: $\mu \neq 32.5$ (the mean age has changed)

The test is performed at the conventional level of significance, which is $\alpha = .05$. From the standard normal distribution, it is determined that $P(z > 1.96) = .025$ and therefore $P(z < -1.96) = .025$. The rejection and nonrejection regions are therefore as follows:

rejection region	nonrejection region	rejection region	z
	-1.96	1.96	

The standard deviation of the policyholder ages is known to remain fairly constant, and therefore the standard error is $\sigma_{\bar{x}} = \dfrac{5.5}{\sqrt{50}} = .778$. The mean age of the sample of 50 policyholders is determined to be 34.4 years. The computed test statistic is $z = \dfrac{\bar{x} - \mu_0}{\sigma_{\bar{x}}} = \dfrac{34.4 - 32.5}{.778} = 2.44$. That is, assuming the mean age of all policyholders has not changed from two years ago, the mean of the current sample is 2.44 standard errors above the population mean. Since this exceeds 1.96, the null hypothesis is rejected and it is concluded that the current mean age exceeds 32.5 years. Notice that the four steps in Table 9.2 were followed in performing the test of hypothesis.

Table 9.2 gives a set of steps that may be used to perform a test of hypothesis about the mean of a population.

Table 9.2

Steps for Testing a Hypothesis Concerning a Population Mean: Large Sample
Step 1: State the null and research hypothesis. The null hypothesis is represented symbolically by H_0: $\mu = \mu_0$, and the research hypothesis is of the form H_a: $\mu \neq \mu_0$ or H_a: $\mu < \mu_0$ or H_a: $\mu > \mu_0$.
Step 2: Use the standard normal distribution table and the level of significance, α, to determine the rejection region.
Step 3: Compute the value of the test statistic as follows: $z = \dfrac{\bar{x} - \mu_0}{\sigma_{\bar{x}}}$, where $\bar{x}$ is the mean of the sample, μ_0 is given in the null hypothesis, and $\sigma_{\bar{x}}$ is computed by dividing σ by $\sqrt{n}$. If σ is unknown, estimate it by using the sample standard deviation, s, when computing $\sigma_{\bar{x}}$.
Step 4: State your conclusion. The null hypothesis is rejected if the computed value of the test statistic falls in the rejection region. Otherwise, the null hypothesis is not rejected.

EXAMPLE 9.10 The police department in a large city claims that the mean 911 response time for domestic disturbance calls is 10 minutes. A "watchdog group" believes that the mean response time is greater than 10 minutes. If μ represents the mean response time for all such calls, the watchdog group wishes to test the research hypothesis that $\mu > 10$ at level of significance $\alpha = .01$. The null and research hypothesis are:

H_0: $\mu = 10$ (the police department claim is correct)
H_a: $\mu > 10$ (the police department claim is not correct)

From the standard normal distribution table, it is found that $P(z > 2.33) = .01$. The rejection and nonrejection regions are as follows:

nonrejection region	rejection region	z
	2.33	

A sample of 35 response times for domestic disturbance calls is obtained and the mean response time is found to be 11.5 minutes and the standard deviation of the 35 response times is 6.0 minutes. The standard error is estimated to be $\dfrac{s}{\sqrt{n}} = \dfrac{6.0}{\sqrt{35}} = 1.01$ minutes. The computed test statistic is $z = \dfrac{\bar{x} - \mu_0}{\sigma_{\bar{x}}} = \dfrac{11.5 - 10.0}{1.01} = 1.49$.

Since the computed test statistic does not exceed 2.33, the police department claim is not rejected. Note that the study has not proved the claim to be true. However, the results of the study are not strong enough to refute the claim at the 1% level of significance.

EXAMPLE 9.11 A sociologist wishes to test the null hypothesis that the mean age of gang members in a large city is 14 years vs. the alternative that the mean is less than 14 years at level of significance $\alpha = .10$. A sample of 40 ages from the police gang unit records in the city found that $\overline{x} = 13.1$ years and $s = 2.5$ years. The estimated standard error is $\frac{s}{\sqrt{n}} = \frac{2.5}{\sqrt{40}} = .40$ years. Assuming the null hypothesis to be true, the sample mean is 2.25 standard errors below the population mean, since $z = \frac{13.1 - 14}{.40} = -2.25$. For $\alpha = .10$, the rejection region is $z < -1.28$. Since the computed test statistic is less than -1.28, the null hypothesis is rejected and it is concluded that the mean age of the gang members is less than 14 years.

The process of testing a statistical hypothesis is similar to the proceedings in a courtroom in the United States. An individual, charged with a crime, is assumed not guilty. However, the prosecution believes the individual is guilty and provides sample evidence to try and prove that the person is guilty. The null and alternative hypothesis may be stated as follows:

H_0: the individual charged with the crime is not guilty
H_a: the individual charged with the crime is guilty

If the evidence is strong, the null hypothesis is rejected and the person is declared guilty. If the evidence is circumstantial and not strong enough, the person is declared not guilty. Notice that the person is not usually declared innocent, but is found not guilty. The evidence usually does not prove the null hypothesis to be true, but is not strong enough to reject it in favor of the alternative.

CALCULATING TYPE II ERRORS

In testing a statistical hypothesis, a Type II error occurs when the test results in not rejecting the null hypothesis when the research hypothesis is true. The probability of a Type II error is represented by the Greek letter β and is defined in formula (9.2):

$$\beta = P(\text{not rejecting the null hypothesis when the research hypothesis is true}) \qquad (9.2)$$

Consider once again the mail order company example discussed in previous sections. The null and research hypothesis are:

H_0: $\mu = \$155$ (the mean sales per order this year is \$155)
H_a: $\mu \neq \$155$ (the mean sales per order this year is not \$155)

The rejection and nonrejection regions are:

rejection region	nonrejection region	rejection region	$\overline{x}$
145		165	

The level of significance is $\alpha = .0456$. The level of significance is computed under the assumption that the null hypothesis is true, that is, $\mu = 155$. The probability of a Type II error is calculated under the assumption that the research hypothesis is true. However, the research hypothesis is true whenever $\mu \neq \$155$. Suppose we wish to calculate β when $\mu = 157.5$. The sequence of steps to compute β are as follows:

$$\beta = P(\text{not rejecting the null hypothesis when the research hypothesis is true})$$

$$\beta = P(145 < \overline{x} < 165 \text{ when } \mu = 157.5)$$

The event $145 < \overline{x} < 165$ when $\mu = 157.5$ is equivalent to $\dfrac{145 - 157.5}{5} < \dfrac{\overline{x} - 157.5}{5} < \dfrac{165 - 157.5}{5}$ or $-2.5 < z < 1.5$. The calculation of β is therefore given as follows:

$$\beta = P(-2.5 < z < 1.5) = .4332 + .4938 = .9270$$

Suppose we wish to compute β when $\mu = 160$. The sequence of steps to compute β are as follows:

$$\beta = P(\text{not rejecting the null hypothesis when the research hypothesis is true})$$

$$\beta = P(145 < \overline{x} < 165 \text{ when } \mu = 160.0)$$

The event $145 < \overline{x} < 165$ when $\mu = 160.0$ is equivalent to $\dfrac{145 - 160}{5} < \dfrac{\overline{x} - 160}{5} < \dfrac{165 - 160}{5}$ or $-3 < z < 1$. The calculation of β is therefore given as follows:

$$\beta = P(-3 < z < 1) = .4986 + .3413 = .8399$$

Notice that the value of β is not constant, but depends on the alternative value assumed for μ. Table 9.3 gives the values of β for several different values of μ.

Table 9.3

μ	β
145.0	.5000
147.5	.6915
150.0	.8399
152.5	.9270
155.0	.9544
157.5	.9270
160.0	.8399
162.5	.6915
165.0	.5000

A plot of β vs. μ is called an *operating characteristic curve*. An operating characteristic curve for Table 9.3 is shown in Fig. 9-10.

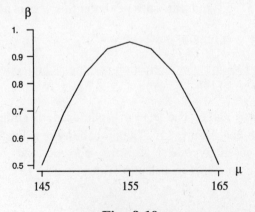

Fig. 9-10

EXAMPLE 9.12 Suppose we wish to construct an operating characteristic curve for the hypothesis concerning the tire manufacturer's claim discussed in Example 9.3. Recall that the null and research hypothesis were as follows:

$$H_0: \mu = 60{,}000 \text{ miles (the tire manufacturer's claim is correct)}$$
$$H_a: \mu < 60{,}000 \text{ miles (the tire manufacturer's claim is false)}$$

The rejection and nonrejection regions were as follows:

rejection region	nonrejection region	$\overline{x}$
58,000		

The level of significance is $\alpha = .0228$. To illustrate the computation of β, suppose we wish to compute the probability of committing a Type II error when $\mu = 59{,}000$.

$$\beta = P(\text{not rejecting the null hypothesis when the research hypothesis is true})$$

$$\beta = P(\overline{x} > 58{,}000 \text{ when } \mu = 59{,}000)$$

The event $\overline{x} > 58{,}000$ when $\mu = 59{,}000$ is equivalent to $\dfrac{\overline{x} - 59{,}000}{1{,}000} > \dfrac{58{,}000 - 59{,}000}{1{,}000}$ or $z > -1$. Therefore,

$$\beta = P(z > -1) = .5000 + .3413 = .8413.$$

Table 9.4 gives the β values for several different μ values.

Table 9.4

μ	β
56,000	.0228
56,500	.0668
57,000	.1587
57,500	.3085
58,000	.5000
58,500	.6915
59,000	.8413
59,500	.9332
60,000	.9772

The operating characteristic curve corresponding to Table 9.4 is shown in Fig. 9-11.

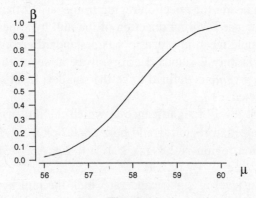

Fig. 9-11

P VALUES

Consider once again the mail order company example. The null and research hypotheses are:

H_0: μ = \$155 (the mean sales per order this year is \$155)
H_a: $\mu \neq$ \$155 (the mean sales per order this year is not \$155)

For level of significance α = .05, the rejection region is shown in Fig. 9-12.

rejection region	nonrejection region	rejection region	z
−1.96		1.96	

Fig. 9-12

To understand the concept of p value, consider the following four scenarios. Also recall that the standard error is $\sigma_{\bar{x}} = \dfrac{50}{\sqrt{100}} = 5$.

Scenario 1: The mean of the 100 sampled accounts is found to equal \$149. The computed test statistic is $z = \dfrac{149 - 155}{5} = -1.2$, and since the computed test statistic falls in the nonrejection region, the null hypothesis is not rejected.

Scenario 2: The mean of the 100 sampled accounts is found to equal \$164. The computed test statistic is z = 1.8, and since the computed test statistic falls in the nonrejection region, the null hypothesis is not rejected.

Scenario 3: The mean of the 100 sampled accounts is found to equal \$165. The computed test statistic is z = 2.0, and since the computed test statistic falls in the rejection region, the null hypothesis is rejected.

Scenario 4: The mean of the 100 sampled accounts is found to equal \$140. The computed test statistic is z = −3.0, and since the computed test statistic falls in the rejection region, the null hypothesis is rejected.

Notice in scenarios 1 and 2 that even though the evidence is not strong enough to reject the null hypothesis, the test statistic is nearer the rejection region in scenario 2 than scenario 1. Also notice in scenarios 3 and 4 that the evidence favoring rejection of the null hypothesis is stronger in scenario 4 than scenario 3, since the sample mean in scenario 4 is 3 standard errors away from the population mean, and the sample mean is only 2 standard errors away in scenario 3. The *p value* is used to reflect these differences. The *p value* is defined to be the smallest level of significance at which the null hypothesis would be rejected.

In scenario 1, the smallest level of significance at which the null hypothesis would be rejected would be one in which the rejection region would be $z \leq -1.2$ or $z \geq 1.2$. The level of significance corresponding to this rejection region is $2 \times P(z \geq 1.2) = 2 \times (.5 - .3849) = .2302$. The p value corresponding to the test statistic $z = -1.2$ is .2302.

In scenario 2, the smallest level of significance at which the null hypothesis would be rejected would be one in which the rejection region would be $z \leq -1.8$ or $z \geq 1.8$. The level of significance corresponding to this rejection region is $2 \times P(z \geq 1.8) = 2 \times (.5 - .4641) = .0718$. The p value corresponding to the test statistic $z = 1.8$ is .0718.

In scenario 3, the smallest level of significance at which the null hypothesis would be rejected would be one in which the rejection region would be $z \leq -2.0$ or $z \geq 2.0$. The level of significance corresponding to this rejection region is $2 \times P(z \geq 2.0) = 2 \times (.5 - .4772) = .0456$. The p value corresponding to the test statistic $z = 2.0$ is .0456.

In scenario 4, the smallest level of significance at which the null hypothesis would be rejected would be one in which the rejection region would be $z \leq -3.0$ or $z \geq 3.0$. The level of significance corresponding to this rejection region is $2 \times P(z \geq 3.0) = 2 \times (.5 - .4987) = .0026$. The p value corresponding to the test statistic $z = 3.0$ is .0026.

To summarize the procedure for computing the p value for a two-tailed test, suppose z^* represents the computed test statistic when testing $H_0: \mu = \mu_o$ vs. $H_a: \mu \neq \mu_0$. The p value is given by formula (9.3):

$$\text{p value} = P(|z| > |z^*|) \tag{9.3}$$

When using the p value approach to testing hypothesis, the null hypothesis is rejected if the p value is less than α. In addition, the p value gives an idea about the strength of the evidence for rejecting the null hypothesis. The smaller the p value, the stronger the evidence for rejecting the null hypothesis.

EXAMPLE 9.13 In Example 9.10, the following statistical hypothesis was tested for $\alpha = .01$:

$$H_0: \mu = 10 \text{ (the police department claim is correct)}$$
$$H_a: \mu > 10 \text{ (the police department claim is not correct)}$$

The computed test statistic is $z^* = 1.49$. The smallest level of significance at which the null hypothesis would be rejected is one in which the rejection region is $z > 1.49$. The p value is therefore equal to $P(z > 1.49)$. From the standard normal distribution table, we find the p value $= .5 - .4319 = .0681$. Using the p value approach to testing, the null hypothesis is not rejected since the p value $> \alpha$.

To summarize the procedure for computing the p value for an upper-tailed test, suppose z^* represents the computed test statistic when testing $H_0: \mu = \mu_o$ vs. $H_a : \mu > \mu_0$. The p value is given by

$$\text{p value} = P(z > z^*) \tag{9.4}$$

EXAMPLE 9.14 In Example 9.11, the following statistical hypothesis was tested for $\alpha = .10$.

$$H_0: \mu = 14 \text{ (the mean age of gang members in the city is 14)}$$
$$H_a: \mu < 14 \text{ (the mean age of gang members in the city is less than 14)}$$

The computed test statistic is $z^* = -2.25$. The smallest level of significance at which the null hypothesis would be rejected is one in which the rejection region is $z < -2.25$. The p value is therefore equal to $P(z < -2.25)$. Using the standard normal distribution table, we find the p value $= .5 - .4878 = .0122$. Using the p value approach to testing, the null hypothesis is rejected since the p value $< .10$.

To summarize the procedure for computing the p value for a lower-tailed test, suppose z^* represents the computed test statistic when testing $H_0: \mu = \mu_0$ vs. $H_a: \mu < \mu_0$. The p value is given by

$$\text{p value} = P(z < z^*) \tag{9.5}$$

EXAMPLE 9.15 A random sample of 40 community college tuition costs was selected to test the null hypothesis that the mean cost for all community colleges equals $1500 vs. the alternative that the mean does not equal $1500. The level of significance is .05. The data are shown in Table 9.5.

Table 9.5

1200	850	1750	930
850	3000	1650	1640
1700	2100	900	1320
1500	500	2050	1750
700	500	1780	2500
1200	1950	675	2310
1500	1000	1080	2900
2000	950	680	1875
1950	560	900	1450
750	500	1500	950

The Minitab analysis for this test of hypothesis is shown below. The command **set c1** is used to put the data in column 1. The name cost is given to column 1. The command **standard deviation c1** is used to find s. The command **ztest mean = 1500 sigma = 655.44 data in c1;** is used to specify the value of μ_0, the estimate of sigma, and the column where the data are located. The subcommand **alternative 0.** indicates a two-tailed test. Since the p value is .32 and exceeds .05, the null hypothesis is not rejected.

```
MTB > set c1
DATA >1200   850  1700  1500   700  1200  1500  2000  1950   750
DATA > 850  3000  2100   500   500  1950  1000   950   560   500
DATA >1750  1650   900  2050  1780   675  1080   680   900  1500
DATA > 930  1640  1320  1750  2500  2310  2900  1875  1450   950
DATA > end
MTB > name c1 'cost'
MTB > standard deviation c1
```

Column Standard Deviation

```
Standard deviation of cost = 655.44
MTB > ztest mean = 1500 sigma = 655.44 data in c1;
SUBC> alternative 0.
```

Z-Test
Test of mu = 1500 vs mu not = 1500
The assumed sigma = 655

Variable	N	Mean	St Dev	SE Mean	Z	P
cost	40	1396	655	104	−1.00	.32

To test the research hypothesis that $\mu < 1500$, the subcommand **alternative −1.** is used. The p value is now equal to .16.

```
MTB > ztest mean = 1500 sigma = 655.44 data in c1;
SUBC > alternative −1.
```

Z-Test
Test of mu = 1500 vs mu < 1500
The assumed sigma = 655

Variable	N	Mean	St Dev	SE Mean	Z	P
cost	40	1396	655	104	−1.00	.16

To test the research hypothesis that $\mu > 1500$, the subcommand **alternative 1.** is used. The p value is now equal to .84.

```
MTB > ztest mean = 1500 sigma = 655.44 data in c1;
SUBC > alternative 1.
```

Z-Test
Test of mu = 1500 vs mu > 1500
The assumed sigma = 655

Variable	N	Mean	St Dev	SE Mean	Z	P
cost	40	1396	655	104	−1.00	.84

HYPOTHESIS TESTS ABOUT A POPULATION MEAN: SMALL SAMPLES

The procedure given in Table 9.2, for testing statistical hypothesis about the population mean when the sample size is large (n ≥ 30), is valid for all types of population distributions. The procedures given in this section when the sample size is small (n < 30) require the assumption that the population have a normal distribution. If the population standard deviation is known, the standard normal distribution table is used to determine the rejection and nonrejection regions. If the population standard deviation is unknown, the t distribution table is used to determine the rejection and nonrejection regions. Since the population standard deviation is usually unknown in practice, we shall discuss this case only. Table 9.6 summarizes the procedure for small samples from a normally distributed population with unknown σ.

Table 9.6

Steps for Testing a Hypothesis Concerning a Population Mean: Small Sample Normally Distributed Population with Unknown σ
Step 1: State the null and research hypothesis. The null hypothesis is represented symbolically by H_0: $\mu = \mu_0$, and the research hypothesis is of the form H_a: $\mu \neq \mu_0$ or H_a: $\mu < \mu_0$ or H_a: $\mu > \mu_0$.
Step 2: Use the t distribution table, with degrees of freedom = n – 1 and the level of significance, α, to determine the rejection region.
Step 3: Compute the value of the test statistic as follows: $t = \dfrac{\bar{x} - \mu_0}{s_{\bar{x}}}$, where $\bar{x}$ is the mean of the sample, μ_0 is given in the null hypothesis, and $s_{\bar{x}}$ is computed by dividing s by $\sqrt{n}$.
Step 4: State your conclusion. The null hypothesis is rejected if the computed value of the test statistic falls in the rejection region. Otherwise, the null hypothesis is not rejected.

EXAMPLE 9.16 In order to test the hypothesis that the mean age of all commercial airplanes is 10 years vs. the alternative that the mean is not 10 years, a sample of 25 airplanes is selected from several airlines. A histogram of the 25 ages is mound-shaped and it is therefore assumed that the distribution of all airplane ages is normally distributed. The sample mean is found to equal 11.5 years and the sample standard deviation is equal to 5.5 years. The level of significance is chosen to be $\alpha = .05$. From the t distribution tables with df = 24, it is determined that the area to the right of 2.064 is .025. By symmetry, the area to the left of −2.064 is also .025. The rejection regions as well as the corresponding areas under the t distribution curve are shown in Fig. 9-13.

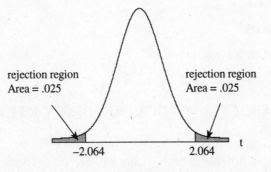

Fig. 9-13

The estimated standard error is $s_{\bar{x}} = \dfrac{5.5}{\sqrt{25}} = 1.1$ years. The computed test statistic is $t = \dfrac{\bar{x} - \mu_0}{s_{\bar{x}}} = \dfrac{11.5 - 10}{1.1} =$ 1.36, and since this value does not fall in the rejection region, the null hypothesis is not rejected.

EXAMPLE 9.17 A social worker wishes to test the hypothesis that the mean monthly household payment for food stamps in Shelby county is \$225 vs. the alternative that it is greater than that amount. The payments for 20 randomly selected households are given in Table 9.7.

Table 9.7

300	280	190	250
250	220	255	272
200	210	245	228
225	290	265	213
275	310	235	277

A histogram of the payments is shown in Fig. 9-14. The histogram indicates that it is reasonable to assume that the payments are normally distributed.

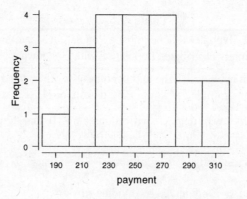

Fig. 9-14

The Minitab analysis for the data is shown below. The command **ttest mean = 225 data in c1;** gives the value for μ_0 and the location of the sample data in column 1. The subcommand **alternative = 1.** Identifies the alternative as being upper-tailed. The p value indicates that the results are significant for any level of significance greater than .0023.

```
MTB > set c1
DATA > 300    250   200   225   275   280   220   210   290   310
DATA > 190    255   245   265   235   250   272   228   213   277
DATA > end
MTB > ttest mean = 225 data in c1;
SUBC > alternative = 1.
T-Test of the Mean
Test of mu = 225.00 vs mu > 225.00
```

Variable	N	Mean	St Dev	SE Mean	T	P
Payment	20	249.50	34.04	7.61	3.22	0.0023

HYPOTHESIS TESTS ABOUT A POPULATION PROPORTION: LARGE SAMPLES

In Chapter 7 the sampling distribution of $\bar{p}$, the sample proportion, was discussed. When the sample size satisfies the inequalities $np > 5$ and $nq > 5$, the sampling distribution of the sample proportion is normally distributed with mean equal to p and standard error $\sigma_{\bar{p}} = \sqrt{\dfrac{pq}{n}}$. This result is sometimes referred to as the *central limit theorem for the sample proportion*. Furthermore, it was shown that $z = \dfrac{\bar{p} - \mu_{\bar{p}}}{\sigma_{\bar{p}}}$ has a standard normal distribution. These results form the theoretical underpinnings for tests of hypothesis concerning p, the population proportion. Table 9.8 gives the steps that may be followed when testing a hypothesis about the population proportion.

Table 9.8

Steps for Testing a Hypothesis Concerning a Population Proportion: Large Samples (np > 5 and nq > 5)
Step 1: State the null and research hypothesis. The null hypothesis is represented symbolically by H_0: p = p_0. and the research hypothesis is of the form H_a: p ≠ p_0 or H_a: p < p_0 or H_a: p > p_0.
Step 2: Use the standard normal distribution table to determine the rejection and nonrejection regions.
Step 3: Compute the value of the test statistic as follows: $z = \dfrac{\bar{p} - p_0}{\sigma_{\bar{p}}}$, where $\bar{p}$ is the sample proportion, p_0 is specified in the null hypothesis, and $\sigma_{\bar{p}} = \sqrt{\dfrac{p_0 \times q_0}{n}}$.
Step 4: State your conclusion. The null hypothesis is rejected if the computed value of the test statistic falls in the rejection region. Otherwise, the null hypothesis is not rejected.

EXAMPLE 9.18 Health-care coverage for employees varies with company size. It is reported that 30% of all companies with fewer than 10 employees provide health benefits for their employees. A sample of 50 companies with fewer than 10 employees is selected to test H_0: p = .3 vs. H_a: p ≠ .3 at α = .01. It is found that 19 of the 50 companies surveyed provide health benefits for their employees. Using the standard normal table, it is found that the area under the curve to the right of 2.58 is .005 and by symmetry, the area to the left of –2.58 is also .005. The rejection regions as well as the associated areas under the standard normal curve are shown in Fig. 9-15.

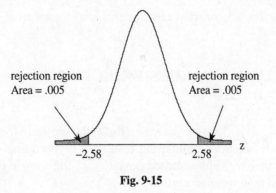

rejection region
Area = .005

rejection region
Area = .005

−2.58   2.58   z

Fig. 9-15

Assuming the null hypothesis to be true, the standard error of the proportion is $\sigma_{\bar{p}} = \sqrt{\dfrac{p_0 \times q_0}{n}} = \sqrt{\dfrac{.3 \times .7}{50}} =$ .065. The sample proportion is $\bar{p} = \dfrac{19}{50} = .38$. The computed value of the test statistic is $z^* = \dfrac{.38 - .30}{.065} = 1.23$. Since the computed value of the test statistic is not in the rejection region, the null hypothesis is not rejected. The p value is $P(|z| > 1.23) = P(z < -1.23) + P(z > 1.23) = 2 \times P(z > 1.23) = 2 \times (.5 - .3907) = .2186$.

EXAMPLE 9.19 A national survey asked the following question of 2500 registered voters: "Is the character of a candidate for president important to you when deciding for whom to vote?" Two thousand of the responses were yes. Let p represent the percent of all registered voters who believe the character of the president is important when deciding for whom to vote. The results of the survey were used to test the null hypothesis H_0: $p = 90\%$ vs. H_a: $p < 90\%$ at level of significance $\alpha = .05$. The sample percent is $\overline{p} = \frac{2000}{2500} \times 100\% = 80\%$.

Assuming the null hypothesis to be true, the standard error is $\sigma_{\overline{p}} = \sqrt{\dfrac{p_0 \times q_0}{n}} = \sqrt{\dfrac{90 \times 10}{2500}} = .60\%$. The computed test statistic is $z^* = \dfrac{80 - 90}{.60} = -16.7$. The null hypothesis would be rejected at all practical levels of significance.

Solved Problems

NULL HYPOTHESIS AND ALTERNATIVE HYPOTHESIS

9.1　A study in 1992 established the mean commuting distance for workers in a certain city to be 15 miles. Because of the westward spread of the city, it is hypothesized that the current mean commuting distance exceeds 15 miles. A traffic engineer wishes to test the hypothesis that the mean commuting distance for workers in this city is greater than 15 miles. Give the null and alternative hypothesis for this scenario.

Ans.　H_0: $\mu = 15$, H_a: $\mu > 15$, where μ represents the current mean commuting distance in this city.

9.2　The mean number of sick days used per year nationally is reported to be 5.5 days. A study is undertaken to determine if the mean number of sick days used for nonunion members in Kansas differs from the national mean. Give the null and alternative hypothesis for this scenario.

Ans.　H_0: $\mu = 5.5$, H_a: $\mu \neq 5.5$, where μ represents the mean number of sick days used for nonunion members in Kansas.

TEST STATISTIC, CRITICAL VALUES, REJECTION AND NONREJECTION REGIONS

9.3　Refer to problem 9.1. The decision is made to reject the null hypothesis if the mean of a sample of 49 randomly selected commuting distances is more than 2.5 standard errors above the mean established in 1992. The standard deviation of the sample is found to equal 3.5 miles. Give the rejection and nonrejection regions.

Ans.　The estimated standard error of the mean is $s_{\overline{x}} = \dfrac{s}{\sqrt{n}} = \dfrac{3.5}{\sqrt{49}} = .5$ miles. The null hypothesis is to be rejected if the sample mean of the 49 commuting distances exceeds the 1992 population mean by 2.5 standard errors. That is, reject H_0 if $\overline{x} > 15 + 2.5 \times .5 = 16.25$.

nonrejection region | rejection region　　　　　$\overline{x}$

16.25

9.4 Refer to problem 9.2. The null hypothesis is to be rejected if the mean of a sample of 250 nonunion members differs from the national mean by more than 2 standard errors. The standard deviation of the sample is found to equal 2.1 days. Give the rejection and nonrejection regions.

Ans. The estimated standard error of the mean is $s_{\bar{x}} = \dfrac{s}{\sqrt{n}} = \dfrac{2.1}{\sqrt{250}} = .13$ days. The null hypothesis is to be rejected if the mean of the sample differs from the national mean by more than 2 standard errors. That is, reject H_0 if $\bar{x} < 5.5 - 2 \times .13 = 5.24$ or if $\bar{x} > 5.5 + 2 \times .13 = 5.76$.

rejection region	nonrejection region	rejection region	$\bar{x}$
5.24		5.76	

TYPE I AND TYPE II ERRORS

9.5 Refer to problems 9.1 and 9.2. Describe, in words, how a Type I and a Type II error for both problems might occur.

Ans. In problem 9.1, a Type I error is made if it is concluded that the current mean commuting distance exceeds 15 miles when in fact the current mean is equal to 15 miles. A Type II error is made if it is concluded that the current mean commuting distance is 15 miles when in fact the mean exceeds 15 miles.

In problem 9.2, a Type I error is made if it is concluded that the mean number of sick days used for the nonunion members in Kansas is different than the national mean when in fact it is not different. A Type II error is made if it is concluded that the mean number of sick days used for nonunion members in Kansas is equal to the national mean when in fact it differs from the national mean.

9.6 Refer to problems 9.3 and 9.4. Find the level of significance for the test procedures described in both problems.

Ans. In problem 9.3, α = P(rejecting the null hypothesis when the null hypothesis is true) or α = P($\bar{x} >$ 16.25 when $\mu = 15$) = P(z > 2.5), since 16.25 is 2.5 standard errors above $\mu = 15$. α = P(z > 2.5) = .5 − .4938 = .0062.

In problem 9.4, α = P(rejecting the null hypothesis when the null hypothesis is true) or α = P($\bar{x} <$ 5.24 or $\bar{x} > 5.76$ when $\mu = 5.5$) = P(z < −2 or z > 2), since 5.24 is 2 standard errors below $\mu = 5.5$ and 5.76 is 2 standard errors above $\mu = 5.5$. α = P(z < −2) + P(z > 2) = 2 × (.5 − .4772) = .0456.

HYPOTHESIS TESTS ABOUT A POPULATION MEAN: LARGE SAMPLES

9.7 A sample of size 50 is used to test the following hypothesis: H_0: $\mu = 17.5$ vs. H_a: $\mu \neq 17.5$ and the level of significance is .01. Give the critical values, the rejection and nonrejection regions, the computed test statistic, and your conclusion if: (*a*) $\bar{x} = 21.5$ and s = 5.5 (σ unknown); (*b*) $\bar{x} = 21.5$ and $\sigma = 5.0$

Ans. The critical values and therefore the rejection and nonrejection regions are the same in both parts. Since P(z < −2.58) = P(z > 2.58) = .005, the critical values are −2.58 and 2.58. The rejection and nonrejection regions are as follows:

rejection region	nonrejection region	rejection region	z
−2.58		2.58	

In part (*a*) the standard error is estimated by $s_{\bar{x}} = \dfrac{s}{\sqrt{n}} = \dfrac{5.5}{\sqrt{50}} = .78$, since the population standard deviation is unknown. The computed test statistic is $z = \dfrac{21.5 - 17.5}{.78} = 5.13$, and the null hypothesis is rejected since this value falls in the rejection region.

In part (*b*) the standard error is computed by $\sigma_{\bar{x}} = \dfrac{\sigma}{\sqrt{n}} = \dfrac{5}{\sqrt{50}} = .71$, since the population standard deviation is known. The computed test statistic is $z = \dfrac{21.5 - 17.5}{.71} = 5.63$, and the null hypothesis is rejected since this value falls in the rejection region.

9.8 A state environmental study concerning the number of scrap-tires accumulated per tire dealership during the past year was conducted. The null hypothesis is $H_0: \mu = 2500$ and the research hypothesis is $H : \mu \neq 2500$, where μ represents the mean number of scrap-tires per dealership in the state. For a random sample of 85 dealerships, the mean is 2750 and the standard deviation is 950. Conduct the hypothesis test at the 5% level of significance.

Ans. The null hypothesis is rejected if $|z*| > 1.96$, where $z*$ is the computed value of the test statistic. The estimate standard error is $s_{\bar{x}} = \dfrac{s}{\sqrt{n}} = \dfrac{950}{\sqrt{85}} = 103$ miles. The computed value of the test statistic is $z* = \dfrac{2750 - 2500}{103} = 2.43$. It is concluded that the mean number of scrap-tires per dealership exceeded 2500 last year.

CALCULATING TYPE II ERRORS

9.9 In problems 9.1 and 9.3, the hypothesis system $H_0: \mu = 15$ and $H_a: \mu > 15$ was tested by using a sample of size 49. The null hypothesis was to be rejected if the sample mean exceeded 16.25. Suppose the true value of μ is 16.5. What is the probability that this test procedure will not result in the rejection of the null hypothesis?

Ans. $\beta = P(\text{not rejecting the null hypothesis when } \mu = 16.5) = P(\bar{x} < 16.25 \text{ when } \mu = 16.5)$. In problem 9.3, the estimated standard error was found to equal .5. The event $\bar{x} < 16.25$ is equivalent to the event $z = \dfrac{\bar{x} - 16.5}{.5} < \dfrac{16.25 - 16.5}{.5} = -.5$ when $\mu = 16.5$. Therefore, $\beta = P(z < -.5) = P(z > .5) = .5 - .1915 = .3085$. That is, there is a 30.85% chance that, even though the mean commuting distance has increased from the 15 miles figure of 1992 to the current figure of 16.5 miles, the hypothesis test will result in concluding that the current mean is the same as the 1992 mean.

9.10 In problems 9.2 and 9.4, the hypothesis system $H_0: \mu = 5.5$ and $H_a: \mu \neq 5.5$ was tested by using a sample of size 250. The null hypothesis was to be rejected if the sample mean was less than 5.24 or exceeded 5.76. Suppose the true value of μ is 5.0. What is the probability that this test procedure will not result in the rejection of the null hypothesis?

Ans. $\beta = P(\text{not rejecting the null hypothesis when } \mu = 5.0) = P(5.24 < \bar{x} < 5.76 \text{ when } \mu \text{ is } 5.0)$. In problem 9.4, the estimated standard error was found to equal .13. The event $5.24 < \bar{x} < 5.76$ when μ is 5.0 is equivalent to the event $\dfrac{5.24 - 5.0}{.13} < \dfrac{x - 5.0}{.13} < \dfrac{5.76 - 5.0}{.13}$ which is the same as the event

$1.85 < z < 5.85$. Therefore $\beta = P(1.85 < z < 5.85)$. Since there is practically zero probability to the right of 5.85, $\beta = P(z > 1.85) = .5 - .4678 = .0322$.

P VALUES

9.11 Use the p value approach to test the hypothesis in problem 9.8.

Ans.   The computed value of the test statistic is $z^* = 2.43$. Since the hypothesis is two-tailed, the p value is given by p value = $P(|\,z\,| > |\,z^*\,|)$ or equivalently $P(z < -|\,z^*\,|) + P(z > |\,z^*\,|) = P(z < -2.43) + P(z > 2.43)$. Since these two probabilities are equal, we have the p value = $2 \times P(z > 2.43) = 2 \times (.5 - .4925) = .015$. Using the p value approach, the null hypothesis is rejected since the p value $< \alpha$.

9.12 Find the p values for the following hypothesis systems and z^* values.
(*a*) $H_0: \mu = 22.5$ vs. $H_a: \mu \neq 22.5$ and $z^* = 1.76$
(*b*) $H_0: \mu = 37.1$ vs. $H_a: \mu \neq 37.1$ and $z^* = -2.38$
(*c*) $H_0: \mu = 12.5$ vs. $H_a: \mu < 12.5$ and $z^* = -.98$
(*d*) $H_0: \mu = 22.5$ vs. $H_a: \mu < 22.5$ and $z^* = 0.35$
(*e*) $H_0: \mu = 100$ vs. $H_a: \mu > 100$ and $z^* = 2.76$
(*f*) $H_0: \mu = 72.5$ vs. $H_a: \mu > 72.5$ and $z^* = -0.25$

Ans.   (*a*)   The p value is given by p value = $P(|\,z\,| > |\,z^*\,|) = P(|\,z\,| > 1.76) = 2 \times P(z > 1.76) = 2 \times (.5 - .4608) = .0784$.
(*b*)   The p value is given by p value = $P(|\,z\,| > |\,z^*\,|) = P(|\,z\,| > 2.38)$, since the absolute value of -2.38 is 2.38. Therefore, p value = $P(|\,z\,| > 2.38) = 2 \times P(z > 2.38) = 2 \times (.5 - .4913) = .0174$.
(*c*)   The p value is given by p value = $P(z < z^*) = P(z < -.98) = P(z > .98) = (.5 - .3365) = .1635$.
(*d*)   The p value is given by p value = $P(z < z^*) = P(z < .35) = .5 + .1368 = .6368$.
(*e*)   The p value is given by p value = $P(z > z^*) = P(z > 2.76) = .5 - .4971 = .0029$.
(*f*)   The p value is given by p value = $P(z > z^*) = P(z > -.25) = .5 + .0987 = .5987$.

HYPOTHESIS TESTS ABOUT A POPULATION MEAN: SMALL SAMPLES

9.13 A psychological test, used to measure the level of hostility, is known to produce scores which are normally distributed, with mean = 35 and standard deviation = 5. The test is used to test the null hypothesis that $\mu = 35$ vs. the research hypothesis that $\mu > 35$, where μ represents the mean score for criminal defense lawyers. Sixteen criminal defense lawyers are administered the test and the sample mean is found to equal 39.5. The sample standard deviation is found to equal 10. The level of significance is selected to be $\alpha = .01$.
(*a*) Perform the test assuming that $\sigma = 5$. This amounts to assuming that the standard deviation of test scores for criminal defense lawyers is the same as the general population.
(*b*) Perform the test assuming σ is unknown.

Ans.   (*a*)   Since the test scores are normally distributed and σ is known, the test statistic, $z = \dfrac{\bar{x} - 35}{\sigma_{\bar{x}}}$ is

   normally distributed. The rejection region is found using the standard normal distribution tables. The null hypothesis is rejected if the test statistic exceeds 2.33.The calculated value of the test statistic is $z^* = \dfrac{39.5 - 35}{1.25} = 3.6$, where the standard error is computed as

   $\sigma_{\bar{x}} = \dfrac{\sigma}{\sqrt{n}} = \dfrac{5}{\sqrt{16}} = 1.25$. The null hypothesis is rejected since 3.6 falls in the rejection region.

(b) Because σ is unknown and the sample size is small the test statistic, $t = \dfrac{\bar{x} - 35}{s_{\bar{x}}}$ has a t distribution with df = 16 − 1 = 15. The rejection region is found by using the t distribution with df = 15. The null hypothesis is rejected if the test statistic exceeds 2.602. The calculated value of the test statistic is $t^* = \dfrac{39.5 - 35}{2.5} = 1.80$, where the estimated standard error is computed as $s_{\bar{x}} = \dfrac{s}{\sqrt{n}} = \dfrac{10}{\sqrt{16}} = 2.5$. The null hypothesis is not rejected, since 1.80 does not fall in the rejection region.

9.14 The weights in pounds of weekly garbage for 25 households is shown in Table 9.9. Use these data to test the hypothesis that the weekly mean for all households in the city is 10 pounds. The alternative hypothesis is that the mean differs from 10 pounds. The level of significance is $\alpha = .05$.

Table 9.9

5.5	13.2	15.5	16.5	18.0
7.5	4.0	3.3	9.0	17.6
12.5	2.5	7.5	8.5	4.9
10.0	14.0	10.0	16.6	2.9
15.3	13.3	6.6	9.7	13.4

Ans.   The Minitab solution is shown below.

```
MTB > set c1
DATA >   5.5   7.5   12.5   10.0   15.5   13.2   4.0   2.5   14.0   13.3
DATA >  15.5   3.3   7.5   10.0   6.6   16.5   9.0   8.5   16.6   9.7
DATA >  18.0   17.6   4.9   2.9   13.4
DATA > end
MTB > ttest mean = 10 data in c1;
SUBC > alt = 0.
```

T-Test of the Mean

Test of mu = 10.000 vs mu not = 10.000

Variable	N	Mean	St Dev	SE Mean	T	P
C1	25	10.320	4.923	0.985	0.33	0.75

Since the p value > .05, the null hypothesis is not rejected.

HYPOTHESIS TESTS ABOUT A POPULATION PROPORTION: LARGE SAMPLES

9.15 A survey of 2500 women between the ages of 15 and 50 found that 28% of those surveyed relied on the pill for birth control. Use these sample results to test the null hypothesis that p = 25% vs. the research hypothesis that p $\neq$ 25%, where p represents the population percentage using the pill for birth control. Conduct the test at $\alpha = .05$ by giving the critical values, the computed value of the test statistic, and your conclusion.

Ans.   The critical values are ± 1.96. The test statistic is given by $z = \dfrac{\bar{p} - p_0}{\sigma_{\bar{p}}}$, where $\sigma_{\bar{p}} = \sqrt{\dfrac{p_0 \times q_0}{n}}$.

The standard error of the proportion is $\sigma_{\bar{p}} = \sqrt{\dfrac{25 \times 75}{2500}} = .87\%$, and the computed test statistic is

$z = \dfrac{28 - 25}{.87} = 3.45$. The null hypothesis is rejected since the computed test statistic exceeds the critical value, 1.96.

9.16 A poll taken just prior to election day finds that 389 of 700 registered voters intend to vote for Jake Konvalina for mayor of a large Midwestern city. Test the null hypothesis that $p = .5$ vs. the alternative that $p > .5$ at $\alpha = .05$, where p represents the proportion of all voters who intend to vote for Jake. Use the p value approach to do the test.

Ans.   The sample proportion is $\bar{p} = \dfrac{389}{700} = .556$, and the standard error of the proportion is

$\sigma_{\bar{p}} = \sqrt{\dfrac{p_0 \times q_0}{n}} = \sqrt{\dfrac{.5 \times .5}{700}} = .019$. The computed test statistic is $z^* = \dfrac{\bar{p} - p_0}{\sigma_{\bar{p}}} = \dfrac{.556 - .5}{.019} = 2.95$.

The p value is given by $P(z > 2.95) = .5 - .4984 = .0016$. The null hypothesis is rejected since the p value $< .05$.

Supplementary Problems

NULL HYPOTHESIS AND ALTERNATIVE HYPOTHESIS

9.17 Classify each of the following as a lower-tailed, upper-tailed, or two-tailed test:
(*a*) $H_0: \mu = 35$, $H_a: \mu < 35$ (*b*) $H_0: \mu = 1.2$, $H_a: \mu \neq 1.2$ (*c*) $H_0: \mu = \$1050$, $H_a: \mu > \$1050$

Ans.   (*a*) lower-tailed test   (*b*) two-tailed test   (*c*) upper-tailed test

9.18 Suppose the current mean cost to incarcerate a prisoner for one year is \$18,000. Consider the following scenarios.
(*a*) A prison reform plan is implemented which prison authorities feel will reduce prison costs.
(*b*) It is uncertain how a new prison reform plan will affect costs.
(*c*) A prison reform plan is implemented which prison authorities feel will increase prison costs.

Let μ represent the mean cost per prisoner per year after the reform plan is implemented. Give the research hypothesis in symbolic form for each of the above cases.

Ans.   (*a*) $H_a: \mu < \$18,000$ (*b*) $H_a: \mu \neq \$18,000$ (*c*) $H_a: \mu > \$18,000$

TEST STATISTIC, CRITICAL VALUES, REJECTION AND NONREJECTION REGIONS

9.19 A coin is tossed 10 times to determine if it is balanced. The coin will be declared "fair," i.e., balanced, if between 2 and 8 heads, inclusive, are obtained. Otherwise, the coin will be declared "unfair." The null hypothesis corresponding to this test is H_0: (the coin is fair), and the alternative hypothesis is H_a: (the coin is unfair). Identify the test statistic, the critical values, and the rejection and nonrejection regions.

Ans.   The test statistic is the number of heads to occur in the 10 tosses of the coin. The critical values are 1 and 9. The rejection region is $x = 0, 1, 9$, or 10, where x represents the number of heads to occur in the 10 tosses. The nonrejection region is $x = 2, 3, 4, 5, 6, 7$, or 8.

9.20 A die is tossed 6 times to determine if it is balanced. The die will be declared "fair," i.e., balanced, if the face "6" turns up 3 or fewer times. Otherwise, the die will be declared "unfair." The null hypothesis corresponding to this test is H_0: (the die is fair), and the alternative hypothesis is H_a: (the die is unfair). Identify the test statistic, the critical values, and the rejection and nonrejection regions.

Ans. The test statistic is the number of times the face "6" turns up in 6 tosses. The critical value is 4. The rejection region is x = 4, 5, or 6, where x represents the number of times the face "6" turns up in the 6 tosses of the die. The nonrejection region is x = 0, 1, 2, or 3.

TYPE I AND TYPE II ERRORS

9.21 In problems 9.19 and 9.20, describe how a Type I and a Type II error would occur.

Ans. In problem 9.19, if the coin is fair, but you happen to be unlucky and get 9 or 10 heads or 9 or 10 tails and therefore conclude that the coin is unfair, you commit a Type I error. If the coin is actually bent so that it favors heads more often than tails or vice versa, but you obtain between 2 and 8 heads, then you commit a Type II error.

In problem 9.20, if the die is fair, but you happen to be unlucky and obtain an unusual happening such as 4 or more sixes in the 6 tosses and therefore conclude that the die is unfair, you commit a Type I error. If the die is loaded so that the face "6" comes up more often than the other faces, but when you toss it, you actually obtain 3 or fewer sixes in the 6 tosses, then you commit a Type II error.

9.22 Find the level of significance in problems 9.19 and 9.20.

Ans. The level of significance is given by α = P(rejecting the null hypothesis when the null hypothesis is true). In problem 9.19, if the null hypothesis is true, then x, the number of heads in 10 tosses of a fair coin, has a binomial distribution and the level of significance is α = P(x = 0, 1, 9, or 10 when p = .5). Using the binomial tables, we find that the level of significance is α = .0010 + .0098 + .0010 + .0098 = .0216. In problem 9.20, if the null hypothesis is true, then x, the number of times the face "6" turns up in 6 tosses, has a binomial distribution with n = 6 and p = $\frac{1}{6}$ = .167, and the level of significance is α = P(x = 4, 5, or 6 when p = .167). Using the binomial probability formula, we find α = .008096 + .000649 + .000022 = .008767.

HYPOTHESIS TESTS ABOUT A POPULATION MEAN: LARGE SAMPLES

9.23 Home Videos Inc. surveys 450 households and finds that the mean amount spent for renting or buying videos is $13.50 per month and the standard deviation of the sample is $7.25. Is this evidence sufficient to conclude that μ > $12.75 per month at level of significance α = .01?

Ans. The computed test statistic is z* = 2.19 and the critical value is 2.33. The null hypothesis is not rejected.

9.24 A survey of 700 hourly wages in the U.S. is taken and it is found that: $\bar{x}$ = $17.65 and s = $7.55. Is this evidence sufficient to conclude that μ > $17.20, the stated current mean hourly wage in the U.S.? Test at α = .05.

Ans. The computed test statistic is z* = 1.58 and the critical value is 1.65. The null hypothesis is not rejected.

CALCULATING TYPE II ERRORS

9.25 In problem 9.23, find the probability of a Type II error if $\mu = \$13.25$. Use the sample standard deviation given in problem 9.23 as your estimate of σ.

   Ans. $\beta = $ P(not rejecting the null hypothesis when $\mu = \$13.25$) = P($\overline{x} < 13.55$ when $\mu = \$13.25$) = P($z < .87$) = .8078.

9.26 In problem 9.24, find the probability of a Type II error if $\mu = \$18.00$. Use the sample standard deviation given in problem 9.24 as your estimate of σ.

   Ans. $\beta = $ P(not rejecting the null hypothesis when $\mu = \$18.00$) = P($\overline{x} < 17.67$ when $\mu = \$18.00$) = P($z < -1.16$) = .123.

P VALUES

9.27 Find the p value for the sample results given in problem 9.23. Use this computed p value to test the hypothesis given in the problem.

   Ans.   The p value = P($Z > 2.19$) = .0143. Since the p value exceeds the preset α, the null hypothesis is not rejected.

9.28 Find the p value for the sample results given in problem 9.24. Use this computed p value to test the hypothesis given in the problem.

   Ans.   The p value = P($Z > 1.58$) = .0571. Since the p value exceeds the preset α, the null hypothesis is not rejected.

HYPOTHESIS TESTS ABOUT A POPULATION MEAN: SMALL SAMPLES

9.29 The mean score on the KSW Computer Science Aptitude Test is equal to 13.5. This test consists of 25 problems and the mean score given above was obtained from data supplied by many colleges and universities. Metropolitan College administers the test to 25 randomly selected students and obtains a mean equal to 12.0 and a standard deviation equal to 4.1. Can it be concluded that the mean for all Metropolitan students is less than the mean reported nationally? Assume the scores at Metropolitan are normally distributed and use $\alpha = .05$.

   Ans.   The null hypothesis is rejected since $t^* = -1.83 < -1.711$. The mean score for Metropolitan students is less than the national mean.

9.30 The claim is made that nationally, the mean payout per \$1 waged is 90 cents for casinos. Twenty randomly selected casinos are selected from across the country. The data and a Minitab analysis are shown below and on the next page. If the null hypothesis is H_0: $\mu = 90$, the alternative hypothesis is H_a: $\mu < 90$, and the level of significance is $\alpha = .01$, what is your conclusion?

Data Display
payout

85	82	85	83	91	78	90
86	95	82	86	74	89	92
96	81	90	86	92	90	

MTB > ttest mean = 90 data in c1;
SUBC > alt = −1.

T-Test of the Mean
Test of mu = 90.00 vs mu < 90.00

Variable	N	Mean	St Dev	SE Mean	T	P
Payout	20	86.65	5.63	1.26	−2.66	0.0077

Ans. Since the p value is less than $\alpha = .01$, reject the claim and conclude that the mean payout is less than 90 cents per dollar.

HYPOTHESIS TESTS ABOUT A POPULATION PROPORTION: LARGE SAMPLES

9.31 A survey of 300 gun owners is taken and 40% of those surveyed say they have a gun for protection for self/family. Use these results to test H_0: p = 32% vs. H_a: p > 32%, where p is the percent of all gun owners who say they have a gun for protection of self/family. Test at $\alpha = .025$.

Ans. The null hypothesis is rejected, since $z^* = 2.97 > 1.96$. It is concluded that the percent is greater than 32%.

9.32 Perform the following hypothesis about p.
   (a) H_0: p = .35 vs. H_a: p ≠ .35, n = 100, $\overline{p}$ = .38, α = .10
   (b) H_0: p = .75 vs. H_a: p < .75, n = 700, $\overline{p}$ = .71, α = .05
   (c) H_0: p = .55 vs. H_a: p > .55, n = 390, $\overline{p}$ = .57, α = .01

Ans. (a) $z^* = 0.63$, p value = .5286, Do not reject the null hypothesis.
   (b) $z^* = -2.44$, p value = .0073, Reject the null hypothesis.
   (c) $z^* = 0.79$, p value = .2148, Do not reject the null hypothesis.

Chapter 10

Inferences for Two Populations

SAMPLING DISTRIBUTION OF $\overline{X}_1 - \overline{X}_2$ FOR LARGE INDEPENDENT SAMPLES

Two samples drawn from two populations are *independent samples* if the selection of the sample from population 1 does not affect the selection of the sample from population 2. The following notation will be used for the sample and population measurements:

μ_1 and μ_2 = means of populations 1 and 2

σ_1 and σ_2 = standard deviations of populations 1 and 2

n_1 and n_2 = sizes of the samples drawn from populations 1 and 2 ($n_1 \geq 30$, $n_2 \geq 30$)

$\overline{x}_1$ and $\overline{x}_2$ = means of the samples selected from populations 1 and 2

s_1 and s_2 = standard deviations of the samples selected from populations 1 and 2

When two large samples ($n_1 \geq 30$, $n_2 \geq 30$) are selected from two populations, the sampling distribution of $\overline{x}_1 - \overline{x}_2$ is normal. The mean or expected value of the random variable $\overline{x}_1 - \overline{x}_2$ is given by formula (*10.1*):

$$E(\overline{x}_1 - \overline{x}_2) = \mu_{\overline{x}_1 - \overline{x}_2} = \mu_1 - \mu_2 \qquad (10.1)$$

The standard error of $\overline{x}_1 - \overline{x}_2$ is given by

$$\sigma_{\overline{x}_1 - \overline{x}_2} = \sqrt{\frac{\sigma_1^2}{n_1} + \frac{\sigma_2^2}{n_2}} \qquad (10.2)$$

Figure 10-1 illustrates that $\overline{x}_1 - \overline{x}_2$ is normally distributed and centers at $\mu_1 - \mu_2$. The standard error of the curve is given by formula (*10.2*).

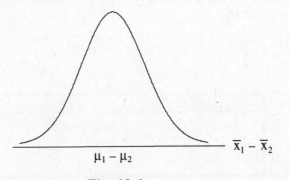

$$\overline{x}_1 - \overline{x}_2$$
$$\mu_1 - \mu_2$$

Fig. 10-1

Formula (10.3) is used to transform the distribution of $\overline{x}_1 - \overline{x}_2$ to a standard normal distribution.

$$z = \frac{\overline{x}_1 - \overline{x}_2 - (\mu_1 - \mu_2)}{\sigma_{\overline{x}_1 - \overline{x}_2}} \qquad (10.3)$$

EXAMPLE 10.1 The mean height of adult males is 69 inches and the standard deviation is 2.5 inches. The mean height of adult females is 65 inches and the standard deviation is 2.5 inches. Let population 1 be the population of male heights, and population 2 the population of female heights. Suppose samples of 50 each are selected from both populations. Then $\overline{x}_1 - \overline{x}_2$ is normally distributed with mean $\mu_{\overline{x}_1 - \overline{x}_2} = \mu_1 - \mu_2 = 69 - 65 = 4$ inches and standard error equal to $\sigma_{\overline{x}_1 - \overline{x}_2} = \sqrt{\dfrac{\sigma_1^2}{n_1} + \dfrac{\sigma_2^2}{n_2}} = \sqrt{\dfrac{6.25}{50} + \dfrac{6.25}{50}} = .5$ inches. The probability that the mean height of the sample of males exceeds the mean height of the sample of females by more than 5 inches is represented by $P(\overline{x}_1 - \overline{x}_2 > 5)$. The event $\overline{x}_1 - \overline{x}_2 > 5$ is converted to an equivalent event involving z by using formula (10.3). The z value corresponding to $\overline{x}_1 - \overline{x}_2 = 5$ is $z = \dfrac{\overline{x}_1 - \overline{x}_2 - (\mu_1 - \mu_2)}{\sigma_{\overline{x}_1 - \overline{x}_2}} = \dfrac{5 - 4}{.5} = 2$. From the standard normal distribution table the area to the right of 2 is $.5 - .4772 = .0228$. Only 2.28% of the time would the mean height of a sample of 50 male heights exceed the mean height of a sample of 50 female heights by 5 or more inches.

ESTIMATION OF $\mu_1 - \mu_2$ USING LARGE INDEPENDENT SAMPLES

The difference in sample means, $\overline{x}_1 - \overline{x}_2$, is a *point estimate of* $\mu_1 - \mu_2$. An *interval estimate for* $\mu_1 - \mu_2$ is obtained by using formula (10.3). Since 95% of the area under the standard normal curve is between -1.96 and 1.96, we have the following probability statement:

$$P(-1.96 < \frac{\overline{x}_1 - \overline{x}_2 - (\mu_1 - \mu_2)}{\sigma_{\overline{x}_1 - \overline{x}_2}} < 1.96) = .95$$

Solving the inequality inside the parenthesis for $\mu_1 - \mu_2$, we obtain the following 95% confidence interval for $\mu_1 - \mu_2$.

$$(\overline{x}_1 - \overline{x}_2) \pm 1.96 \times \sigma_{\overline{x}_1 - \overline{x}_2}$$

The general form of a *confidence interval for* $\mu_1 - \mu_2$ is given by formula (10.4), where z is determined by the specified level of confidence. The z values for the most common levels of confidence are given in Table 10.1.

$$(\overline{x}_1 - \overline{x}_2) \pm z \times \sigma_{\overline{x}_1 - \overline{x}_2} \qquad (10.4)$$

Table 10.1

Confidence level	Z value
80	1.28
90	1.65
95	1.96
99	2.58

EXAMPLE 10.2 Keyhole heart bypass was performed on 48 patients and conventional surgery was performed on 55 patients. The time spent on breathing tubes was recorded for all patients and is summarized in Table 10.2.

The difference in sample means is $\bar{x}_1 - \bar{x}_2 = 3.0 - 15.0 = -12.0$. Since population standard deviations are unknown, they are estimated with the sample standard deviations. The *estimated standard error of the difference in sample means* is represented by $S_{\bar{x}_1 - \bar{x}_2} = \sqrt{\dfrac{s_1^2}{n_1} + \dfrac{s_2^2}{n_2}} = \sqrt{\dfrac{2.25}{48} + \dfrac{13.69}{55}} = .544$.

Table 10.2

Sample	Sample size	Mean	Standard deviation
1. Keyhole	48	3.0 hours	1.5 hours
2. Conventional	55	15.0 hours	3.7 hours

A 90% confidence interval for $\mu_1 - \mu_2$ is found by using formula (*10.4*). The z value is determined to be 1.65 from Table 10.1. The interval is $-12.0 \pm 1.65 \times .544$, or -12.0 ± 0.9. The interval extends from $-12.0 - 0.9 = -12.9$ to $-12.0 + 0.9 = -11.1$. That is, we are 95% confident that the keyhole procedure requires on the average from 11.1 to 12.9 hours less time on breathing tubes. Note that when population standard deviations are unknown and the samples are large, that the sample standard deviations are substituted for the population standard deviations.

TESTING HYPOTHESIS ABOUT $\mu_1 - \mu_2$ USING LARGE INDEPENDENT SAMPLES

The procedure for testing hypothesis about the difference in population means when using two large independent samples is given in Table 10.3.

Table 10.3

Steps for Testing a Hypothesis Concerning $\mu_1 - \mu_2$: Large Independent Samples
Step 1: State the null and research hypothesis. The null hypothesis is represented symbolically by H_0: $\mu_1 - \mu_2 = D_0$, and the research hypothesis is of the form H_a: $\mu_1 - \mu_2 \neq D_0$ or H_a: $\mu_1 - \mu_2 < D_0$ or H_a: $\mu_1 - \mu_2 > D_0$.
Step 2: Use the standard normal distribution table and the level of significance, α, to determine the rejection region.
Step 3: Compute the value of the test statistic as follows: $z^* = \dfrac{\bar{x}_1 - \bar{x}_2 - D_0}{\sigma_{\bar{x}_1 - \bar{x}_2}}$, where $\bar{x}_1 - \bar{x}_2$ is the computed difference in the sample means, D_0 is the hypothesized difference in the population means as given in the null hypothesis, and $\sigma_{\bar{x}_1 - \bar{x}_2} = \sqrt{\dfrac{\sigma_1^2}{n_1} + \dfrac{\sigma_2^2}{n_2}}$ or if the population standard deviations are unknown, $S_{\bar{x}_1 - \bar{x}_2} = \sqrt{\dfrac{s_1^2}{n_1} + \dfrac{s_2^2}{n_2}}$ is used as an estimate of $\sigma_{\bar{x}_1 - \bar{x}_2}$.
Step 4: State your conclusion. The null hypothesis is rejected if the computed value of the test statistic falls in the rejection region. Otherwise, the null hypothesis is not rejected.

EXAMPLE 10.3 Keyhole heart bypass was performed on 48 patients and conventional surgery was performed on 55 patients. The length of hospital stay in days was recorded for each patient. A summary of the data are given in Table 10.4.

Table 10.4

Sample	Sample size	Mean	Standard deviation
1. Keyhole	48	3.5 days	1.5 days
2. Conventional	55	8.0 days	2.0 days

These results are used to test the research hypothesis that the mean hospital stay for keyhole patients is less than the mean hospital stay for conventional patients with level of significance $\alpha = .01$. The null hypothesis is H_0: $\mu_1 - \mu_2 = 0$ and the research hypothesis is H_a: $\mu_1 - \mu_2 < 0$. This is a lower-tailed test, and the critical value is -2.33, since $P(z < -2.33) = .01$. This critical value is determined in exactly the same way as it was in chapter 9. The null hypothesis is rejected if the computed test statistic is less than the critical value. The following quantities are needed to compute the value of the test statistic: $D_0 = 0$, $\overline{x}_1 - \overline{x}_2 = 3.5 - 8.0 = -4.5$, and the estimated standard error is $S_{\overline{x}_1 - \overline{x}_2} = \sqrt{\dfrac{s_1^2}{n_1} + \dfrac{s_2^2}{n_2}} = \sqrt{\dfrac{2.25}{48} + \dfrac{4}{55}} = .346$. The computed test statistic is:

$$z^* = \frac{\overline{x}_1 - \overline{x}_2 - D_0}{\sigma_{\overline{x}_1 - \overline{x}_2}} = \frac{-4.5 - 0}{.346} = -13.0$$

The null hypothesis is rejected and it is concluded that the mean hospital stay for the keyhole procedure is less than that for the conventional procedure.

SAMPLING DISTRIBUTION OF $\overline{X}_1 - \overline{X}_2$ FOR SMALL INDEPENDENT SAMPLES FROM NORMAL POPULATIONS WITH EQUAL (BUT UNKNOWN) STANDARD DEVIATIONS

When the sample sizes are small (one or both less than 30) and the population standard deviations are unknown, the substitution of sample standard deviations for the population standard deviations, as was done in Examples 10.2 and 10.3, is not appropriate. The use of the standard normal table for confidence intervals and testing hypothesis is not valid in this case. Statisticians have developed two different procedures for the case where the sample sizes are small and the population standard deviations are unknown. Both cases require the assumption that both populations have normal distributions. In this section it will also be assumed that the populations have equal standard deviations.

A statistical test for deciding whether to assume $\sigma_1 = \sigma_2$ or $\sigma_1 \neq \sigma_2$ utilizes the two sample standard deviations and a statistical distribution called the F distribution. *However, a rule of thumb used by some statisticians states that if* $.5 \leq \dfrac{S_1}{S_2} \leq 2$, *then assume* $\sigma_1 = \sigma_2$. *Otherwise, assume that* $\sigma_1 \neq \sigma_2$.

Suppose σ is the common population standard deviation, i. e., $\sigma_1 = \sigma_2 = \sigma$. The standard error of $\overline{x}_1 - \overline{x}_2$ is given by formula (*10.2*) as $\sigma_{\overline{x}_1 - \overline{x}_2} = \sqrt{\dfrac{\sigma_1^2}{n_1} + \dfrac{\sigma_2^2}{n_2}}$. Replacing the two population standard deviations by σ and factoring it out, we get formula (*10.5*):

$$\sigma_{\overline{x}_1 - \overline{x}_2} = \sqrt{\sigma^2 \left(\frac{1}{n_1} + \frac{1}{n_2} \right)} \tag{10.5}$$

Now, σ^2, the common population variance, is estimated by pooling the sample variances as a weighted average. The *pooled estimator of* σ^2 is represented by S^2 and is given by

$$S^2 = \frac{(n_1 - 1)s_1^2 + (n_2 - 1)s_2^2}{n_1 + n_2 - 2} \tag{10.6}$$

Replacing σ^2 by S^2 in formula (*10.5*), the *estimated standard error of the difference in the sample means* is obtained and is given in formula (*10.7*)

$$S_{\bar{x}_1-\bar{x}_2} = \sqrt{S^2\left(\frac{1}{n_1}+\frac{1}{n_2}\right)} \qquad (10.7)$$

Replacing $\sigma_{\bar{x}_1-\bar{x}_2}$ by $S_{\bar{x}_1-\bar{x}_2}$ in formula (10.3), we obtain

$$t = \frac{\bar{x}_1 - \bar{x}_2 - (\mu_1 - \mu_2)}{S_{\bar{x}_1-\bar{x}_2}} \qquad (10.8)$$

When sampling from two normally distributed populations having equal population standard deviations, the statistic given in formula (10.8) has a t distribution with degrees of freedom given by

$$df = n_1 + n_2 - 2 \qquad (10.9)$$

EXAMPLE 10.4 A sample of size 10 is randomly selected from a normally distributed population and a second sample of size 15 is selected from another normally distributed population. The standard deviation of the first sample is equal to 13.24 and the standard deviation of the second sample is equal to 24.25. The rule of thumb, given above, suggests that the populations may be assumed to have a common standard deviation since $.5 \leq \dfrac{S_1}{S_2} = .55 \leq 2$. A pooled estimate of the common variance is given as follows:

$$S^2 = \frac{(n_1-1)s_1^2 + (n_2-1)s_2^2}{n_1+n_2-2} = \frac{9 \times 13.24^2 + 14 \times 24.25^2}{10+15-2} = 426.5458$$

and a pooled estimate of the common standard deviation is $S = \sqrt{426.5458} = 20.65$. The estimated standard error of the difference in sample means is as follows:

$$S_{\bar{x}_1-\bar{x}_2} = \sqrt{S^2\left(\frac{1}{n_1}+\frac{1}{n_2}\right)} = \sqrt{426.5458 \times \left(\frac{1}{10}+\frac{1}{15}\right)} = 8.43$$

The computations of S and $S_{\bar{x}_1-\bar{x}_2}$ are required for setting confidence intervals on $\mu_1 - \mu_2$ and for testing hypothesis concerning $\mu_1 - \mu_2$ when the two samples are small, and the population standard deviations are unknown but assumed equal.

ESTIMATION OF $\mu_1 - \mu_2$ USING SMALL INDEPENDENT SAMPLES FROM NORMAL POPULATIONS WITH EQUAL (BUT UNKNOWN) STANDARD DEVIATIONS

The difference in sample means, $\bar{x}_1 - \bar{x}_2$, is a *point estimate of* $\mu_1 - \mu_2$. An *interval estimate for* $\mu_1 - \mu_2$ is obtained by using formula (10.8). When independent small samples are selected from two normal populations having unknown but equal standard deviations, the general form of a confidence interval for $\mu_1 - \mu_2$ is given by formula (10.10), where t is obtained from the t distribution table and is determined by the level of confidence and the degrees of freedom which is given by df = $n_1 + n_2 - 2$. The standard error of the difference in sample means, $S_{\bar{x}_1-\bar{x}_2}$, is given by formula (10.7).

$$(\bar{x}_1 - \bar{x}_2) \pm t \times S_{\bar{x}_1-\bar{x}_2} \qquad (10.10)$$

EXAMPLE 10.5 Keyhole heart bypass was performed on 8 patients and conventional surgery was performed on 10 patients. The time spent on breathing tubes was recorded for all patients and is summarized in Table 10.5. The difference in sample means is $\overline{x}_1 - \overline{x}_2 = 3.0 - 7.0 = -4.0$ hours. The pooled estimate of the common population variance is obtained as follows:

$$S^2 = \frac{(n_1 - 1)s_1^2 + (n_2 - 1)s_2^2}{n_1 + n_2 - 2} = \frac{7 \times 2.25 + 9 \times 7.29}{8 + 10 - 2} = 5.085$$

The standard error of the difference in sample means is:

$$S_{\overline{x}_1 - \overline{x}_2} = \sqrt{S^2 \left(\frac{1}{n_1} + \frac{1}{n_2} \right)} = \sqrt{5.085 \times \left(\frac{1}{8} + \frac{1}{10} \right)} = 1.070$$

Table 10.5

Sample	Sample size	Mean	Standard deviation
1. Keyhole	8	3.0 hours	1.5 hours
2. Conventional	10	7.0 hours	2.7 hours

To find a 95% confidence interval for $\mu_1 - \mu_2$, we need to find the t value for use in formula (*10.10*). The degrees of freedom is df $= n_1 + n_2 - 2 = 8 + 10 - 2 = 16$. From the t distribution table having 16 degrees of freedom, we find the following: $P(-2.120 < t < 2.120) = .95$. The proper value of t is therefore 2.120. The margin of error is t $\times$ $S_{\overline{x}_1 - \overline{x}_2}$ = 2.120 $\times$ 1.070 = 2.2684 or 2.27 to 2 decimal places. The 95% confidence interval is -4.0 ± 2.27. Assuming that the times spent on breathing tubes are normally distributed for both procedures, the difference $\mu_1 - \mu_2$ is between -6.27 and -1.73 hours with 95% confidence.

TESTING HYPOTHESIS ABOUT $\mu_1 - \mu_2$ USING SMALL INDEPENDENT SAMPLES FROM NORMAL POPULATIONS WITH EQUAL (BUT UNKNOWN) STANDARD DEVIATIONS

The procedure for testing a hypothesis about the difference in population means when using two small independent samples from normal populations with equal (but unknown) standard deviations is given in Table 10.6.

Table 10.6

Steps for Testing a Hypothesis Concerning $\mu_1 - \mu_2$: Small Independent Samples from Normal Populations with Equal (but Unknown) Standard Deviations
Step 1: State the null and research hypothesis. The null hypothesis is represented symbolically by H_0: $\mu_1 - \mu_2 = D_0$ and the research hypothesis is of the form H_a: $\mu_1 - \mu_2 \neq D_0$ or H_a: $\mu_1 - \mu_2 < D_0$ or H_a: $\mu_1 - \mu_2 > D_0$.
Step 2: Use the t distribution table with degrees of freedom equal to $n_1 + n_2 - 2$ and the level of significance, α, to determine the rejection region.
Step 3: Compute the value of the test statistic as follows: $t^* = \dfrac{\overline{x}_1 - \overline{x}_2 - D_0}{S_{\overline{x}_1 - \overline{x}_2}}$, where $\overline{x}_1 - \overline{x}_2$ is the computed difference in the sample means, D_0 is the hypothesized difference in the population means as stated in the null hypothesis, and $S_{\overline{x}_1 - \overline{x}_2} = \sqrt{S^2 \left(\frac{1}{n_1} + \frac{1}{n_2} \right)}$, where $S^2 = \dfrac{(n_1 - 1)s_1^2 + (n_2 - 1)s_2^2}{n_1 + n_2 - 2}$.
Step 4: State your conclusion. The null hypothesis is rejected if the computed value of the test statistic falls in the rejection region. Otherwise, the null hypothesis is not rejected.

EXAMPLE 10.6 Keyhole heart bypass was performed on 8 patients and conventional surgery was performed on 10 patients. The length of hospital stay in days was recorded for each patient. A summary of the data is given in Table 10.6.

Table 10.6

Sample	Sample size	Mean	Standard deviation
1. Keyhole	8	3.5 days	1.5 days
2. Conventional	10	8.0 days	2.0 days

These results are used to test the research hypothesis that the mean hospital stay for keyhole patients is less than the mean hospital stay for conventional patients with level of significance $\alpha = .01$. The null hypothesis is H_0: $\mu_1 - \mu_2 = 0$ and the research hypothesis is H_a: $\mu_1 - \mu_2 < 0$. To determine the rejection region, note that the test is a lower-tailed test and the degrees of freedom is $df = n_1 + n_2 - 2 = 8 + 10 - 2 = 16$. Using the t distribution table, we find that for $df = 16$, $P(t > 2.583) = .01$, and therefore, $P(t < -2.583) = .01$. The shaded rejection region is shown in Fig. 10-2.

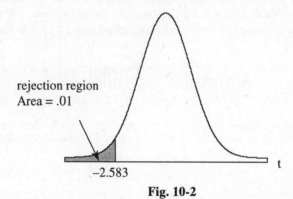

rejection region
Area = .01

−2.583

t

Fig. 10-2

The null hypothesis is to be rejected if the computed test statistic is less than −2.583. The computed test statistic is determined according to step 3 in Table 10.6. The difference in sample means is $\overline{x}_1 - \overline{x}_2 = 3.5 - 8.0 = -4.5$. The value specified for D_0 is 0. The pooled estimate of the common population variance is

$$S^2 = \frac{(n_1-1)s_1^2 + (n_2-1)s_2^2}{n_1 + n_2 - 2} = \frac{7 \times 2.25 + 9 \times 4.00}{8 + 10 - 2} = 3.2344$$

and the standard error of the difference in sample means is

$$S_{\overline{x}_1 - \overline{x}_2} = \sqrt{S^2\left(\frac{1}{n_1} + \frac{1}{n_2}\right)} = \sqrt{3.2344 \times \left(\frac{1}{8} + \frac{1}{10}\right)} = .853$$

The computed test statistic is

$$t^* = \frac{\overline{x}_1 - \overline{x}_2 - D_0}{S_{\overline{x}_1 - \overline{x}_2}} = \frac{-4.5 - 0}{.853} = -5.28.$$

Since t^* is less than the critical value, the null hypothesis is rejected and it is concluded that mean length of hospital stay is smaller for the keyhole procedure than the conventional procedure. This hypothesis test procedure assumes that the two populations are normally distributed.

EXAMPLE 10.7 A sociological study compared the dating practices of high school senior males and high school senior females. The two samples were selected independently of one another. The number of dates per month were recorded for each participant. The data are shown in Table 10.7.

Table 10.7

Males	Females
8	7
11	7
7	12
5	11
13	14
10	10
10	10
8	9
9	4
5	8

The assumptions of equal variances and normal populations are usually checked out with statistical software before the actual test of equality of means is carried out. This has been made much easier because of the widespread availability of statistical software. The checking of assumptions as well as the actual test of equality of population means for the data in Table 10.7 will be illustrated by using Minitab. The commands **% normplot mdates** and **% normplot fdates** produced the following plots, which may be used to check for normality.

Normal Probability Plot

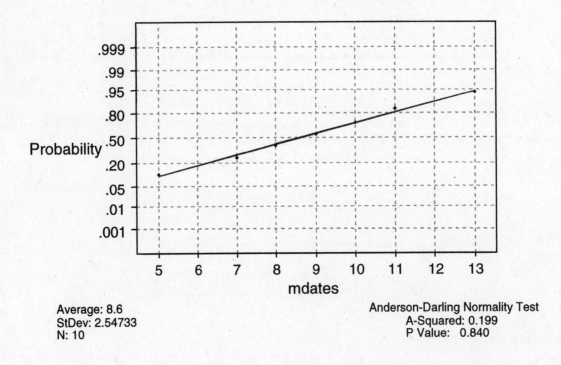

Average: 8.6
StDev: 2.54733
N: 10

Anderson-Darling Normality Test
A-Squared: 0.199
P Value: 0.840

Normal Probability Plot

Average: 9.2
StDev: 2.85968
N: 10

Anderson-Darling Normality Test
A-Squared: 0.147
P Value: 0.948

If the sample data are selected from a normal population, the points on the normal probability plot will tend to fall along a straight line. Normality of the population is usually rejected if the p value is less than $\alpha = .05$. For the above data, the p values are 0.840 and 0.948 and normality of the populations is not rejected.

An edited version of the Minitab procedure to test for equal population variances is shown below. A 1 in column 1 indicates the response in c2 came from the male group and a 2 indicates the response came from the female group. The assumption of equal variances is usually rejected if the p value shown in Bartlett's Test (normal distribution) is less than .05. Since the p value is 0.736, the assumption of equal variances is not rejected.

	C1	C2
1	1	8
2	1	11
3	1	7
4	1	5
5	1	13
6	1	10
7	1	10
8	1	8
9	1	9
10	1	5
11	2	7
12	2	7
13	2	12
14	2	11
15	2	14
16	2	10
17	2	10
18	2	9
19	2	4
20	2	8

MTB > %vartest c2 c1
Homogeneity of Variance
Response   C2
Factors   C1

Bartlett's Test (normal distribution)
Test Statistic: 0.114
P value   : 0.736

Having checked out the normality assumptions and the equal population variances assumption, the test procedure is now conducted.

MTB > twot data in c2, groups in c1;
SUBC > alternative is 0;
SUBC > pooled procedure.

Two Sample T-Test and Confidence Interval
Two sample T for dates

Sample	N	Mean	St Dev	SE Mean
1	10	8.60	2.55	0.81
2	10	9.20	2.86	0.90

95% CI for mu (1) – mu (2): (–3.14, 1.94)
T-Test mu (1) = mu (2) (vs not =): T = –0.50   P = 0.63   DF = 18

To use the TWOT command in Minitab, the responses must be in a column and the group from which the responses came must be in another column. The data are entered in columns 1 and 2 as shown above. The null hypothesis is H_0: $\mu_1 - \mu_2 = 0$. The alternative command indicates the nature of the research hypothesis. A –1 indicates a lower-tailed test, a 0 indicates a two-tailed test, and a 1 indicates an upper-tailed test. The subcommand **pooled procedure** indicates that equal population standard deviations are assumed and a pooled estimated is to be computed.

The output gives the sample size, mean, standard deviation, and standard error of the mean for both groups separately. A 95% confidence interval for $\mu_1 - \mu_2$ is given. The computed test statistic is $t^* = -0.50$. The two-tailed p value is 0.63. The degrees of freedom is 18. The pooled standard deviation estimate is S = 2.71.

SAMPLING DISTRIBUTION OF $\overline{X}_1 - \overline{X}_2$ FOR SMALL INDEPENDENT SAMPLES FROM NORMAL POPULATIONS WITH UNEQUAL (AND UNKNOWN) STANDARD DEVIATIONS

When the population standard deviations are unknown and it is not reasonable to assume they are equal, the sample variances are not pooled as they were in the previous three sections. Statisticians have developed a test statistic for this case, which has a distribution, which is approximately a t distribution. The standard error of the difference in the sample means is approximated by substituting the sample variances for the population variances in formula (10.2) as was done in the large sample case to obtain the estimated standard error given in the following:

$$S_{\overline{x}_1 - \overline{x}_2} = \sqrt{\frac{s_1^2}{n_1} + \frac{s_2^2}{n_2}} \qquad\qquad (10.11)$$

In formula (10.3), we replace $\sigma_{\overline{x}_1 - \overline{x}_2}$ by the expression for $S_{\overline{x}_1 - \overline{x}_2}$ in formula (10.11), and obtain the statistic shown in formula (10.12):

$$t = \frac{\overline{x}_1 - \overline{x}_2 - (\mu_1 - \mu_2)}{S_{\overline{x}_1 - \overline{x}_2}} \qquad (10.12)$$

When sampling from two normally distributed populations having unequal population standard deviations, the statistic given in formula (10.12) has an approximate t distribution and the degrees of freedom is given by

$$df = \text{minimum of } \{(n_1 - 1), (n_2 - 1)\} \qquad (10.13)$$

EXAMPLE 10.8 A sample of size 13 is taken from a normally distributed population, and a sample of size 15 is taken from a second normally distributed population. The mean of population 1 is 70 and the mean of population 2 is 65. The population variances are unknown, but assumed to be unequal. The following statistic has an approximate t distribution.

$$t = \frac{\overline{x}_1 - \overline{x}_2 - 5}{S_{\overline{x}_1 - \overline{x}_2}}$$

The degrees of freedom is df = minimum{12, 14} = 12.

ESTIMATION OF $\mu_1 - \mu_2$ USING SMALL INDEPENDENT SAMPLES FROM NORMAL POPULATIONS WITH UNEQUAL (AND UNKNOWN) STANDARD DEVIATIONS

The difference in sample means, $\overline{x}_1 - \overline{x}_2$, is a *point estimate of* $\mu_1 - \mu_2$. An *interval estimate for* $\mu_1 - \mu_2$ is obtained by using formula (10.12). When independent small samples are selected from two normal populations having unknown and unequal standard deviations, the general form of a confidence interval for $\mu_1 - \mu_2$ is given by formula (10.14), where t is obtained from the t distribution table and is determined by the level of confidence and the degrees of freedom which is given by df = minimum of $\{(n_1 - 1), (n_2 - 1)\}$. The standard error of the difference in sample means, $S_{\overline{x}_1 - \overline{x}_2}$, is given by formula (10.11).

$$(\overline{x}_1 - \overline{x}_2) \pm t \times S_{\overline{x}_1 - \overline{x}_2} \qquad (10.14)$$

EXAMPLE 10.9 Keyhole heart bypass was performed on 8 patients and conventional surgery was performed on 10 patients. The time spent on breathing tubes was recorded for all patients and is summarized in Table 10.8. The difference in sample means is $\overline{x}_1 - \overline{x}_2 = 3.5 - 5.0 = -1.5$ hours. Because the ratio of the sample standard deviation for sample 2 divided by sample 1 is 3.4, it is not reasonable to assume that $\sigma_1 = \sigma_2$.

Table 10.8

Sample	Sample size	Mean	Standard deviation
1. Keyhole	8	3.5 hours	0.5 hour
2. Conventional	10	5.0 hours	1.7 hours

To find a 95% confidence interval for $\mu_1 - \mu_2$, we need to find the t value for use in formula (10.14). The degrees of freedom is df = minimum of $\{(n_1 - 1), (n_2 - 1)\}$ = minimum $\{7, 9\}$ = 7. From the t distribution table having 7 degrees of freedom, we find the following: P(-2.365 < t < 2.365) = .95. The proper value of t is therefore 2.365. The standard error of the difference in sample means is $S_{\overline{x}_1 - \overline{x}_2} = \sqrt{\dfrac{s_1^2}{n_1} + \dfrac{s_2^2}{n_2}}$. Using the sample standard deviations in Table 10.8, we find that $S_{\bar{x}_1 - \bar{x}_2} = \sqrt{\dfrac{.25}{8} + \dfrac{2.89}{10}} = .566$. The margin of error is $t \times S_{\bar{x}_1 - \bar{x}_2} =$ $2.365 \times .566 = 1.3386$ or 1.34 to 2 decimal places. The 95% confidence interval is -1.5 ± 1.34. Assuming that the times spent on breathing tubes are normally distributed for both procedures, the difference $\mu_1 - \mu_2$ is between -2.84 and -0.16 hours with 95% confidence.

EXAMPLE 10.10 Mothers Against Drunk Driving (MADD) have pushed for lowering the limit for driving while intoxicated from .1% to .08%. A study of alcohol-related traffic deaths compared the blood alcohol levels of individuals involved in such accidents in states with a 0.08% limit with those in states with a .1% limit. The results are shown in Table 10.9. Minitab will be used to set a 95% confidence interval on $\mu_1 - \mu_2$, where μ_1 is the mean blood-alcohol level of such individuals in states having a .08% limit and μ_2 is the mean blood-alcohol level in states having a .1% limit.

Table 10.9

0.08% level	.1% level
.12	.09
.05	.16
.07	.03
.09	.21
.07	.16
.04	.12
.12	.09
.11	.09
.03	.14
.06	.19
.06	.08
.09	.09
.10	.11
.10	.16
.12	.24

Normal probability plots, such as those in Example 10.7, indicate that it is reasonable to assume that the samples are taken from normally distributed populations. An edited version of the Minitab test for equal population variances is shown below. Before executing the Minitab command **%vartest c2 c1**, the data in Table 10.9 are set in columns c1 and c2, where c1 contains a 1 or a 2, depending on where the sample value in c2 comes from. The set up for the columns is similar to that shown in Example 10.7. Bartlett's Test is used to test the null hypothesis H_0: The population standard deviations are equal vs. the alternative hypothesis H_a: The population standard deviations are not equal. The hypothesis of equal standard deviations is usually rejected if the p value for Bartlett's Test is less than .05. In this case, the p value is equal to 0.025, and equal population standard deviations is rejected.

MTB > %vartest c2 c1

Homogeneity of Variance
Response   C2
Factors   C1
ConfLvl   95.0000
Bartlett's Test (normal distribution)
Test Statistic: 4.998
P value   : 0.025

Since it is reasonable to assume that the populations have normally distributed populations with unequal population standard deviations, the techniques of this section are appropriate. The command **twot data in c2 groups in c1** produces a 95% confidence interval for $\mu_1 - \mu_2$. If the population standard deviations were assumed to be equal, the subcommand **SUBC> pooled procedure** would be added to the twot command. The Minitab output gives the sample size, mean, standard deviation, and standard error for both groups. The 95% confidence interval for $\mu_1 - \mu_2$ extends from –0.0829 to –0.014. The edited output is as follows:

MTB > twot data in c2 groups in c1

Two Sample T-Test and Confidence Interval

Two sample T for C2

C1	N	Mean	St Dev	SE Mean
1	15	0.0820	0.0300	0.0078
2	15	0.1307	0.0562	0.015

95% CI for mu (1) – mu (2): (–0.0829, –0.014)

TESTING HYPOTHESIS ABOUT $\mu_1 - \mu_2$ USING SMALL INDEPENDENT SAMPLES FROM NORMAL POPULATIONS WITH UNEQUAL (AND UNKNOWN) STANDARD DEVIATIONS

The procedure for testing a hypothesis about the difference in population means when using two small independent samples from normal populations with unequal (and unknown) standard deviations is given in Table 10.10.

Table 10.10

Steps for Testing a Hypothesis Concerning $\mu_1 - \mu_2$: Small Independent Samples from Normal Populations with Unequal (and Unknown) Standard Deviations
Step 1: State the null and research hypothesis. The null hypothesis is represented symbolically by H_0: $\mu_1 - \mu_2 = D_0$ and the research hypothesis is of the form H_a: $\mu_1 - \mu_2 \neq D_0$ or H_a: $\mu_1 - \mu_2 < D_0$ or H_a: $\mu_1 - \mu_2 > D_0$.
Step 2: Use the t distribution table with degrees of freedom equal to df = minimum of $\{(n_1 - 1), (n_2 - 1)\}$ and the level of significance, α, to determine the rejection region.
Step 3: Compute the value of the test statistic as follows: $t^* = \dfrac{\overline{x}_1 - \overline{x}_2 - D_0}{S_{\overline{x}_1 - \overline{x}_2}}$, where $\overline{x}_1 - \overline{x}_2$ is the computed difference in the sample means, D_0 is the hypothesized difference in the population means as stated in the null hypothesis, and $S_{\overline{x}_1 - \overline{x}_2} = \sqrt{\dfrac{s_1^2}{n_1} + \dfrac{s_2^2}{n_2}}$.
Step 4: State your conclusion. The null hypothesis is rejected if the computed value of the test statistic falls in the rejection region. Otherwise, the null hypothesis is not rejected.

EXAMPLE 10.11 Keyhole heart bypass was performed on 8 patients and conventional surgery was performed on 10 patients. The length of hospital stay in days was recorded for each patient. A summary of the data is given in Table 10.11. The null hypothesis H_0: $\mu_1 - \mu_2 = -4$ vs. H_a: $\mu_1 - \mu_2 \neq -4$ is of interest to the researchers involved in the study. In words, the null hypothesis states that the mean hospital stay is 4 days less for the keyhole procedure than for the conventional procedure. The research hypothesis states that the difference in means is not equal to 4 days. Since the sample standard deviation for the conventional procedure is 2.5 times the sample standard deviation for the keyhole procedure, it is assumed that population standard deviations are unequal. The sample observations (which are not shown) indicate that it is reasonable to assume that both populations are normally distributed.

Table 10.11

Sample	Sample size	Mean	Standard deviation
1. Keyhole	8	3.5 days	1.2 days
2. Conventional	10	8.0 days	3.0 days

The degrees of freedom for the t distribution is $df = $ minimum of $\{(n_1 - 1), (n_2 - 1)\} = $ minimum of $\{7, 9\} = 7$. For $\alpha = .01$, an area of .005 is allocated to each tail, since the research hypothesis is two-tailed. Using the t distribution with 7 degrees of freedom, it is found that $P(t > 3.499) = .005$. The critical values are ± 3.499. The estimated standard error of the difference in sample means is :

$$S_{\overline{x}_1 - \overline{x}_2} = \sqrt{\frac{s_1^2}{n_1} + \frac{s_2^2}{n_2}} = \sqrt{\frac{1.44}{8} + \frac{9.0}{10}} = 1.039.$$

The computed test statistic is:

$$t^* = \frac{\overline{x}_1 - \overline{x}_2 - D_0}{S_{\overline{x}_1 - \overline{x}_2}} = \frac{3.5 - 8.0 - (-4)}{1.039} = -.48$$

Since t^* is between -3.499 and 3.499, the evidence is not sufficient to reject the null hypothesis.

EXAMPLE 10.12 The data in Table 10.9 are used to test the research hypothesis H_a: $\mu_1 - \mu_2 < 0$. The statement $\mu_1 - \mu_2 < 0$ is equivalent to $\mu_1 < \mu_2$. The statistical statement $\mu_1 < \mu_2$ states that the mean blood-alcohol level for individuals involved in alcohol-related traffic deaths is lower in states with a .08% limit than the mean level in states with a .1% limit. In Example 10.10, it was shown that it was reasonable to assume normal population distributions and unequal population standard deviations. After putting the data in columns c1 and c2 as described in Example 10.10, the following Minitab output was obtained.

MTB > twot data in c2 groups in c1;
SUBC > alternative = –1.

Two Sample T-Test and Confidence Interval

Two sample T for C2

C1	N	Mean	St Dev	SE Mean
1	15	0.0820	0.0300	0.0078
2	15	0.1307	0.0562	0.0150

95% CI for mu (1) – mu (2): (–0.0829, –0.014)
T-Test mu (1) = mu (2) (vs <): T= –2.96 P=0.0038 DF= 21

The subcommand, **SUBC> alternative = –1**, indicates that a lower-tailed test is required. Note that the research hypothesis is supported at the $\alpha = .05$ level, since the p value is 0.0038. The computed test statistic is shown to be $T = -2.96$ and the degrees of freedom is shown as 21. The test statistic is computed as shown in Table 10.10. The degrees of freedom is not computed by $df = $ minimum of $\{(n_1 - 1), (n_2 - 1)\} = $ minimum of $\{14, 14\} = 14$. The degrees of freedom computation is given by a more complicated formula. The use of $df = $ minimum of $\{(n_1 - 1), (n_2 - 1)\}$ gives a more conservative test than that used by Minitab. Both ways of computing the degrees of freedom may be found in various textbooks.

SAMPLING DISTRIBUTION OF $\overline{d}$ FOR NORMALLY DISTRIBUTED DIFFERENCES COMPUTED FOR DEPENDENT SAMPLES

The previous sections in this chapter have dealt with independent samples selected from two populations. When sample values are purposely matched or paired, the samples are called *dependent samples*. The samples are also referred to as *paired* or *matched samples*. Dependent samples include pairs of measurements such as before/after measurements made on the same person or machine, pre- and post-test scores taken on the same person, similar measurements made on twins who have undergone different treatments, and so forth. Table 10.12 illustrates a typical set of paired data. The differences are formed by subtracting the second sample value from the first.

Table 10.12

Pair	Sample 1	Sample 2	Difference
1	x_1	y_1	$d_1 = x_1 - y_1$
2	x_2	y_2	$d_2 = x_2 - y_2$
.	.	.	.
.	.	.	.
.	.	.	.
n	x_n	y_n	$d_n = x_n - y_n$

The estimation and testing procedures for experiments involving paired samples involve the differences shown in Table 10.12. As in previous sections, we shall investigate the sampling distribution of the mean of the sample differences first, and then apply these results to establish confidence intervals and test hypothesis concerning the mean of the population of all possible differences. The following notation will be used for inferences involving dependent samples.

μ_d = the mean of the population of paired differences

σ_d = the standard deviation of the population of paired differences

$\overline{d}$ = the mean of the paired differences which are computed from the samples

S_d = the standard deviation of the paired differences which are computed from the samples

n = the number of paired differences which are computed from the samples

$\sigma_{\overline{d}}$ = the standard error of $\overline{d}$

$S_{\overline{d}}$ = the estimated standard error of $\overline{d}$

The sample mean of paired differences is given by

$$\overline{d} = \frac{\Sigma d}{n}$$ (10.15)

The sample standard deviation of paired differences is given by

$$S_d = \sqrt{\frac{\Sigma d^2 - (\Sigma d)^2 / n}{n - 1}}$$ (10.16)

The estimated standard error of $\overline{d}$ is given by

$$S_{\overline{d}} = \frac{S_d}{\sqrt{n}}$$ (10.17)

When the population of differences is normally distributed, the statistic given in formula (10.18) has a t distribution with $(n - 1)$ degrees of freedom.

$$t = \frac{\overline{d} - \mu_d}{S_{\overline{d}}}$$ (10.18)

This result is perfectly analogous to the previous result given in Chapters 8 and 9; namely the result that the statistic given in formula (10.19) has a t distribution with df = n – 1 when the sample values comprising $\overline{x}$ come from a normally distributed population having mean equal to μ.

$$t = \frac{\overline{x} - \mu}{S_{\overline{x}}} \qquad (10.19)$$

The confidence interval for μ_d as well as the procedure for testing hypothesis about μ_d are analogous to the techniques given in Chapters 8 and 9 for estimating the mean of one population or testing an hypothesis about the mean of a population. The only difference between the techniques is that in the case of dependent samples, the procedures are applied to differences.

EXAMPLE 10.13 A digital Sphygmomanometer gives diastolic blood pressure readings that are 5 units higher on the average than those obtained by the traditional method used by most medical professionals. Suppose 20 individuals have their diastolic blood pressure taken both ways. Let x represent the readings obtained by using the digital Sphygmomanometer, and let y represent the readings obtained by the traditional method. The differences in readings are found by the formula d = x − y. The mean population difference is μ_d = 5 units. Assuming that the population of all differences is normally distributed, the following statistic has a t distribution with df = 20 − 1 = 19:

$$t = \frac{\overline{d} - 5}{S_{\overline{d}}}$$

ESTIMATION OF μ_d USING NORMALLY DISTRIBUTED DIFFERENCES COMPUTED FROM DEPENDENT SAMPLES

When a paired sample of size n is selected and differences are computed as shown in Table 10.12, a confidence interval for the population mean difference, μ_d, can be formed by using the sampling distribution of the statistic given in formula (10.18). The confidence interval is given by formula (10.20), where t is determined by the level of confidence and df = n − 1. The standard error of $\overline{d}$ is computed by using formulas (10.16) and (10.17).

$$\overline{d} \pm t \times S_{\overline{d}} \qquad (10.20)$$

EXAMPLE 10.14 A psychological study compared the amount of manganese in tears that lubricate the eye with the amount in emotional tears. A measurement, which is related to the amount manganese, was taken for each type of tear on 10 different subjects. The results are shown in Table 10.13. The computations needed to set a 99% confidence interval on μ_d are given below the table.

Table 10.13

Subject	Lubricating tears	Emotional tears	Difference
1	10	12	−2
2	13	12	1
3	11	9	2
4	10	11	−1
5	9	7	2
6	8	9	−1
7	10	11	−1
8	10	13	−3
9	8	10	−2
10	12	10	2

The sum of the differences is $\Sigma d = -2 + 1 + 2 - 1 + 2 - 1 - 1 - 3 - 2 + 2 = -3$ and the sum of the squares of the differences is $\Sigma d^2 = 4 + 1 + 4 + 1 + 4 + 1 + 1 + 9 + 4 + 4 = 33$. The mean difference is $\overline{d} = -.3$. The sample standard deviation for the differences is

$$S_d = \sqrt{\frac{\Sigma d^2 - (\Sigma d)^2 / n}{n-1}} = \sqrt{\frac{33 - \frac{9}{10}}{9}} = 1.889$$

The standard error of $\overline{d}$ is

$$S_{\overline{d}} = \frac{S_d}{\sqrt{n}} = \frac{1.889}{\sqrt{10}} = .597$$

The t value for a 99% confidence interval is found as follows: The degrees of freedom is $df = 10 - 1 = 9$. From the t distribution table for 9 degrees of freedom, we find that $P(-3.250 < t < 3.250) = .99$. The 99% confidence interval is found by using formula (10.20) to be: $-.3 \pm 3.25 \times .597$, or the 99% confidence interval is $(-2.24, 1.64)$. The confidence interval is valid provided the population of differences is normally distributed.

EXAMPLE 10.15 A Minitab solution to Example 10.14 is given below. The lubricating tear data is put in column 1 and the emotional tear data is put in column 2. The command **let c3 = c1 – c2** computes the differences and puts them in column 3. The command **name c3 'diff'** assigns the name diff to column 3. The command **tinterval 99% confidence data in c3** computes the 99% confidence interval for μ_d. Note that Minitab computes and prints out the mean difference, the sample standard deviation of differences, the standard error of $\overline{d}$, as well as the 99% confidence interval. These quantities are the same as those computed by hand in Example 10.14.

Data Display
Row lubricate emotion

1	10	12
2	13	12
3	11	9
4	10	11
5	9	7
6	8	9
7	10	11
8	10	13
9	8	10
10	12	10

MTB > let c3 = c1 – c2
MTB > name c3 'diff'
MTB > tinterval 99% confidence data in c3

Confidence Intervals

Variable	N	Mean	St Dev	SE Mean	99.0 % CI
diff	10	−0.300	1.889	0.597	(−2.241, 1.641)

TESTING HYPOTHESIS ABOUT μ_d USING NORMALLY DISTRIBUTED DIFFERENCES COMPUTED FROM DEPENDENT SAMPLES

The procedure for a hypothesis test about the mean difference for paired samples is given in Table 10.14.

Table 10.14

Steps for Testing a Hypothesis Concerning μ_d: Normally Distributed Differences Computed from Dependent Samples
Step 1: State the null and research hypothesis. The null hypothesis is represented symbolically by H_0: $\mu_d = D_0$ and the research hypothesis is of the form H_a: $\mu_d \neq D_0$ or H_a: $\mu_d < D_0$ or H_a: $\mu_d > D_0$.
Step 2: Use the t distribution table with degrees of freedom equal to df = n – 1 and the level of significance, α, to determine the rejection region.
Step 3: Compute the value of the test statistic as follows: $t^* = \dfrac{\bar{d} - D_0}{S_{\bar{d}}}$, where $\bar{d} = \dfrac{\Sigma d}{n}$, D_0 is the hypothesized value of μ_d in the null hypothesis, $S_{\bar{d}} = \dfrac{S_d}{\sqrt{n}}$, and $S_d = \sqrt{\dfrac{\Sigma d^2 - (\Sigma d)^2 / n}{n - 1}}$.
Step 4: State your conclusion. The null hypothesis is rejected if the computed value of the test statistic falls in the rejection region. Otherwise, the null hypothesis is not rejected.

EXAMPLE 10.16 A sociological study concerning marriage and the family was conducted and one of several factors of interest was the educational level of husbands and wives. Table 10.15 gives the number of years of education beyond high school for 15 couples. These data were used to test the null hypothesis that the mean educational levels are the same vs. the research hypothesis that the educational levels are different at level of significance $\alpha = .05$.

Table 10.15

Couple	Husband	Wife	Difference
1	4	2	2
2	0	4	–4
3	7	6	1
4	4	4	0
5	2	3	–1
6	8	4	4
7	4	4	0
8	3	1	2
9	1	0	1
10	2	2	0
11	4	4	0
12	8	2	6
13	0	4	–4
14	4	0	4
15	1	4	–3

The degrees of freedom is df = 15 – 1 = 14, and since the research hypothesis is two-tailed and $\alpha = .05$, a t distribution curve with area equal to .025 in each tail will determine the rejection regions. Consulting the t distribution table with df = 14, we find that $P(t > 2.145) = .025$. Figure 10-3 illustrates the rejection regions.

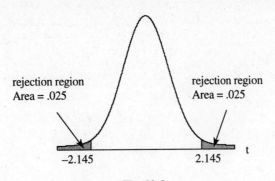

Fig. 10-3

The sum of the differences is $\Sigma d = 8$ and the sum of the squares of the differences is $\Sigma d^2 = 120$. The mean difference is $\overline{d} = \dfrac{\Sigma d}{n} = \dfrac{8}{15} = .533$, and the standard deviation is

$$S_d = \sqrt{\frac{\Sigma d^2 - (\Sigma d)^2 / n}{n-1}} \doteq \sqrt{\frac{120 - 8^2 / 15}{14}} = 2.8752$$

The standard error of the mean difference is

$$S_{\overline{d}} = \frac{S_d}{\sqrt{n}} = \frac{2.8752}{\sqrt{15}} = .7424$$

The computed test statistic is

$$t^* = \frac{\overline{d} - D_0}{S_{\overline{d}}} = \frac{533 - 0}{.7424} = 0.72$$

Since $t^* = 0.72$ does not fall within the rejection region shown in Fig. 10-3, we are unable to conclude that the educational levels of husbands and wives are different.

EXAMPLE 10.17 A Minitab solution for Example 10.16 is shown below. The data for the husbands is put into column c1 and the data for the wives is put into column c2. The differences are put into c3, and a one-sample t test is performed on c3. The subcommand **SUBC> alternative = 0.** indicates a two-tailed alternative hypothesis. Note that the mean, standard deviation, standard error, and computed t values are the same as those computed in Example 10.16. Since the computed p value is 0.48, the null hypothesis is not rejected at the $\alpha = .05$ level.

Data Display

Row	husband	wife
1	4	2
2	0	4
3	7	6
4	4	4
5	2	3
6	8	4
7	4	4
8	3	1
9	1	0
10	2	2
11	4	4
12	8	2
13	0	4
14	4	0
15	1	4

```
MTB > let c3 = c1 – c2
MTB > name c3 'diff'
MTB > ttest mean = 0, data in c3;
SUBC > alternative = 0.
```

T-Test of the Mean
Test of mu = 0.000 vs mu not = 0.000

Variable	N	Mean	St Dev	SE Mean	T	P
diff	15	0.533	2.875	0.742	0.72	0.48

SAMPLING DISTRIBUTION OF $\bar{P}_1 - \bar{P}_2$ FOR LARGE INDEPENDENT SAMPLES

Previous sections in this chapter have discussed inferences concerning the differences in means for two populations. The remaining sections of Chapter 10 are concerned with inferences concerning the differences in population proportions or percents. How does the percent of defectives produced by machine 1 compare with the percent of defectives produced by machine 2? Is there a difference in the percent of males who will vote for a presidential candidate and the percent of females who will vote for the candidate? Is the percentage of smokers the same for African-Americans as it is for Whites? All of these questions involve the comparison of percents or proportions. Sample proportions will be used to estimate the differences in population proportions or to test a hypothesis about the differences in population proportions. The following notation will be used.

p_1 and p_2 = proportions in populations 1 and 2 having the characteristic of interest

n_1 and n_2 = sizes of the independent samples drawn from populations 1 and 2

$\bar{p}_1$ and $\bar{p}_2$ = proportions in samples 1 and 2 having the characteristic of interest

$q_1 = 1 - p_1$ and $q_2 = 1 - p_2$ = proportions in populations 1 and 2 not having the characteristic

$\bar{q}_1 = 1 - \bar{p}_1$ and $\bar{q}_2 = 1 - \bar{p}_2$ = proportions in samples 1 and 2 not having the characteristic

When the samples sizes, n_1 and n_2, are such that $n_1 p_1 > 5$, $n_2 p_2 > 5$, $n_1 q_1 > 5$ and $n_2 q_2 > 5$, the sampling distribution of $\bar{p}_1 - \bar{p}_2$ is normal. The mean or expected value of $\bar{p}_1 - \bar{p}_2$ is given by formula (10.21):

$$E(\bar{p}_1 - \bar{p}_2) = \mu_{\bar{p}_1 - \bar{p}_2} = p_1 - p_2 \qquad (10.21)$$

The standard error of $\bar{p}_1 - \bar{p}_2$ is given by formula (10.22).

$$\sigma_{\bar{p}_1 - \bar{p}_2} = \sqrt{\frac{p_1 \times q_1}{n_1} + \frac{p_2 \times q_2}{n_2}} \qquad (10.22)$$

When the sample sizes are large enough to satisfy the above requirements, the distribution of $\bar{p}_1 - \bar{p}_2$ is as shown in Fig. 10-4. The standard error of the curve is given by formula (10.22).

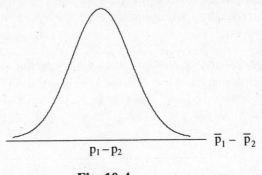

Fig. 10-4

Formula (10.23) is used to transform the distribution of $\bar{p}_1 - \bar{p}_2$ to a standard normal distribution.

$$z = \frac{\bar{p}_1 - \bar{p}_2 - (p_1 - p_2)}{\sigma_{\bar{p}_1 - \bar{p}_2}}$$

(10.23)

EXAMPLE 10.18 Three percent of the items produced by machine 1 have a minor defect and 2% of the same item produced by machine 2 have a minor defect. If samples of size 500 are selected from each machine, the distribution of $\bar{p}_1 - \bar{p}_2$ will be normal since $n_1 p_1 = 500 \times .03 = 15$, $n_2 p_2 = 500 \times .02 = 10$, $n_1 q_1 = 500 \times .97 = 485$ and $n_2 q_2 = 500 \times .98 = 490$. Normality may be assumed, since all four products are greater than 5. The mean value of $\bar{p}_1 - \bar{p}_2$ is $.03 - .02 = .01$, and the standard error of $\bar{p}_1 - \bar{p}_2$ is

$$\sigma_{\bar{p}_1 - \bar{p}_2} = \sqrt{\frac{p_1 \times q_1}{n_1} + \frac{p_2 \times q_2}{n_2}} = \sqrt{\frac{.03 \times .97}{500} + \frac{.02 \times .98}{500}} = .00987.$$

The probability that the percent defective in the sample from machine 1 exceeds the percent defective in the sample from machine 2 by 3% or more is expressed as $P(\bar{p}_1 - \bar{p}_2 > .03)$. The event $\bar{p}_1 - \bar{p}_2 > .03$ is transformed to an equivalent event involving z by the use of formula (10.23). The equivalent event is $\dfrac{\bar{p}_1 - \bar{p}_2 - .01}{.00987} > \dfrac{.03 - .01}{.00987} = 2.03$, or $z > 2.03$. The event $\bar{p}_1 - \bar{p}_2 > .03$ is equivalent to the event $z > 2.03$, and therefore $P(\bar{p}_1 - \bar{p}_2 > .03) = P(z > 2.03) = .5 - .4788 = .0212$. There are only approximately 2 chances out of 100 that the percent defective in sample 1 will exceed the percent defective in sample 2 by 3% or more.

ESTIMATION OF $P_1 - P_2$ USING LARGE INDEPENDENT SAMPLES

The difference in sample proportions, $\bar{p}_1 - \bar{p}_2$, is *a point estimate of* $p_1 - p_2$. An *interval estimate of* $p_1 - p_2$ is obtained by using formula (10.23). However, the standard error as given in formula (10.22) will need to be estimated since p_1 and p_2 are unknown. If the population proportions are estimated by their corresponding sample proportions, we obtain the estimated standard error of $\bar{p}_1 - \bar{p}_2$. The estimated standard error of $\bar{p}_1 - \bar{p}_2$ is represented by $S_{\bar{p}_1 - \bar{p}_2}$ and is given by formula (10.24).

$$S_{\bar{p}_1 - \bar{p}_2} = \sqrt{\frac{\bar{p}_1 \times \bar{q}_1}{n_1} + \frac{\bar{p}_2 \times \bar{q}_2}{n_2}}$$

(10.24)

The confidence interval for $p_1 - p_2$ is given by formula (10.25):

$$(\bar{p}_1 - \bar{p}_2) \pm z \times S_{\bar{p}_1 - \bar{p}_2}$$

(10.25)

The confidence interval is valid provided that $n_1p_1 > 5$, $n_2p_2 > 5$, $n_1q_1 > 5$ and $n_2q_2 > 5$. Since p_1, q_1, p_2, and q_2 are unknown, the corresponding sample quantities are substituted to check the validity of using the confidence interval.

EXAMPLE 10.19 A survey of 2000 ninth-graders found that 32% had used cigarettes in the past week and a survey of 1500 high school seniors found that 35% had used cigarettes in the past week. Suppose p_1 represents the proportion of all ninth-graders who used cigarettes in the past week and p_2 represents the proportion of all seniors who used cigarettes in the past week. A point estimate for $p_1 - p_2$ is –3%. The estimated standard error of $\overline{p}_1 - \overline{p}_2$ is as follows:

$$S_{\overline{p}_1-\overline{p}_2} = \sqrt{\frac{\overline{p}_1 \times \overline{q}_1}{n_1} + \frac{\overline{p}_2 \times \overline{q}_2}{n_2}} = \sqrt{\frac{32 \times 68}{2000} + \frac{35 \times 65}{1500}} = 1.614\%$$

A 95% confidence interval for $p_1 - p_2$ is given by $(\overline{p}_1 - \overline{p}_2) \pm z \times S_{\overline{p}_1-\overline{p}_2}$ or $-3\% \pm 1.96 \times 1.614\%$ or $-3\% \pm 3.2\%$. The 95% confidence interval extends from –6.2% to 0.2%. Note that $n_1 \times \overline{p}_1 = 2000 \times .32 = 640$, $n_1 \times \overline{q}_1 = 2000 \times .68 = 1360$, $n_2 \times \overline{p}_2 = 1500 \times .35 = 525$, and $n_2 \times \overline{q}_2 = 1500 \times .65 = 975$. Since all 4 of these quantities exceed 5, the confidence interval is valid.

TESTING HYPOTHESIS ABOUT $P_1 - P_2$ USING LARGE INDEPENDENT SAMPLES

The most common null hypothesis concerning $p_1 - p_2$ is H_0: $p_1 - p_2 = 0$. Recall that in testing hypothesis, the null hypothesis is always assumed to be true and the test statistic is computed under this assumption. If the test statistic value is judged to be highly unusual, then the assumption that the null hypothesis is true is rejected. When H_0 is assumed to be true, $p_1 - p_2 = 0$ or $p_1 = p_2$. If we let p be the common value of p_1 and p_2, then the standard error of $\overline{p}_1 - \overline{p}_2$ simplifies as follows:

$$\sigma_{\overline{p}_1-\overline{p}_2} = \sqrt{\frac{p_1 \times q_1}{n_1} + \frac{p_2 \times q_2}{n_2}} = \sqrt{\frac{p \times q}{n_1} + \frac{p \times q}{n_2}} = \sqrt{p \times q\left(\frac{1}{n_1} + \frac{1}{n_2}\right)}$$

Since p and q are unknown, they must be estimated from the two samples. Let x_1 be the number in sample 1 with the characteristic of interest and let x_2 be the number in sample 2 with the characteristic of interest. A *pooled estimate of p* is given by

$$\overline{p} = \frac{x_1 + x_2}{n_1 + n_2} \tag{10.26}$$

Substituting $\overline{p}$ for p and $\overline{q} = 1 - \overline{p}$ for q in the above expression for $\sigma_{\overline{p}_1-\overline{p}_2}$, we obtain the estimated standard error of $\overline{p}_1 - \overline{p}_2$ as given in

$$S_{\overline{p}_1-\overline{p}_2} = \sqrt{\overline{p} \times \overline{q}\left(\frac{1}{n_1} + \frac{1}{n_2}\right)} \tag{10.27}$$

The test statistic for testing the null hypothesis H_0: $p_1 - p_2 = 0$ is obtained by using formula (10.23) with $p_1 - p_2$ replaced by 0 and $\sigma_{\overline{p}_1-\overline{p}_2}$ estimated by $S_{\overline{p}_1-\overline{p}_2}$ as given in formula (10.27). The resulting test statistic is given in formula (10.28):

$$z = \frac{\overline{p}_1 - \overline{p}_2}{S_{\overline{p}_1 - \overline{p}_2}}$$

(10.28)

The steps for testing a hypothesis concerning $p_1 - p_2$ are given in Table 10.16.

Table 10.16

Steps for Testing a Hypothesis Concerning $p_1 - p_2$: Large Independent Samples $n_1 p_1 > 5$, $n_2 p_2 > 5$, $n_1 q_1 > 5$ and $n_2 q_2 > 5$
Step 1: State the null and research hypothesis. The null hypothesis is represented symbolically by H_0: $p_1 - p_2 = 0$ and the research hypothesis is of the form H_a: $p_1 - p_2 < 0$ or H_a: $p_1 - p_2 > 0$ or H_a: $p_1 - p_2 \neq 0$.
Step 2: Use the standard normal distribution table and the level of significance, α, to determine the rejection region.
Step 3: Compute the value of the test statistic as follows: $z^* = \dfrac{\overline{p}_1 - \overline{p}_2}{S_{\overline{p}_1 - \overline{p}_2}}$, where $\overline{p}_1 - \overline{p}_2$ is the computed difference in sample proportions, $S_{\overline{p}_1 - \overline{p}_2} = \sqrt{\overline{p} \times \overline{q} \left(\dfrac{1}{n_1} + \dfrac{1}{n_2} \right)}$, $\overline{p} = \dfrac{x_1 + x_2}{n_1 + n_2}$, and $\overline{q} = 1 - \overline{p}$.
Step 4: State your conclusion. The null hypothesis is rejected if the computed value of the test statistic falls in the rejection region. Otherwise, the null hypothesis is not rejected.

EXAMPLE 10.20 A study was conducted to compare teen cigarette use for whites and Hispanics. Suppose p_1 represents the proportion of teenage whites who use cigarettes and p_2 represents the proportion of teenage Hispanics who use cigarettes. The null hypothesis is H_0: $p_1 - p_2 = 0$ and the research hypothesis is H_a: $p_1 - p_2 \neq 0$. The critical values for a level of significance equal to .05 are ± 1.96. The sample results are given in Table 10.17.

Table 10.17

Sample	Sample size	Number of smokers	Sample proportion
1. Whites	$n_1 = 1500$	$x_1 = 555$	$\overline{p}_1 = 0.37$
2. Hispanics	$n_2 = 500$	$x_2 = 175$	$\overline{p}_2 = 0.35$

The pooled proportion is $\overline{p} = \dfrac{x_1 + x_2}{n_1 + n_2} = \dfrac{555 + 175}{1500 + 500} = 0.365$. The estimated standard error for $\overline{p}_1 - \overline{p}_2$ is:

$$S_{\overline{p}_1 - \overline{p}_2} = \sqrt{\overline{p} \times \overline{q} \left(\frac{1}{n_1} + \frac{1}{n_2} \right)} = \sqrt{.365 \times .635 \left(\frac{1}{1500} + \frac{1}{500} \right)} = 0.02486$$

and the computed test statistic is:

$$z^* = \frac{\overline{p}_1 - \overline{p}_2}{S_{\overline{p}_1 - \overline{p}_2}} = \frac{.02}{.02486} = 0.80.$$

Based on this study, we would not be able to conclude that a difference exists between the proportion of smokers within the two groups of teenagers.

Solved Problems

SAMPLING DISTRIBUTION OF $\overline{X}_1 - \overline{X}_2$ FOR LARGE INDEPENDENT SAMPLES

10.1 A sample of size 50 is taken from a population having a mean equal to 90 and a standard deviation equal to 15. A second sample of size 70 and independent of the first sample is selected from another population having mean equal to 75 and standard deviation equal to 10. The mean of the sample of size 50 is represented by $\overline{x}_1$ and the mean of the sample of size 70 is represented as $\overline{x}_2$.
   (a) What type distribution does $\overline{x}_1 - \overline{x}_2$ have?
   (b) What is the expected value of $\overline{x}_1 - \overline{x}_2$?
   (c) What is the standard error of $\overline{x}_1 - \overline{x}_2$?
   (d) Transform $\overline{x}_1 - \overline{x}_2$ to a standard normal.

   Ans.   (a) Because both samples are 30 or more, $\overline{x}_1 - \overline{x}_2$ will have a normal distribution.
   (b) The expected value of $\overline{x}_1 - \overline{x}_2$ is $E(\overline{X}_1 - \overline{X}_2) = \mu_{\overline{x}_1 - \overline{x}_2} = \mu_1 - \mu_2 = 90 - 75 = 15$.

   (c) The standard error of $\overline{x}_1 - \overline{x}_2$ is $\sigma_{\overline{x}_1 - \overline{x}_2} = \sqrt{\dfrac{\sigma_1^2}{n_1} + \dfrac{\sigma_2^2}{n_2}} = \sqrt{\dfrac{225}{50} + \dfrac{100}{70}} = 2.43$.

   (d) $z = \dfrac{\overline{x}_1 - \overline{x}_2 - 15}{2.43}$ is a standard normal variable.

ESTIMATION OF $\mu_1 - \mu_2$ USING LARGE INDEPENDENT SAMPLES

10.2 Table 10.18 gives the summary statistics for the number of years that 250 men and women have spent with their current employers. Use these results to find a 90% confidence interval for $\mu_1 - \mu_2$, the mean difference in years spent with their current employer for men and women.

Table 10.18

Sample	Sample size	Mean	Standard deviation
1. Men	250	5.5 years	2.1 years
2. Women	250	3.3 years	1.8 years

   Ans.   The general form of the confidence interval for $\mu_1 - \mu_2$ is $(\overline{X}_1 - \overline{X}_2) \pm z \times \sigma_{\overline{x}_1 - \overline{x}_2}$. The difference in means is $\overline{X}_1 - \overline{X}_2 = 5.5 - 3.3 = 2.2$. The z value for a 90% confidence interval is 1.65. The estimated standard error of the difference in means is $S_{\overline{x}_1 - \overline{x}_2} = \sqrt{\dfrac{s_1^2}{n_1} + \dfrac{s_2^2}{n_2}} = \sqrt{\dfrac{4.41}{250} + \dfrac{3.24}{250}} = .1749$.
   The 90% margin of error when using $\overline{X}_1 - \overline{X}_2$ as an estimate of $\mu_1 - \mu_2$ is $1.65 \times .1749$ or $.2886$. The confidence interval is $2.2 \pm .3$. The confidence interval extends from 1.9 to 2.5 years.

TESTING HYPOTHESIS ABOUT $\mu_1 - \mu_2$ USING LARGE INDEPENDENT SAMPLES

10.3 Use the data in Table 10.18 to test the null hypothesis that $\mu_1 - \mu_2 = 1.5$ years vs. the research hypothesis that $\mu_1 - \mu_2 > 1.5$ years at level of significance $\alpha = .01$.

Ans.   Since the samples are large, the standard normal distribution table is used to find the critical value. From the standard normal table, we find that $P(z > 2.33) = .01$, and therefore the critical value is 2.33. The computed value of the test statistic is given as follows:

$$z^* = \frac{\overline{x}_1 - \overline{x}_2 - D_0}{\sigma_{\overline{x}_1 - \overline{x}_2}} = \frac{5.5 - 3.3 - 15}{.1749} = 4.00$$

The standard error of the difference in means is replaced by the estimated standard error of the difference in means. It is concluded that the mean exceeds 1.5 years.

SAMPLING DISTRIBUTION OF $\overline{X}_1 - \overline{X}_2$ FOR SMALL INDEPENDENT SAMPLES FROM NORMAL POPULATIONS WITH EQUAL (BUT UNKNOWN) STANDARD DEVIATIONS

10.4 A sample of size 10 is taken from a normal population having a mean equal to 35 and a sample of size 15 is taken from another normal population having mean equal to 40. The two normal populations have equal variances. The mean and variance of the sample of size 10 are represented by $\overline{x}_1$ and s_1^2 and the mean and variance of the sample of size 15 are represented by $\overline{x}_2$ and s_2^2.

(*a*) Give the expression for the pooled estimate of the common population variance.

(*b*) Give the expression for the estimated standard error of the difference in the sample means.

(*c*) Use the results in parts (*a*) and (*b*) to form a statistic that has a t distribution with 23 degrees of freedom.

Ans.   (*a*) $S^2 = \dfrac{(n_1 - 1)s_1^2 + (n_2 - 1)s_2^2}{n_1 + n_2 - 2} = \dfrac{9 \times s_1^2 + 14 \times s_2^2}{23}$

(*b*) $S_{\overline{x}_1 - \overline{x}_2} = \sqrt{S^2\left(\dfrac{1}{n_1} + \dfrac{1}{n_2}\right)} = \sqrt{S^2 \times \left(\dfrac{1}{10} + \dfrac{1}{15}\right)}$, where S^2 is given in part (*a*).

(*c*) $t = \dfrac{\overline{x}_1 - \overline{x}_2 - (\mu_1 - \mu_2)}{S_{\overline{x}_1 - \overline{x}_2}} = \dfrac{\overline{x}_1 - \overline{x}_2 - (35 - 40)}{S_{\overline{x}_1 - \overline{x}_2}} = \dfrac{\overline{x}_1 - \overline{x}_2 + 5}{S_{\overline{x}_1 - \overline{x}_2}}$, where $S_{\overline{x}_1 - \overline{x}_2}$ is given in part (*b*).

ESTIMATION OF $\mu_1 - \mu_2$ USING SMALL INDEPENDENT SAMPLES FROM NORMAL POPULATIONS WITH EQUAL (BUT UNKNOWN) STANDARD DEVIATIONS

10.5 A comparison of motel room rates for single occupancy was made for the cities of Omaha, Nebraska and Kansas City, Missouri. The rates for the two cities are shown in Table 10.19. Using the command **% normplot** of Minitab for both samples, it is found that it is reasonable to assume that both populations are normally distributed. Using the command **%vartest** of Minitab, it is also found that it is reasonable to assume that the populations have equal variability. The Minitab output for setting a 99% confidence interval on $\mu_1 - \mu_2$ is given below. Verify the Pooled standard deviation, the degrees of freedom, and the 99% confidence interval given in the output.

Table 10.19

Omaha	Kansas City
75	80
70	75
70	75
65	85
85	85
85	100
80	60
90	65
90	95
60	105
60	70

MTB > twot 99% confidence data in c2, groups in c1;
SUBC > pooled.

Two Sample T-Test and Confidence Interval
Two sample T for rate

city	N	Mean	St Dev	SE Mean
1	11	75.5	11.3	3.4
2	11	81.4	14.3	4.3

99% CI for mu (1) – mu (2): (–21.6, 9.7)
T-Test mu (1) = mu (2) (vs not =): T = –1.07 P = 0.30 DF = 20
Both use Pooled St Dev = 12.9

Ans. The difference in sample mean is $\bar{x}_1 - \bar{x}_2 = 75.5 - 81.4 = -5.9$.

The pooled variance is $S^2 = \dfrac{(n_1-1)s_1^2 + (n_2-1)s_2^2}{n_1+n_2-2} = \dfrac{10 \times 127.69 + 10 \times 204.49}{11+11-2} = 166.09$, and S = $\sqrt{166.09} = 12.9$.

The standard error of the difference in the sample means is $S_{\bar{x}_1-\bar{x}_2} = \sqrt{s^2\left(\dfrac{1}{n_1}+\dfrac{1}{n_2}\right)} = \sqrt{166.09 \times \left(\dfrac{1}{11}+\dfrac{1}{11}\right)} = 5.50$.

Using the t distribution table with df = 20 and right-hand tail area equal to .005, we find the t value is 2.845. The 99% margin of error is 2.845 × 5.50 = 15.6. The 99% confidence interval extends from –5.9 – 15.6 = –21.5 to –5.9 + 15.6 = 9.7.

TESTING HYPOTHESIS ABOUT $\mu_1 - \mu_2$ USING SMALL INDEPENDENT SAMPLES FROM NORMAL POPULATIONS WITH EQUAL (BUT UNKNOWN) STANDARD DEVIATIONS

10.6 Use the motel rate data in Table 10.19 and the Minitab output given in problem 10.5 to test the null hypothesis $H_0: \mu_1 - \mu_2 = 0$ vs. $H_a: \mu_1 - \mu_2 \neq 0$ at significance level $\alpha = .01$.
  (a) Give the critical values for performing the test.
  (b) Give the computed value of the test statistic, and your conclusion based upon this value and the critical value in part (a).
  (c) Give the p value and your conclusion based on this value.

Ans. (*a*) The degrees of freedom for the t distribution is 20. Since the research hypothesis is two-tailed, the significance level is divided by 2 and .005 is put into each tail of the distribution. By consulting the t distribution table, we find that for df = 20 and right-tail area = .005, the t value is 2.845. The critical values are ±2.845.

(*b*) The computed t value, from the Minitab output in problem 10.5, is t* = −1.07. Since this value does not fall in the rejection region, the null hypothesis is not rejected. It cannot be concluded that the mean motel rates differ for the two cities.

(*c*) The p value, from the Minitab output in problem 10.5, is equal to 0.30. Since this exceeds the preset level of significance, the null hypothesis is not rejected. The same conclusion is reached as in part (*b*).

SAMPLING DISTRIBUTION OF $\overline{X}_1 - \overline{X}_2$ FOR SMALL INDEPENDENT SAMPLES FROM NORMAL POPULATIONS WITH UNEQUAL (AND UNKNOWN) STANDARD DEVIATIONS

10.7 Refer to problem 10.4. Suppose the two populations have unequal population variances. What changes are needed to the answers given in the problem?

Ans. Since the population variances are unequal, the sample variances are not pooled together to estimate a common population variance. The standard error of the difference in the sample means is given by $S_{\overline{x}_1 - \overline{x}_2} = \sqrt{\dfrac{s_1^2}{n_1} + \dfrac{s_2^2}{n_2}} = \sqrt{\dfrac{s_1^2}{10} + \dfrac{s_2^2}{15}}$. The statistic $t = \dfrac{\overline{x}_1 - \overline{x}_2 + 5}{S_{\overline{x}_1 - \overline{x}_2}}$ has a t distribution and the degrees of freedom is given by df = minimum of $\{(n_1 - 1), (n_2 - 1)\}$ = minimum of $\{9, 14\}$ = 9. Note that the degrees of freedom is reduced from 23 to 9. This problem illustrates the importance of checking the assumptions underlying the estimation and testing procedures. The computation of the test statistic as well as the degrees of freedom is determined by the assumption concerning the variances of the two populations.

ESTIMATION OF $\mu_1 - \mu_2$ USING SMALL INDEPENDENT SAMPLES FROM NORMAL POPULATIONS WITH UNEQUAL (AND UNKNOWN) STANDARD DEVIATIONS

10.8 Refer to problem 10.5. Compute the standard error of the difference in means, the degrees of freedom, and the 99% confidence interval assuming unequal population variances. Compare the results with those in problem 10.5.

Ans. The standard error of the difference in sample means is $S_{\overline{x}_1 - \overline{x}_2} = \sqrt{\dfrac{s_1^2}{n_1} + \dfrac{s_2^2}{n_2}} = \sqrt{\dfrac{11.3^2}{11} + \dfrac{14.3^2}{11}}$ which equals 5.50. This is the same answer obtained in the equal variances case. This will always occur if the sample sizes are equal.

The degrees of freedom is df = minimum of $\{(n_1 - 1), (n_2 - 1)\}$ = minimum of $\{10, 10\}$ = 10. Using the t distribution table with df = 10 and right-hand tail area equal to .005, we find the t value is 3.169. The 99% margin of error is 3.169 × 5.50 = 17.4. The 99% confidence interval extends from −5.9 − 17.4 = −23.3 to −5.9 + 17.4 = 11.5.

The 99% confidence interval in the equal variances case is (−21.5, 9.7). The 99% confidence interval for the unequal variances case is (−23.3, 11.5). Note that the interval in the unequal variances case is wider.

TESTING HYPOTHESIS ABOUT $\mu_1 - \mu_2$ USING SMALL INDEPENDENT SAMPLES FROM NORMAL POPULATIONS WITH UNEQUAL (AND UNKNOWN) STANDARD DEVIATIONS

10.9 In a study of internet users, the average time spent online per week was determined for a group of college graduates as well as a group of non-college graduates. The results of the study are shown in Table 10.20. Test the research hypothesis that $\mu_1 > \mu_2$ at level of significance $\alpha = .05$. Give the critical value, the computed test statistic, and your conclusion. Assume that the times are normally distributed for both populations.

Table 10.20

Sample	Sample size	Mean	Standard deviation
1. College graduate	14	8.6 hours	1.1 hours
2. Non-college graduate	12	6.3 hours	2.7 hours

Ans. Some statisticians use the following rule to decide whether population variances are equal or not:

If $.5 \le \dfrac{S_1}{S_2} \le 2$, *then assume* $\sigma_1 = \sigma_2$. *Otherwise, assume that* $\sigma_1 \neq \sigma_2$. Since the ratio of the sample is less than .5, we assume that the populations have unequal standard deviations. The degrees of freedom for this model is df = minimum of $\{(n_1 - 1), (n_2 - 1)\}$ = minimum of $\{13, 11\}= 11$. The critical value is determined by using the t distribution table with df = 11 and right-tail area equal to .05. This value is found to equal 1.796.

The standard error of the difference in sample means is $S_{\bar{x}_1 - \bar{x}_2} = \sqrt{\dfrac{s_1^2}{n_1} + \dfrac{s_2^2}{n_2}} = \sqrt{\dfrac{1.21}{14} + \dfrac{7.29}{12}} =$

.833. The computed test statistic is $t^* = \dfrac{\bar{x}_1 - \bar{x}_2 - (\mu_1 - \mu_2)}{S_{\bar{x}_1 - \bar{x}_2}} = \dfrac{8.6 - 6.3 - 0}{.833} = 2.76$. The research

hypothesis is supported since $t^* = 2.76$ exceeds the critical value, 1.796.

SAMPLING DISTRIBUTION OF $\bar{d}$ FOR NORMALLY DISTRIBUTED DIFFERENCES COMPUTED FOR DEPENDENT SAMPLES

10.10 Table 10.21 gives a set of paired data, along with the differences for the pairs. Answer the following questions concerning these paired data.

(a) Find the following: $\bar{d} = \dfrac{\Sigma d}{n}$, $S_d = \sqrt{\dfrac{\Sigma d^2 - (\Sigma d)^2 / n}{n - 1}}$, and $S_{\bar{d}} = \dfrac{S_d}{\sqrt{n}}$.

(b) What parameters do each of the statistics in part (a) estimate?

(c) What assumption is needed in order that the statistic $t = \dfrac{\bar{d} - \mu_d}{S_{\bar{d}}}$ have a t distribution

with $(n - 1) = (6 - 1) = 5$ degrees of freedom?

Table 10.21

Pair	Sample 1	Sample 2	Difference
1	18	15	$d_1 = 18 - 15 = 3$
2	23	22	$d_2 = 23 - 22 = 1$
3	27	25	$d_3 = 27 - 25 = 2$
4	20	22	$d_4 = 20 - 22 = -2$
5	19	19	$d_5 = 19 - 19 = 0$
6	24	22	$d_6 = 24 - 22 = 2$

Ans. (*a*) $\Sigma d = 3 + 1 + 2 - 2 + 0 + 2 = 6$, $\Sigma d^2 = 9 + 1 + 4 + 4 + 0 + 4 = 22$, $\overline{d} = \frac{6}{6} = 1$, $S_d = \sqrt{\dfrac{22 - \frac{36}{6}}{5}}$

$= 1.789$, and $S_{\overline{d}} = \dfrac{S_d}{\sqrt{n}} = \dfrac{1.789}{\sqrt{6}} = 0.73$.

(*b*) The statistic, $\overline{d}$, estimates μ_d, the mean of the population of paired differences. The statistic, S_d, estimates σ_d, the standard deviation of the population of paired differences. The statistic, $S_{\overline{d}}$, estimates $\sigma_{\overline{d}}$, the standard error of the population of paired differences.

(*c*) It is assumed that the population of paired differences is normally distributed.

ESTIMATION OF μ_d USING NORMALLY DISTRIBUTED DIFFERENCES COMPUTED FROM DEPENDENT SAMPLES

10.11 A sociological study compared the salaries of 10 professional African-American women with the salaries of 10 corresponding professional White-American women. The women were paired according to certain salient characteristics and the 10 pairs were chosen from ten different professions. The salaries (in thousands) are shown in Table 10.22.

Table 10.22

Pair	African-American	White	Difference
1	65	60	5
2	50	55	–5
3	75	70	5
4	80	75	5
5	105	95	10
6	90	100	–10
7	65	70	–5
8	60	50	10
9	115	105	10
10	80	90	–10

The Minitab command **tinterval 90 percent confidence data in c3 is** used to produce the following output. The differences are computed and put into column c3. The command requests a 90% confidence interval on the mean difference.

MTB > let c3 = c1 – c2
MTB > print c1 – c3

Data Display

Row	Afamer	white	diff
1	65	60	5
2	50	55	–5
3	75	70	5
4	80	75	5
5	105	95	10
6	90	100	–10
7	65	70	–5
8	60	50	10
9	115	105	10
10	80	90	–10

MTB > **tinterval 90 percent confidence data in c3**

Variable	N	Mean	St Dev	SE Mean	90.0 % CI
diff	10	1.50	8.18	2.59	(−3.24, 6.24)

Verify that $\overline{d} = 1.50$, $S_d = 8.18$, $S_{\overline{d}} = 2.59$, and that the confidence interval is as given in the output.

Ans.   Using the differences given above, we find that $\Sigma d = 15$, and $\overline{d} = 1.5$. We also find that $\Sigma d^2 = 625$, and $S_d = \sqrt{\dfrac{\Sigma d^2 - (\Sigma d)^2 / n}{n-1}} = \sqrt{\dfrac{625 - 22.5}{9}} = 8.18$. The estimated standard error of $\overline{d}$ is $S_{\overline{d}} = \dfrac{S_d}{\sqrt{n}} = \dfrac{8.18196}{\sqrt{10}} = 2.59$.

The confidence interval is given by $\overline{d} \pm t \times S_{\overline{d}}$. The t value for a 90% confidence interval is found as follows. The degrees of freedom is df = 10 − 1 = 9. For a 90% confidence interval 5% is put into each tail of the t distribution. Using the t distribution table with 9 degrees of freedom and a right-hand tail area equal to .05, we find the t value to be 1.833. The 90% margin of error is $t \times S_{\overline{d}} = 1.833 \times 2.59 = 4.75$. The 90% confidence interval goes from 1.50 − 4.75 = −3.25 to 1.50 + 4.75 = 6.25.

TESTING HYPOTHESIS ABOUT μ_d USING NORMALLY DISTRIBUTED DIFFERENCES COMPUTED FROM DEPENDENT SAMPLES

10.12 Use the data in Table 10.22 to test the research hypothesis $\mu_d \neq 0$ at level of significance $\alpha = .01$.

Ans.   The degrees of freedom is one less than the number of pairs, i.e., df = 9. Since the research hypothesis is two-tailed, tail areas equal to .005 are put into both tails of the t distribution. From the t distribution table the critical value is determined by using a right-tail area equal to .005. The critical value is equal to 3.250.

From problem 10.11, the following are found: $\overline{d} = 1.50$, $S_{\overline{d}} = 2.59$, and $D_0 = 0$. The computed test statistic is $t^* = \dfrac{\overline{d} - D_0}{S_{\overline{d}}} = \dfrac{1.50 - 0}{2.59} = 0.58$. Since this value falls between −3.250 and 3.250, the null hypothesis is not rejected.

SAMPLING DISTRIBUTION OF $\overline{P}_1 - \overline{P}_2$ FOR LARGE INDEPENDENT SAMPLES

10.13 Population 1 has $p_1 = .010$ and population 2 has $p_2 = .005$. Independent samples of size 5000 each are selected from both populations. Find the probability that $\overline{p}_1 - \overline{p}_2$ exceeds .015?

Ans.   The mean value of $\overline{p}_1 - \overline{p}_2$ is .010 − .005 = .005. The standard error of $\overline{p}_1 - \overline{p}_2$ is as follows:

$$\sigma_{\overline{p}_1 - \overline{p}_2} = \sqrt{\dfrac{p_1 \times q_1}{n_1} + \dfrac{p_2 \times q_2}{n_2}} = \sqrt{\dfrac{.010 \times .990}{5000} + \dfrac{.005 \times .995}{5000}} = .001725.$$

Since $n_1 p_1 > 5$, $n_2 p_2 > 5$, $n_1 q_1 > 5$ and $n_2 q_2 > 5$, $\overline{p}_1 - \overline{p}_2$ has a normal distribution. We are asked to find $P(\overline{p}_1 - \overline{p}_2 > .015)$. The transformation $z = \dfrac{\overline{p}_1 - \overline{p}_2 - (p_1 - p_2)}{\sigma_{\overline{p}_1 - \overline{p}_2}} = \dfrac{\overline{p}_1 - \overline{p}_2 - .005}{.001725}$ is used to transform the statistic $\overline{p}_1 - \overline{p}_2$ to a standard normal variable. The same transformation on .015

gives the value $\dfrac{.015 - .005}{.001725} = 5.80$. We have the following: $P(\overline{p}_1 - \overline{p}_2 > .015) = P(z > 5.80)$, which is approximately 0. It is highly unlikely that $\overline{p}_1 - \overline{p}_2$ exceeds .015.

ESTIMATION OF $P_1 - P_2$ USING LARGE SAMPLES

10.14 Three thousand commuters in both New York and Chicago were surveyed and the percentage of commuters who took more than 60 minutes to get to work were determined for both groups. It was found that 16.5% in New York and 10.7% in Chicago required more than 60 minutes. Set a 90% confidence interval on $p_1 - p_2$, where p_1 corresponds to New York.

Ans. The confidence interval is given by $(\overline{p}_1 - \overline{p}_2) \pm z \times S_{\overline{p}_1-\overline{p}_2}$, where $S_{\overline{p}_1-\overline{p}_2} = \sqrt{\dfrac{\overline{p}_1 \times \overline{q}_1}{n_1} + \dfrac{\overline{p}_2 \times \overline{q}_2}{n_2}}$.

The z value for 90% confidence is 1.65. The difference in sample percentages is 16.5% − 10.7% = 5.8%. The standard error is $S_{\overline{p}_1-\overline{p}_2} = \sqrt{\dfrac{16.5 \times 83.5}{3000} + \dfrac{10.7 \times 89.3}{3000}} = 0.88\%$. The 90% margin of error is $1.65 \times 0.88 = 1.5\%$. The 90% confidence interval extends from $5.8 - 1.5 = 4.3\%$ to $5.8 + 1.5 = 7.3\%$.

TESTING HYPOTHESIS ABOUT $P_1 - P_2$ USING LARGE INDEPENDENT SAMPLES

10.15 A survey of 50 men and 50 women was conducted and it was found that 16 of the men and 10 of the women used hotel room minibars. Test the hypothesis that $p_1 = p_2$ vs. $p_1 \neq p_2$ at level of significance $\alpha = .05$.

Ans. The critical values are ± 1.96. The pooled estimate of the common proportion is $\overline{p} = \dfrac{x_1 + x_2}{n_1 + n_2} = \dfrac{16 + 10}{50 + 50} = .26$. The standard error of the difference in proportions is $S_{\overline{p}_1-\overline{p}_2} = \sqrt{\overline{p} \times \overline{q}\left(\dfrac{1}{n_1} + \dfrac{1}{n_2}\right)} = \sqrt{.26 \times .74\left(\dfrac{1}{50} + \dfrac{1}{50}\right)} = .0877$. The difference in sample proportions is $\overline{p}_1 - \overline{p}_2 = .32 - .20 = .16$.

The computed test statistic is $z^* = \dfrac{\overline{p}_1 - \overline{p}_2}{S_{\overline{p}_1-\overline{p}_2}} = \dfrac{.16}{.0877} = 1.82$. Since the computed test statistic does not exceed the critical value, we cannot conclude that there is a difference between men and women users of hotel room minibars.

Supplementary Problems

SAMPLING DISTRIBUTION OF $\overline{X}_1 - \overline{X}_2$ FOR LARGE INDEPENDENT SAMPLES

10.16 A sample of size 100 is selected from a population having mean 75 and standard deviation 3. Another independent sample of size 100 is selected from a population having mean 50 and standard deviation 4. Verify that $\overline{x}_1 - \overline{x}_2$ has mean 25 and standard deviation 0.5.

(a) What percent of the time will $\overline{x}_1 - \overline{x}_2$ fall within 0.5 of 25?

(b) What percent of the time will $\overline{x}_1 - \overline{x}_2$ fall within 1.0 of 25?

(c) What percent of the time will $\overline{x}_1 - \overline{x}_2$ fall within 1.5 of 25?

Ans. (*a*) 68%   (*b*) 95%   (*c*) 99.7%

ESTIMATION OF $\mu_1 - \mu_2$ USING LARGE INDEPENDENT SAMPLES

10.17 The information shown in Table 10.23 was obtained from two independent samples selected from two populations.
(*a*) Give a point estimate for $\mu_1 - \mu_2$.
(*b*) Find a 99% confidence interval for $\mu_1 - \mu_2$.

Table 10.23

Sample	Sample size	Mean	Standard deviation
1	50	$9,500	$1,250
2	75	$9,125	$950

Ans.   (*a*) $375   (*b*) $375 \pm 2.58 \times 208.05$ or –$161.77 to $911.7

TESTING HYPOTHESIS ABOUT $\mu_1 - \mu_2$ USING LARGE INDEPENDENT SAMPLES

10.18 Use the data given in problem 10.17 to test the hypothesis $H_0: \mu_1 - \mu_2 = 0$ vs. $H_a: \mu_1 - \mu_2 > 0$ at level of significance $\alpha = .01$.
(*a*) Give the computed test statistic.
(*b*) Give the p value.
(*c*) Give your conclusion.

Ans. (*a*) $z^* = \dfrac{375 - 0}{208.05} = 1.80$ (*b*) p value = $.5 - .4641 = .0359$
(*c*) Do not reject H_0 since p value > α.

SAMPLING DISTRIBUTION OF $\overline{X}_1 - \overline{X}_2$ FOR SMALL INDEPENDENT SAMPLES FROM NORMAL POPULATIONS WITH EQUAL (BUT UNKNOWN) STANDARD DEVIATIONS

10.19 A psychological study compared the language skills and mental development of two groups of two-year-olds. One group consisted of chatty toddlers and the other group consisted of quiet children. The scores on a test which measured language skills are shown for the two groups in Table 10.24. Use Minitab to determine whether it is reasonable to assume the populations of test scores are normally distributed and also determine if it is reasonable to assume that two populations have equal standard deviations.

Table 10.24

Chatty toddlers	Quiet toddlers
75	80
70	75
70	65
65	70
85	90
85	90
80	75
90	85
90	90
60	75
60	80

Ans. The following normal probability plots were produced by Minitab. Beneath the graph, the mean, standard deviation, and sample size are shown as well as a p value for the Anderson-Darlington normality test. The p value corresponds to a null hypothesis, which states that the sample data were selected from a normally distributed population. This hypothesis is rejected if the p value is less than $\alpha = .05$. Otherwise, normality is usually assumed. In this case, it is safe to assume that both samples were obtained from normally distributed populations since the p value for the chatty sample is 0.433 and the p value for the quiet sample is 0.368.

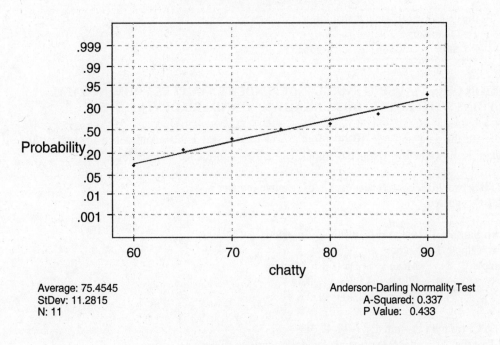

Average: 75.4545
StDev: 11.2815
N: 11

Anderson-Darling Normality Test
A-Squared: 0.337
P Value: 0.433

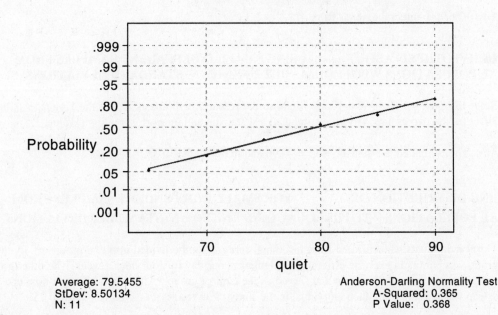

Average: 79.5455
StDev: 8.50134
N: 11

Anderson-Darling Normality Test
A-Squared: 0.365
P Value: 0.368

The following Minitab output may be used to test for equal population standard deviations. The null hypothesis of equal population standard deviations is rejected if the p value corresponding to Bartlett's test is less than $\alpha = .05$. In this case, the p value equals 0.386, and is reasonable to assume that $\sigma_1 = \sigma_2$.

MTB > %vartest c2 c1
Homogeneity of Variance

Response   score
Factors   sample
ConfLvl   95.0000

Bartlett's Test (normal distribution)
Test Statistic: 0.752
P value   : 0.386

ESTIMATION OF $\mu_1 - \mu_2$ USING SMALL INDEPENDENT SAMPLES FROM NORMAL POPULATIONS WITH EQUAL (BUT UNKNOWN) STANDARD DEVIATIONS

10.20   Refer to the data in Table 10.24. Minitab was used to set a 90% confidence interval on $\mu_1 - \mu_2$. The output is shown below. Using the output, give a 90% confidence interval for $\mu_1 - \mu_2$.

MTB > twot 90% confidence, data in c2, sample number in c1;
SUBC > pooled.

Two Sample T-Test and Confidence Interval
Two sample T for score

sample	N	Mean	St Dev	SE Mean
1	11	75.50	11.30	3.4
2	11	79.55	8.50	2.6

90% CI for mu (1) – mu (2): (–11.4, 3.3)
T-Test mu (1) = mu (2) (vs not =): T= –0.96   P=0.35   DF= 20
Both use Pooled St Dev = 9.99

Ans. The 90% interval extends from –11.4 to 3.3.

TESTING HYPOTHESIS ABOUT $\mu_1 - \mu_2$ USING SMALL INDEPENDENT SAMPLES FROM NORMAL POPULATIONS WITH EQUAL (BUT UNKNOWN) STANDARD DEVIATIONS

10.21   Refer to problems 10.19 and 10.20. Suppose the research hypothesis is that the language skills scores differ for the two groups. Give the computed test statistic and the corresponding p value.

Ans. The computed test statistic is $t^* = -0.96$ and the p value is 0.35.

SAMPLING DISTRIBUTION OF   FOR SMALL INDEPENDENT SAMPLES FROM NORMAL POPULATIONS WITH UNEQUAL (AND UNKNOWN) STANDARD DEVIATIONS

10.22   Thirty individuals who suffered from insomnia were randomly divided into two groups of 15 each. One group was put on an exercise program of 40 minutes per day for four days a week. The other group was not put on the exercise program and served as the control group. After six weeks, the time taken to fall asleep was measured for each individual in the study. The results are given in Table 10.25.

Table 10.25

Exercise group	Control group
15	19
17	22
15	25
15	30
16	31
16	15
17	16
14	19
14	19
15	30
15	27
17	28
13	22
14	13
18	32

The Minitab output for the test of equal standard deviations is shown below. Would you assume equal or unequal population standard deviations?

MTB > %vartest c2 c1
Homogeneity of Variance

Response   C2
Factors   C1
ConfLvl   95.0000

Bartlett's Test (normal distribution)
Test Statistic: 23.039
P value   :   0.000

Ans. The Minitab procedure is used to test H_0: $\sigma_1 = \sigma_2$ vs. H_a: $\sigma_1 \neq \sigma_2$. Since the p value = 0.000, the
     null hypothesis should be rejected. Assume that the standard deviations are not equal for the two
     groups.

ESTIMATION OF $\mu_1 - \mu_2$ USING SMALL INDEPENDENT SAMPLES FROM NORMAL POPULATIONS WITH UNEQUAL (AND UNKNOWN) STANDARD DEVIATIONS

10.23 Refer to the data in Table 10.25. The Minitab analysis for setting a 99% confidence interval on $\mu_1 - \mu_2$
     is shown below. This analysis assumes unequal population standard deviations.

MTB > twot 99% data in c2 groups in c1

Two Sample T-Test and Confidence Interval
Two sample T for C2

C1	N	Mean	St Dev	SE Mean
1	15	15.40	1.40	0.36
2	15	23.20	6.27	1.60

99% CI for mu (1) − mu (2): (−12.69, −2.9)
T-Test mu (1) = mu (2) (vs not =): T = −4.70 P = 0.0003 DF = 15

Rather than finding the degrees of freedom by using df = minimum of $\{n_1 - 1, n_2 - 1\}$, Minitab uses a different formula. If the degrees of freedom are found using df = minimum of $\{14, 14\} = 14$, the t value is 2.977. The t value, using df = 15, is 2.947. The confidence interval will be approximately the same for either value. Using df = 14, find a 90% confidence interval for $\mu_1 - \mu_2$.

Ans. $(15.40 - 23.20) \pm 1.761 \times \sqrt{\dfrac{1.40^2}{15} + \dfrac{6.27^2}{15}}$ or -7.8 ± 2.9 or $(-10.7, -4.9)$

TESTING HYPOTHESIS ABOUT $\mu_1 - \mu_2$ USING SMALL INDEPENDENT SAMPLES FROM NORMAL POPULATIONS WITH UNEQUAL (AND UNKNOWN) STANDARD DEVIATIONS

10.24   Refer to problems 10.22 and 10.23. Is there a difference in the time to go to sleep for the two groups?

Ans. Yes, the computed test statistic is $t^* = -4.70$, and the p value is 0.0003.

SAMPLING DISTRIBUTION OF $\bar{d}$ FOR NORMALLY DISTRIBUTED DIFFERENCES COMPUTED FOR DEPENDENT SAMPLES

10.25   Table 10.26 gives the diastolic blood pressure before treatment and six weeks after treatment is started for 10 hypertensive patients. The statistic, $t = \dfrac{\bar{d} - \mu_d}{S_{\bar{d}}}$, has a t distribution with df = n − 1, provided the differences have a normal distribution. The basic normality assumption needs to be verified before setting a confidence interval on μ_d or testing a hypothesis concerning μ_d. The normal probability plot in Minitab can be used to test the following hypothesis: H_0: The differences are normally distributed vs. H_a: The differences are not normally distributed. If the level of significance is set at the conventional level of significance $\alpha = .05$, then the null hypothesis is rejected if the p value < α. Using the Minitab output shown below and on the next page, what decision do you reach concerning the assumption of normality for the differences?

Table 10.26

Patient	Before	After	Difference
1	90	80	10
2	100	85	15
3	95	83	12
4	85	75	10
5	99	85	14
6	105	85	20
7	90	80	10
8	97	79	18
9	99	85	14
10	110	90	20

Ans. The p value for the Anderson-Darling Normality test is 0.233. The null hypothesis is not rejected, and it is assumed that the differences are normally distributed.

Normal Probability Plot

Average: 14.3
StDev: 3.94546
N: 10

Anderson-Darling Normality Test
A-Squared: 0.438
P Value: 0.233

ESTIMATION OF μ_d USING NORMALLY DISTRIBUTED DIFFERENCES COMPUTED FROM DEPENDENT SAMPLES

10.26 Verify the output shown in the following Minitab output for a 90% confidence interval for the mean difference in the blood pressures given in problem 10.25.

MTB > tinterval 90 percent confidence, data in c1
Confidence Intervals

Variable	N	Mean	St Dev	SE Mean	90.0 % CI
diff	10	14.30	3.95	1.25	(12.01, 16.59)

Ans. $\Sigma d = 143$ $\Sigma d^2 = 2185$ $\bar{d} = 14.3$ $S_d = 3.9455$ $S_{\bar{d}} = 1.2477$ $t = 1.833$
$\bar{d} \pm t \times S_{\bar{d}}$ is found to be (12.01, 16.59).

TESTING HYPOTHESIS ABOUT μ_d USING NORMALLY DISTRIBUTED DIFFERENCES COMPUTED FROM DEPENDENT SAMPLES

10.27 Verify the computed test statistic in the below Minitab output for the differences in problem 10.25.

MTB > ttest mu = 0 data in c1

T-Test of the Mean
Test of mu = 0.00 vs mu not = 0.00

Variable	N	Mean	St Dev	SE Mean	T	P
diff	10	14.30	3.95	1.25	11.46	0.0000

Ans. $t^* = \dfrac{\bar{d} - D_0}{S_{\bar{d}}} = \dfrac{14.30 - 0}{1.25} = 11.44$

SAMPLING DISTRIBUTION OF $\overline{P}_1 - \overline{P}_2$ FOR LARGE INDEPENDENT SAMPLES

10.28 In a study of 100 surgery patients, 50 were kept warm with blankets after surgery and 50 were kept cool. Eight of the warm group developed wound infections and 14 of the cool group developed wound infections. Let p_1 represent the proportion of all surgery patients kept warm after surgery who develop wound infections and let p_2 represent the proportion for the cool group. The sample proportions for the two groups are $\overline{p}_1 = .16$ and $\overline{p}_2 = .28$. The difference in sample proportions, $\overline{p}_1 - \overline{p}_2$, will have a normal distribution provided that $n_1 p_1 > 5$, $n_2 p_2 > 5$, $n_1 q_1 > 5$, and $n_2 q_2 > 5$. Since the population proportions are unknown, these conditions cannot be checked directly. In practice, the conditions are checked out by substituting the sample proportions for the population proportions. Use the sample proportions to check out the requirements for assuming normality.

Ans. $n_1 \overline{p}_1 = 8$, $n_2 \overline{p}_2 = 16$, $n_1 \overline{q}_1 = 42$, and $n_2 \overline{q}_2 = 34$ $\overline{p}_1 - \overline{p}_2$ has a normal distribution.

ESTIMATION OF $P_1 - P_2$ USING LARGE SAMPLES

10.29 Refer to problem 10.28. Find a 90% confidence interval for $p_1 - p_2$.

Ans. The 90% confidence interval is $(\overline{p}_1 - \overline{p}_2) \pm z \times S_{\overline{p}_1 - \overline{p}_2}$, where $S_{\overline{p}_1 - \overline{p}_2} = \sqrt{\dfrac{\overline{p}_1 \times \overline{q}_1}{n_1} + \dfrac{\overline{p}_2 \times \overline{q}_2}{n_2}}$.

$(.16 - .28) \pm 1.65 \times .0820$ or $-.12 \pm .14$ or $(-.26 , .02)$

TESTING HYPOTHESIS ABOUT $P_1 - P_2$ USING LARGE INDEPENDENT SAMPLES

10.30 Refer to problem 10.28. Test H_0: $p_1 - p_2 = 0$ vs. H_a: $p_1 - p_2 < 0$ at $\alpha = .05$.

Ans. The pooled estimate of the common proportion is $\overline{p} = \dfrac{x_1 + x_2}{n_1 + n_2} = 0.22$. The standard error is

$S_{\overline{p}_1 - \overline{p}_2} = \sqrt{\overline{p} \times \overline{q} \left(\dfrac{1}{n_1} + \dfrac{1}{n_2} \right)} = .0828$. The computed test statistic is $z^* = \dfrac{\overline{p}_1 - \overline{p}_2}{S_{\overline{p}_1 - \overline{p}_2}} = -1.45$. The critical value is -1.65. D_0 not reject the null hypothesis.

Chapter 11

Chi-square Procedures

CHI-SQUARE DISTRIBUTION

The Chi-square procedures discussed in this chapter utilize a distribution called the Chi-square distribution. The symbol χ^2 is often used rather than the term Chi-square. The Greek letter χ is pronounced Chi. Like the t distribution, the shape of the χ^2 distribution curve is determined by the degrees of freedom (df) associated with the distribution. Figure 11-1 shows χ^2 distributions for 5, 10, and 15 degrees of freedom.

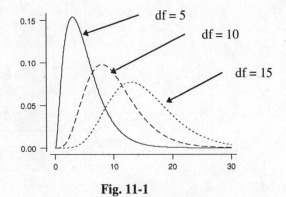

Fig. 11-1

Table 11.1 gives some of the basic properties of χ^2 distribution curves.

Table 11.1

Properties of the χ^2 Distribution
1. The total area under a χ^2 curve is equal to one.
2. A χ^2 curve starts at 0 on the horizontal axis and extends indefinitely to the right, approaching, but never touching the horizontal axis.
3. A χ^2 curve is always skewed to the right.
4. As the number of degrees of freedom becomes larger, the χ^2 curves look more and more like normal curves.
5. The mean of a χ^2 distribution is df and the variance is 2df.
6. When the degrees of freedom is 3 or more, the peak of the χ^2 curve occurs at df −2. This value is the mode of the distribution.

EXAMPLE 11.1 Table 11.2 gives the mean, mode, and standard deviation for each of the three χ^2 curves shown in Fig. 11-1. The mean, mode, and standard deviation are determined by using properties 5 and 6 from Table 11.1.

Table 11.2

df	Mean	Mode	Standard deviation
5	5	3	3.16
10	10	8	4.47
15	15	13	5.48

CHI-SQUARE TABLES

The area in the right tail under the χ^2 distribution curve for various degrees of freedom is given in the Chi-square tables found in Appendix 4. Example 11.2 illustrates how to read this table.

EXAMPLE 11.2 Table 11.3 contains the row corresponding to df = 5 from the Chi-square distribution table. Figures 11-2 and 11-3 give Chi-square curves having df = 5. The shaded area to the right of 11.070 in Fig 11-2 is .050. The shaded area to the right of 1.610 in Fig 11-3 is equal to .900. These areas and Chi-square values are shown in bold print in Table 11.3.

Table 11.3

df	\multicolumn{10}{c}{Area in the right tail under the Chi-square distribution curve}
	.995	.990	.975	.950	**.900**	.100	**.050**	.025	.010	.005
5	0.412	0.554	0.831	1.145	**1.610**	9.236	**11.070**	12.833	15.086	16.750

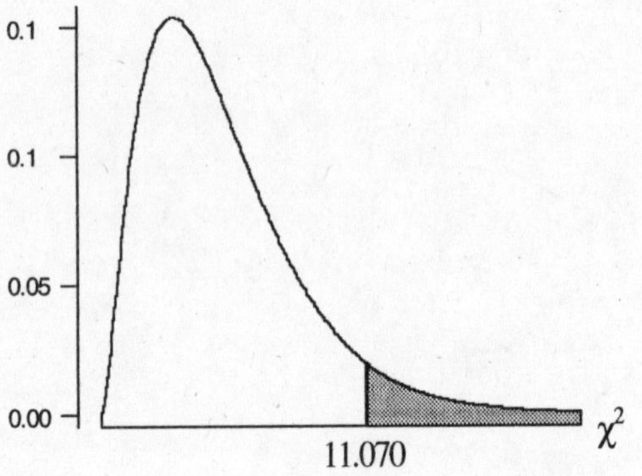

Fig. 11-2

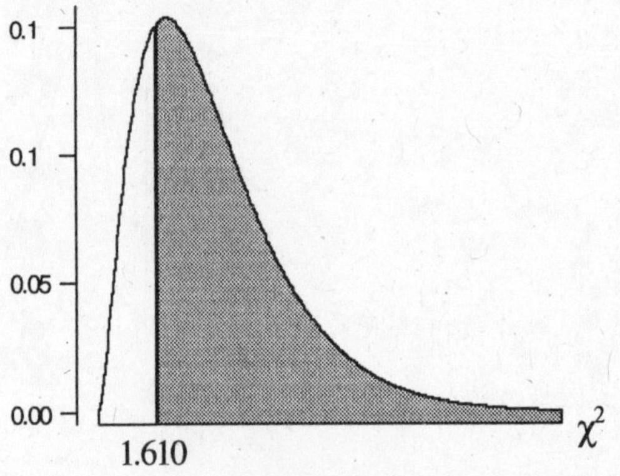

Fig. 11-3

GOODNESS-OF-FIT TEST

In many situations, each element of a population is assigned to one and only one of k categories or classes. Such a population is described by a *multinomial probability distribution*. Example 11.3 describes such a population and illustrates the structure of the null and alternative hypotheses for a *goodness-of-fit test.*

EXAMPLE 11.3 Consider the population of Americans who've dieted. A survey reported that 85% were most likely to go off their diet on the weekend, 10% were most likely to go off on a weekday, and 5% didn't know. This population is divided into three categories: category 1: most likely to go off their diet on the weekend, category 2: most likely to go off their diet on a weekday, and category 3: did not know when they were most likely to go off their diet. The multinomial probability distribution is: $p_1 = .85$, $p_2 = .10$, and $p_3 = .05$. The Delta Health fitness club is interested in whether their members follow this same multinomial probability distribution. A goodness-of-fit test is used to test the null hypothesis: H_0: $p_1 = .85$, $p_2 = .10$, and $p_3 = .05$ vs. the following alternative hypothesis: H_a: The population proportions are not $p_1 = .85$, $p_2 = .10$, and $p_3 = .05$. The probabilities p_1, p_2, and p_3 in the hypotheses statements represent the proportions for the categories as applied to the health fitness club members. The next two sections will describe the steps for performing a goodness-of-fit test.

EXAMPLE 11.4 Table 11.4 gives the age distribution of part-time college students as determined five years ago. If p_1, p_2, p_3, p_4, and p_5 represent the current percentages for the five groups, then the null hypothesis that the current distribution is the same as five years ago is stated as follows: H_0: $p_1 = .25$, $p_2 = .35$, $p_3 = .25$, $p_4 = .10$, and $p_5 = .05$. The research hypothesis is stated as: H_a: The current proportions are not as stated in the null hypothesis. A goodness-of-fit test is used to test this hypothesis system.

Table 11.4

Age	18–24	25–34	35–44	45–54	55 or over
Percent	25%	35%	25%	10%	5%

OBSERVED AND EXPECTED FREQUENCIES

The first step in performing a goodness-of-fit test is the selection of a sample of size n from the population and the determination of the *observed frequencies* for k classes. Recall that in testing hypothesis, the null hypothesis is assumed to be true, and is rejected if a highly unlikely value is obtained for the test statistic. The *expected frequencies* are computed assuming the null hypothesis to be true.

EXAMPLE 11.5 In Example 11.3, 200 members of Delta Health fitness club were surveyed and it was found that 160 were most likely to go off the diet on the weekend, 22 were most likely to go off the diet on a weekday, and 18 did not know. If the null hypothesis is true and the club members follow the nationwide distribution, then the expected numbers in the three categories are: $np_1 = 200 \times .85 = 170$, $np_2 = 200 \times .10 = 20$, and $np_3 = 200 \times .05 = 10$. The observed frequencies are 160, 22, and 18 and the expected frequencies are 170, 20, and 10.

EXAMPLE 11.6 In Example 11.4, 1500 part-time students were surveyed across the country. It was observed that 352 were in the age group 18–24, 501 were in the age group 25–34, 371 were in the age group 35–44, 126 were in the age group 45–54, and the remainder were in the age group 55 or over. If the null hypothesis is true and the distribution is the same as it was five years ago, the expected numbers in the five categories are as follows: $np_1 = 1500 \times .25 = 375$, $np_2 = 1500 \times .35 = 525$, $np_3 = 1500 \times .25 = 375$, $np_4 = 1500 \times .10 = 150$, and $np_5 = 1500 \times .05 = 75$. The observed frequencies are 352, 501, 371, 126, and 150 and the expected frequencies are 375, 525, 375, 150, and 75.

SAMPLING DISTRIBUTION OF THE GOODNESS-OF-FIT TEST STATISTIC

The goodness-of-fit test statistic is given in formula (*11.1*), where o represents an observed frequency and e represents an expected frequency, and the sum is over all k categories.

$$\chi^2 = \sum \frac{(o-e)^2}{e} \qquad (11.1)$$

The test statistic given in formula (*11.1*) has a Chi-square distribution with df = k − 1, provided all the expected frequencies are 5 or more. Some statisticians use a less restrictive requirement, namely, that all expected frequencies are at least one and that at most 20% of the expected frequencies are less than 5. We shall use the requirement that all expected frequencies are 5 or more. This requirement means that a minimum sample size is needed to use this procedure. If the observed and expected frequencies are close, then the computed value for χ^2 will be close to zero since the differences (o − e) will all be near zero. If the observed and expected values differ considerably, then the computed value of χ^2 will be large, supporting the research hypothesis. Since only large values of χ^2 indicate that the null hypothesis should be rejected and the research hypothesis supported, this is always a one-tailed test. That is, the null hypothesis is rejected only for large values of the computed test statistic. Table 11.5 summarizes the procedure for performing a goodness-of-fit test.

Table 11.5

Steps for Performing a Goodness-of-Fit Test
Step 1: State the null and research hypothesis concerning the hypothesized distribution for the k categories.
Step 2: Use the χ^2 table and the level of significance, α, to determine the rejection region.
Step 3: Compute the value of the test statistic as follows: $\chi^2 = \sum \frac{(o-e)^2}{e}$, where o represents the observed frequencies and e represents the expected frequencies. Check to make sure that all expected frequencies are 5 or more.
Step 4: State your conclusion. The null hypothesis is rejected if the computed value of the test statistic falls in the rejection region. Otherwise, the null hypothesis is not rejected.

EXAMPLE 11.7 Refer to Examples 11.3 and 11.5. The null hypothesis may be stated in either of two ways: H_0: The distribution of categories for Delta Health fitness club members is the same as the national distribution or H_0: $p_1 = .85$, $p_2 = .10$, and $p_3 = .05$. The research hypothesis may also be stated in either of two ways: H_a: The distribution of categories for Delta Health fitness club members is not the same as the national distribution or H_a: The population proportions are not $p_1 = .85$, $p_2 = .10$, and $p_3 = .05$. Table 11.6 illustrates the computation of the test statistic. The computed value of the test statistic is $\chi^2* = 7.188$.

Table 11.6

Category	o	e	o − e	$(o-e)^2$	$\frac{(o-e)^2}{e}$
1	160	170	−10	100	.588
2	22	20	2	4	.200
3	18	10	8	64	6.4
Sum	200	200	0		7.188

For level of significance $\alpha = .05$, the critical value is 5.991. The row corresponding to df = 3 − 1 = 2 from the Chi-square table in Appendix 4 is shown in Table 11.7.

Table 11.7

	Area in the right tail under the Chi-square distribution curve									
df	.995	.990	.975	.950	.900	.100	**.050**	.025	.010	.005
2	0.010	0.020	0.051	0.103	0.211	4.605	**5.991**	7.378	9.210	10.597

Since the computed test statistic exceeds 5.991, the null hypothesis is rejected. There are almost twice as many in the Don't Know category at Delta Health fitness club as would be expected if the distributions were the same.

EXAMPLE 11.8 Refer to Examples 11.4 and 11.6. The null hypothesis is H_0: $p_1 = .25$, $p_2 = .35$, $p_3 = .25$, $p_4 = .10$, and $p_5 = .05$. The research hypothesis is stated as: H_a: The current proportions are not as stated in the null hypothesis. The observed and expected frequencies are given in Example 11.6. Table 11.8 illustrates the computation of the goodness-of-fit test statistic. The computed value of the test statistic is $\chi^2* = 81.391$.

Table 11.8

Age category	o	e	o − e	$(o-e)^2$	$\frac{(o-e)^2}{e}$
18–24	352	375	−23	529	1.411
25–34	501	525	−24	576	1.097
35–44	371	375	−4	16	.043
45–54	126	150	−24	576	3.84
55 or over	150	75	75	5625	75.000
Sum	1500	1500	0		81.391

The row corresponding to df $= 5 - 1 = 4$ from the Chi-square table in Appendix 4 is shown in Table 11.9. For level of significance $\alpha = .01$, the critical value is 13.277 and is shown in bold type.

Table 11.9

df	\multicolumn{10}{c}{Area in the right tail under the Chi-square distribution curve}
	.995	.990	.975	.950	.900	.100	.050	.025	**.010**	.005
4	0.207	0.297	0.484	0.711	1.064	7.779	9.488	11.143	**13.277**	14.860

Since the computed test statistic exceeds 13.277, the null hypothesis is rejected. There appears to have been an increase in the 55 or over group from 5 years ago.

EXAMPLE 11.9 The Minitab solutions to Examples 11.7 and 11.8 are shown in Figures 11-4 and 11-5.

```
MTB > set the observed values in c1
DATA > 160  22  18
DATA > end
MTB > set the expected values in c2
DATA > 170  20  10
DATA > end
MTB > let k1 = sum((c1 – c2)**2/c2)
MTB > print k1

k1   7.18824

MTB > cdf k1 k2;
SUBC > Chisquare 2.
MTB > let k3 = 1 – k2
MTB > print k3

k3   0.0274849
```

Fig. 11-4

```
MTB > set the observed values in c1
DATA > 352  501  371  126  150
DATA > end
MTB > set the expected values in c2
DATA > 375  525  375  150  75
DATA > end
MTB > let k1 = sum((c1 – c2)**2/k2)
MTB > print k1

k1   81.3905

MTB > cdf k1 k2;
SUBC > Chisquare 4.
MTB > let k3 = 1 – k2
MTB > print k3

k3   0
```

Fig. 11-5

The upper part of both figures illustrates the computation of the test statistic for Examples 11.7 and 11.8. The computed values, 7.18824 and 81.3905, are the same as shown in Examples 11.7 and 11.8. The portion of the output shown in bold illustrates the computation of the p value for the two Examples. The p value is shown next to k3. The p value for Example 11.7 is 0.027 and the p value for Example 11.8 is 0.

CHI-SQUARE INDEPENDENCE TEST

Consider a survey of 100 males and 100 females concerning their opinion toward capital punishment. Tables 11.10 and 11.11 gives two different sets of results. In Table 11.10, the distribution of opinions is exactly the same for males and females. In this case, we say that the opinion concerning capital punishment is *independent* of the sex of the respondent.

Table 11.10

	Supports	Opposes	Undecided	Row total
Male	70	20	10	100
Female	70	20	10	100
Column total	140	40	20	200

In Table 11.11, the distribution of opinions is clearly different for males and females. In this case, we say that the opinion concerning capital punishment is *dependent* on the sex of the respondent.

Table 11.11

	Supports	Opposes	Undecided	Row total
Male	70	20	10	100
Female	40	50	10	100
Column total	110	70	20	200

Tables 11.10 and 11.11 are called *contingency tables*. In a *Chi-square independence test*, we are interested in using results such as those shown in these tables, to test for the independence of two characteristics on the elements of a population. In the above discussion, the two characteristics are sex of the individual and opinion of the individual concerning capital punishment. The conclusions regarding independence would be clear in the two tables given above. But suppose the results of the survey are not as clear cut as above. How do we decide from the results given in a contingency table if two characteristics are independent? Suppose the results of the survey were as shown in Table 11.12.

Table 11.12

	Supports	Opposes	Undecided	Row total
Male	80	15	5	100
Female	70	25	5	100
Column total	150	40	10	200

In a Chi-square test of independence, the null hypothesis is that the two characteristics are independent. The *observed frequencies* are shown in Table 11.12. A table of *expected frequencies* is also determined assuming the null hypothesis to be true. In Table 11.12, note that 150 out of 200, or 75% of those surveyed, supported capital punishment. If the two characteristics are independent, we would expect 75% of the 100 males and 75% of the 100 females to support capital punishment. From this discussion, note that if e_{11} represents the expected frequency in the first row, first column cell, then

$$e_{11} = 100 \times .75 = \frac{100 \times 150}{200} = 75 = \frac{(\text{row 1 total}) \times (\text{column 1 total})}{\text{sample size}}$$

In general, the expected frequency in row i column j is given by formula (*11.2*):

$$e_{ij} = \frac{(\text{row i total}) \times (\text{column j total})}{\text{samplesize}} \qquad (11.2)$$

Table 11.13 shows the computation of the expected frequencies for Table 11.12.

Table 11.13

	Supports	Opposes	Undecided	Row total
Male	$\frac{100 \times 150}{200} = 75$	$\frac{100 \times 40}{200} = 20$	$\frac{100 \times 10}{200} = 5$	100
Female	$\frac{100 \times 150}{200} = 75$	$\frac{100 \times 40}{200} = 20$	$\frac{100 \times 10}{200} = 5$	100
Column total	150	40	10	200

EXAMPLE 11.10 The table of expected frequencies for Table 11.10 is given in Table 11.14. Notice that when the contingency table indicates independence, the observed and expected frequencies are exactly the same.

Table 11.14

	Supports	Opposes	Undecided	Row total
Male	$\frac{100 \times 140}{200} = 70$	$\frac{100 \times 40}{200} = 20$	$\frac{100 \times 20}{200} = 10$	100
Female	$\frac{100 \times 140}{200} = 70$	$\frac{100 \times 40}{200} = 20$	$\frac{100 \times 20}{200} = 10$	100
Column total	140	40	20	200

EXAMPLE 11.11 The table of expected frequencies for Table 11.11 is given in Table 11.15. Notice that when the contingency table indicates strong dependence, that the observed and expected frequencies are very different.

Table 11.15

	Supports	Opposes	Undecided	Row total
Male	$\frac{100 \times 110}{200} = 55$	$\frac{100 \times 70}{200} = 35$	$\frac{100 \times 20}{200} = 10$	100
Female	$\frac{100 \times 110}{200} = 55$	$\frac{100 \times 70}{200} = 35$	$\frac{100 \times 20}{200} = 10$	100
Column total	110	70	20	200

How different do the observed frequencies and those you would expect when the characteristics are independent need to be before you would reject independence? The test statistic given in the next section will answer this question.

SAMPLING DISTRIBUTION OF THE TEST STATISTIC FOR THE CHI-SQUARE INDEPENDENCE TEST

Table 11.16 gives the observed frequencies and the expected frequencies in parenthesis for the data on Table 11.12. The null hypothesis is: H_0: Opinion concerning capital punishment is independent of the sex of the individual. The research hypothesis is: H_a: Opinion concerning capital punishment differs for males and females. The test statistic for testing this hypothesis is given by formula (*11.3*), where the sum is over all 6 cells. If there are r rows and c columns, there will be r × c cells in the contingency table. The test statistic has a Chi-square distribution with $(r - 1) \times (c - 1)$ degrees of freedom.

$$\chi^2 = \sum \frac{(o - e)^2}{e} \qquad (11.3)$$

Table 11.16

	Supports	Opposes	Undecided	Row total
Male	80 (75)	15 (20)	5 (5)	100
Female	70 (75)	25 (20)	5 (5)	100
Column total	150	40	10	200

The computed value of the test statistic is found as follows:

$$\chi^{2*} = \sum \frac{(o - e)^2}{e} = \frac{(80-75)^2}{75} + \frac{(15-20)^2}{20} + \frac{(5-5)^2}{5} + \frac{(70-75)^2}{75} + \frac{(25-20)^2}{20} + \frac{(5-5)^2}{5} = 3.167$$

The degrees of freedom is equal to df = $(r - 1) \times (c - 1) = (2 - 1) \times (3 - 1) = 2$. The critical value from the Chi-square distribution table is 5.991 for $\alpha = .05$. Since the computed test statistic does not exceed the critical value, the null hypothesis is not rejected. We cannot reject that the characteristics are independent. A Minitab analysis of the same data is shown below. The data are read first. The command **Chisquare c1 – c3** produces the expected frequencies, the computed value of the test statistic, and the p value. The p value is equal to 0.205.

MTB > read c1 – c3
DATA > 80 15 5
DATA > 70 25 5
DATA > end
  2 rows read.
MTB > **Chisquare c1 – c3**
Expected counts are printed below observed counts

	C1	C2	C3	Total
1	80	15	5	100
	75.00	20.00	5.00	
2	70	25	5	100
	75.00	20.00	5.00	
Total	150	40	10	200

Chi-Sq = $0.333 + 1.250 + 0.000 + 0.333 + 1.250 + 0.000 = 3.167$
DF = 2, P value = 0.205

EXAMPLE 11.12 The Minitab output for the data in Tables 11.10 and 11.11 are shown in Figures 11-6 and 11-7.

MTB > read c1 – c3
DATA > 70 20 10
DATA > 70 20 10
DATA > end
  2 rows read.
MTB > Chisquare c1 – c3

Expected counts are printed below observed counts

	C1	C2	C3	Total
1	70	20	10	100
	70.00	20.00	10.00	
2	70	20	10	100
	70.00	20.00	10.00	
Total	140	40	20	200

Chi-Sq = $0.000 + 0.000 + 0.000 + 0.000 + 0.000 +$
        $0.000 = 0.000$
DF = 2, P value = 1.000

MTB > read c1 – c3
DATA > 70 20 10
DATA > 40 50 10
DATA > end
  2 rows read.
MTB > Chisquare c1 – c3

Expected counts are printed below observed counts

	C1	C2	C3	Total
1	70	20	10	100
	55.00	35.00	10.00	
2	40	50	10	100
	55.00	35.00	10.00	
Total	110	70	20	200

Chi-Sq = $4.091 + 6.429 + 0.000 + 4.091 + 6.429 +$
        $0.000 = 21.039$
DF = 2, P value = 0.000

Fig. 11-6   **Fig. 11-7**

SAMPLING DISTRIBUTION OF THE SAMPLE VARIANCE

The population and sample variance was defined and illustrated in Chapter 3. The population variance is given by formula (11.4):

$$\sigma^2 = \frac{\sum (x - \mu)^2}{N} \qquad (11.4)$$

The sample variance is given by

$$s^2 = \frac{\sum (x - \bar{x})^2}{n-1} \qquad (11.5)$$

The concept of the sampling distribution of the sample variance is established in Example 11.13.

EXAMPLE 11.13 Consider the small finite population consisting of the times required for six individuals to open a "Child proof" aspirin bottle. The required times (in seconds) are shown in Table 11.17. The population mean is $\mu = 30$ seconds. The population variance is found as follows:

$$\sigma^2 = \frac{\sum (x - \mu)^2}{N} = \frac{400 + 100 + 0 + 0 + 100 + 400}{6} = 166.67$$

The standard deviation is $\sigma = \sqrt{166.67} = 12.9$.

Table 11.17

Individual	Required time
A	10
B	20
C	30
D	30
E	40
F	50

There are 20 different samples of size 3 possible and they are listed along with the sample variance and sampling error for each sample in Table 11.18.

Table 11.18

Sample number	Sample	Sample variance, s^2	Sampling error, $\lvert s^2 - \sigma^2 \rvert$
1	A, B, C = 10, 20, 30	100.00	66.67
2	A, B, D = 10, 20, 30	100.00	66.67
3	A, B, E = 10, 20, 40	233.48	66.81
4	A, B, F = 10, 20, 50	432.64	265.97
5	A, C, D = 10, 30, 30	133.40	33.27
6	A, C, E = 10, 30, 40	233.48	66.81
7	A, C, F = 10, 30, 50	400.00	233.33
8	A, D, E = 10, 30, 40	233.48	66.81
9	A, D, F = 10, 30, 50	400.00	233.33
10	A, E, F = 10, 40, 50	432.64	265.97
11	B, C, D = 20, 30, 30	33.29	133.38
12	B, C, E = 20, 30, 40	100.00	66.67
13	B, C, F = 20, 30, 50	233.48	66.81
14	B, D, E = 20, 30, 40	100.00	66.67
15	B, D, F = 20, 30, 50	233.48	66.81
16	B, E, F = 20, 40, 50	233.48	66.81
17	C, D, E = 30, 30, 40	33.29	133.38
18	C, D, F = 30, 30, 50	133.40	33.27
19	C, E, F = 30, 40, 50	100.00	66.67
20	D, E, F = 30, 40, 50	100.00	66.67

The distribution of the sample variance is obtained from Table 11.18 and is given in Table 11.19.

Table 11.19

s^2	33.29	100.0	133.40	233.48	400.00	432.64
$P(s^2)$	.1	.3	.1	.3	.1	.1

The above procedure may be used to find the sampling distribution of the sample variance when sampling from a finite population. However, it is clear that it is a tedious procedure even when using a computer. When sampling from an infinite population, the result given in Table 11.20 is used to determine the sampling distribution for a function of the sample variance, namely, $\dfrac{(n-1)S^2}{\sigma^2}$. The proof of this result is beyond the scope of this text. In the next section, this result is utilized to set a confidence interval on σ^2 as well as test hypotheses about σ^2.

Table 11.20

Sampling Distribution of $\dfrac{(n-1)S^2}{\sigma^2}$
When a simple random sample of size n is selected from a normally distributed population having population variance, σ^2, $\dfrac{(n-1)S^2}{\sigma^2}$ has a Chi-square distribution with (n – 1) degrees of freedom, where S^2 is the sample variance.

INFERENCES CONCERNING THE POPULATION VARIANCE

The result given in Table 11.20 will be used to find confidence intervals and test hypotheses about a population variance or standard deviation.

EXAMPLE 11.14 It is important that drug manufacturers control the variation of the dosage in their products. A drug company produces 250 milligram tablets of the antibiotic *amoxicillin*. A sample of size 15 is selected from the production process and the level of *amoxicillin* is determined for each tablet. The results are shown in Table 11.21.

Table 11.21

249.995	249.985	250.000
249.990	249.980	250.000
250.010	250.000	249.999
250.005	250.015	250.001
250.000	250.010	250.000

The mean of the sample is $\bar{x} = 250.000$ mg and the sample variance is $s^2 = .00008538$ mg^2. According to Table 11.20, $\dfrac{(n-1)S^2}{\sigma^2} = \dfrac{14S^2}{\sigma^2}$ has a Chi-square distribution with df = 14. Table 11.22 gives the row corresponding to df = 14 from the Chi-square table in Appendix 4.

Table 11.22

df	\multicolumn{10}{c}{Area in the right tail under the Chi-square distribution curve}
	.995	.990	**.975**	.950	.900	.100	.050	**.025**	.010	.005
14	4.075	4.660	**5.629**	6.571	7.790	21.064	23.685	**26.119**	29.141	31.319

Since $\dfrac{14S^2}{\sigma^2}$ has a Chi-square distribution with df = 14, we know that $\dfrac{14S^2}{\sigma^2}$ is between 5.629 and 26.119 with probability .95. This is true because according to Table 11.22, there is 97.5% of the area under the curve to the right of 5.629 and 2.5% of the area under the curve to the right of 26.119, and therefore 95% of the area under the curve is between 5.629 and 26.119. This may be expressed as follows:

$$P\left(5.629 < \frac{14S^2}{\sigma^2} < 26.119\right) = .95$$

The following notation is often used for the tabled Chi-square values: $\chi^2_{.975} = 5.629$ and $\chi^2_{.025} = 26.119$. The probability statement may be expressed as follows:

$$P\left(\chi^2_{.975} < \frac{14S^2}{\sigma^2} < \chi^2_{.025}\right) = .95$$

Now, the if inequality $\chi^2_{.975} < \dfrac{14S^2}{\sigma^2} < \chi^2_{.025}$ is solved for σ^2, we obtain $\dfrac{14S^2}{\chi^2_{.025}} < \sigma^2 < \dfrac{14S^2}{\chi^2_{.975}}$ as a 95% confidence interval for σ^2. The numerical confidence interval is obtained by replacing S^2 by .00008538 mg^2 and using the values obtained from the Chi-square table. The lower confidence limit is:

$$\frac{14S^2}{\chi^2_{.025}} = \frac{14 \times .00008538}{26.119} = .00004576$$

The upper confidence interval is:

$$\frac{14S^2}{\chi^2_{.975}} = \frac{14 \times .00008538}{5.629} = .0002124$$

The 95% confidence interval for σ^2 goes from .00004576 to .0002124. The 95% confidence interval for σ is obtained by taking the square root of the limits for σ^2. The lower limit is $\sqrt{.00004576} = .006765$ and the upper limit is $\sqrt{.0002124} = .014574$. The 95% confidence interval for the standard deviation (to three decimal places) is (.007, .015).

The general form for a $(1 - \alpha) \times 100\%$ confidence interval for the population variance is given by formula (11.6), where the values of χ^2 are based on the Chi-square distribution with df = n – 1.

$$\frac{(n-1)S^2}{\chi^2_{\alpha/2}} < \sigma^2 < \frac{(n-1)S^2}{\chi^2_{1-\alpha/2}} \tag{11.6}$$

The $(1 - \alpha) \times 100\%$ confidence interval for the population standard deviation is given by formula (11.7).

$$\sqrt{\frac{(n-1)S^2}{\chi^2_{\alpha/2}}} < \sigma < \sqrt{\frac{(n-1)S^2}{\chi^2_{1-\alpha/2}}} \tag{11.7}$$

Both of the confidence intervals assume that the sample was obtained from a normally distributed population.

EXAMPLE 11.15 A random sample of the lengths of bolts produced by Fastners, Inc. was taken. The sample results were as follows: n = 30, $\overline{x}$ = 5.000 cm, and s = 0.055 cm. A 99% confidence interval for σ is determined as follows: For a 99% confidence interval, $1 - \alpha = .99$, and therefore $\alpha = .01$, or $\dfrac{\alpha}{2} = .005$ and $1 - \dfrac{\alpha}{2} = .995$. The degrees of freedom is df = n − 1 = 30 − 1 = 29. Table 11.23 gives that portion of the Chi-square table needed to determine the values for use in formula (*11.7*). The following values are shown in bold print in the table: $\chi^2_{.995} = 13.121$ and $\chi^2_{.005} = 52.336$. The sample variance is $s^2 = (0.055)^2 = .003025$. Substituting into formula (*11.7*) we obtain the following 99% confidence interval for the population standard deviation.

$$\sqrt{\frac{(n-1)S^2}{\chi^2_{\alpha/2}}} < \sigma < \sqrt{\frac{(n-1)S^2}{\chi^2_{1-\alpha/2}}} \text{ or } \sqrt{\frac{29 \times .003025}{52.336}} < \sigma < \sqrt{\frac{29 \times .003025}{13.121}} \text{ or } .041 < \sigma < .082$$

The population standard deviation of bolt lengths is between .041 cm and .082 cm with 99% confidence.

Table 11.23

df	Area in the right tail under the Chi-square distribution curve									
	.995	.990	.975	.950	.900	.100	.050	.025	.010	**.005**
29	**13.121**	14.256	16.047	17.708	19.768	39.087	42.557	45.722	49.588	**52.336**

The sampling distribution of $\dfrac{(n-1)S^2}{\sigma^2}$, given in Table 11.20, is also utilized to test hypotheses concerning the population variance or standard deviation. The steps for testing a population variance are given in Table 11.24.

Table 11.24

Steps for Testing a Hypothesis Concerning σ^2: Sampling from a Normal Population
Step 1: State the null and research hypothesis. The null hypothesis is represented symbolically by H_0: $\sigma^2 = \sigma_0^2$ and the research hypothesis is of the form H_a: $\sigma^2 \neq \sigma_0^2$ or H_a: $\sigma^2 < \sigma_0^2$ or H_a: $\sigma^2 > \sigma_0^2$.
Step 2: For level of significance α, the critical values for H_a: $\sigma^2 \neq \sigma_0^2$ are $\chi^2_{1-\alpha/2}$ and $\chi^2_{\alpha/2}$, the critical value for H_a: $\sigma^2 < \sigma_0^2$ is $\chi^2_{1-\alpha}$, and the critical value for H_a: $\sigma^2 > \sigma_0^2$ is χ^2_α.
Step 3: Compute the value of the test statistic as follows: $\chi^{2*} = \dfrac{(n-1)S^2}{\sigma_0^2}$, where σ_0^2 is given in the null hypothesis and S^2 is computed from your sample.
Step 4: State your conclusion. The null hypothesis is rejected if the computed value of the test statistic falls in the rejection region. Otherwise, the null hypothesis is not rejected.

EXAMPLE 11.16 The ratio of potassium to sodium in an individual's diet is sometimes referred to as the K factor. A sample of 15 Yanomamo Indians from Brazil was obtained and their K factors were determined. The standard deviation of the sample of K factor values was found to equal 0.15. These sample results were used to test the hypothesis H_0: $\sigma = 0.25$ vs. H_a: $\sigma \neq 0.25$ at level of significance $\alpha = .05$. The critical values are shown in bold in Table 11.25. From this table, we have $\chi^2_{.975} = 5.629$ and $\chi^2_{.025} = 26.119$. The computed value of the test statistic is:

$$\chi^{2*} = \frac{(n-1)S^2}{\sigma_0^2} = \frac{14 \times .15^2}{.25^2} = \frac{14 \times .0225}{.0625} = 5.04$$

Since the computed value of the test statistic is less than 5.629, the null hypothesis is rejected and it is concluded that the standard deviation for Yanomamo Indians is less than 0.25.

Table 11.25

Df	\~ Area in the right tail under the Chi-square distribution curve									
	.995	.990	**.975**	.950	.900	.100	.050	**.025**	.010	.005
14	4.075	4.660	**5.629**	6.571	7.790	21.064	23.685	**26.119**	29.141	31.319

Solved Problems

CHI-SQUARE DISTRIBUTION

11.1 What happens to the peak of the Chi-square curve as the degrees of freedom is increased?

Ans.   The peak of the Chi-square curve shifts to the right as the degrees of freedom is increased.

11.2 The area under the Chi-square curve corresponding to χ^2 values greater than 118.498 is .10. Find $P(0 < \chi^2 < 118.498)$.

Ans.   Since the total area under the Chi-square curve is equal to 1, $P(0 < \chi^2 < 118.498)$ is equal to $1 - P(\chi^2 > 118.498) = 1 - .10 = .90$.

CHI-SQUARE TABLES

11.3 Find the value of χ^2 for 5 degrees of freedom and
(*a*) .005 area in the right tail of the Chi-square distribution curve.
(*b*) .005 area in the left tail of the Chi-square distribution curve.

Ans.   Table 11.26 gives the row corresponding to df = 5 from the Chi-square table. (*a*) The table indicates that the area to the right of 16.750 is .005. (*b*) The table indicates that the area to the right of 0.412 is .995. Therefore, the area to the left of 0.412 is 1 – .995 = .005.

Table 11.26

Df	Area in the right tail under the Chi-square distribution curve									
	.995	.990	.975	.950	.900	.100	.050	.025	.010	**.005**
5	**0.412**	0.554	0.831	1.145	1.610	9.236	11.070	12.833	15.086	**16.750**

11.4 Table 11.27 gives the row corresponding to df = 10 from the Chi-square table. Find the area under the Chi-square distribution curve having df = 10 corresponding to χ^2 between 4.865 and 23.209.

Table 11.27

Df	Area in the right tail under the Chi-square distribution curve									
	.995	.990	.975	.950	**.900**	.100	.050	.025	**.010**	.005
10	2.156	2.558	3.247	3.940	**4.865**	15.987	18.307	20.483	**23.209**	25.188

Ans.   According to Table 11.27, the area to the right of 23.209 is .010 and the area to the right of 4.865 is .900. In Fig. 11-8, the shaded area equals .010 and corresponds to χ^2 values exceeding 23.209. The shaded area in Fig. 11-9 equals .900 and corresponds to χ^2 exceeding 4.865. The area corresponding to χ^2 between 4.865 and 23.209 is the difference in the two shaded areas, or .900 – .010 = .890.

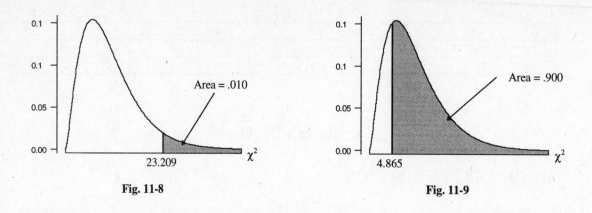

Fig. 11-8   Fig. 11-9

GOODNESS-OF-FIT TEST

11.5 A magazine reported that 75% of the population oppose same sex marriages, 20% approve, and 5% are undecided. A survey is conducted to test the research hypothesis that the distribution is different from that reported by the magazine. State the null and alternative hypotheses in terms of p_1, p_2, and p_3 where p_1 = the population proportion opposed to same sex marriage, p_2 = the population proportion who approve of same sex marriage, and p_3 = the proportion who are undecided.

Ans. The null hypothesis is H_0: The population proportions are $p_1 = .75$, $p_2 = .20$, and $p_3 = .05$. The research hypothesis is H_a: The population proportions are not $p_1 = .75$, $p_2 = .20$, and $p_3 = .05$.

11.6 The fairness of a die may be tested as a goodness-of-fit test. Let p_i represent the proportion of the time that face i turns up when the die is tossed, where i is 1, 2, 3, 4, 5, or 6. If the null hypothesis is that the die is fair and the research hypothesis is that the die is unfair, state H_0 and H_a in terms of the p_i.

Ans. The null hypothesis is H_0: $p_1 = p_2 = p_3 = p_4 = p_5 = p_6 = \frac{1}{6}$. The research hypothesis is H_a: Not all p_i equal $\frac{1}{6}$.

OBSERVED AND EXPECTED FREQUENCIES

11.7 Refer to problem 11.5. A poll was conducted and 585 opposed same sex marriage, 195 approved, 120 were undecided. What results would you expect if the null hypothesis were true?

Ans. The sample size is n = 900. The expected frequencies are: $e_1 = np_1 = 900 \times .75 = 675$, $e_2 = np_2 = 900 \times .20 = 180$, and $e_3 = np_3 = 900 \times .05 = 45$.

11.8 Refer to problem 11.6. The die in question was rolled and face 1 turned up 89 times, face 2 turned up 93 times, face 3 turned up 103 times, face 4 turned up 111 times, face 5 turned up 100 times, and face 6 turned up 104 times. What results would you expect if the null hypothesis is true and the die is fair?

Ans. The sample size is n = 600. If the null hypothesis is true, then each face should occur 100 times in 600 rolls.

SAMPLING DISTRIBUTION OF THE GOODNESS-OF-FIT TEST STATISTIC

11.9 Refer to problems 11.5 and 11.7. Test the hypothesis at $\alpha = .01$.

Ans. The number of categories is k = 3, and the degrees of freedom is df = k − 1 = 2. By referring to the Chi-square distribution table in Appendix 4, we find that the critical value for a 1% level of significance is equal to 9.210. The null hypothesis will be rejected if the computed value of the test statistic exceeds this critical value.

The computation of the test statistic is shown in Table 11.28. Since the computed test statistic is 138.250, the null hypothesis is rejected. There is a much higher number in the undecided group than would be expected if the null hypothesis were true.

Table 11.28

Category	o	e	o − e	$(o-e)^2$	$\frac{(o-e)^2}{e}$
Oppose	585	675	−90	8100	12.000
Support	195	180	15	225	1.250
Undecided	120	45	75	5625	125.000
Sum	900	900	0		138.250

11.10 Refer to problems 11.6 and 11.8. Test the hypothesis at $\alpha = .05$.

Ans. The number of categories is k = 6, and the degrees of freedom is df = k − 1 = 5. Table 11.29 gives the row corresponding to df = 5 from the Chi-square distribution table in Appendix 4. The critical value is 11.070, as shown in bold print.

Table 11.29

	Area in the right tail under the Chi-square distribution curve									
Df	.995	.990	.975	.950	.900	.100	**.050**	.025	.010	.005
5	0.412	0.554	0.831	1.145	1.610	9.236	**11.070**	12.833	15.086	16.750

The computation of the test statistic is shown in Table 11.30. The computed value of the test statistic is 3.16. Based on this experiment, there is no reason to doubt the fairness of the die.

Table 11.30

Turned up face	o	e	o − e	$(o-e)^2$	$\frac{(o-e)^2}{e}$
1	89	100	−11	121	1.21
2	93	100	−7	49	0.49
3	103	100	3	9	0.09
4	111	100	11	121	1.21
5	100	100	0	0	0
6	104	100	4	16	0.16
sum	600	600	0		3.16

CHI-SQUARE INDEPENDENCE TEST

11.11 A study involving several cities from across the country involving crime was conducted. The cities were divided into the categories South, Northeast, North Central, and West. Based on interviews, crime was classified as a major concern, a minor concern, or of no concern for each individual interviewed. The results are shown in Table 11.31. If the level of concern regarding crime is independent of the section of the country, find the expected frequencies for the 12 cells.

Table 11.31

Concern	South	Northeast	North Central	West
Major	25	40	35	40
Minor	65	45	40	40
No concern	15	10	15	20

Ans. The computations of the expected frequencies are shown in Table 11.32.

Table 11.32

Concern	South	Northeast	North Central	West	Row total
Major	$\frac{140\times105}{390}=37.69$	$\frac{140\times95}{390}=34.10$	$\frac{140\times90}{390}=32.31$	$\frac{140\times100}{390}=35.90$	140
Minor	$\frac{190\times105}{390}=51.15$	$\frac{190\times95}{390}=46.28$	$\frac{190\times90}{390}=43.85$	$\frac{190\times100}{390}=48.72$	190
No concern	$\frac{60\times105}{390}=16.15$	$\frac{60\times95}{390}=14.62$	$\frac{60\times90}{390}=13.85$	$\frac{60\times100}{390}=15.38$	60
Column total	105	95	90	100	390

11.12 Table 11.33 gives the results of a study involving marital status and net worth in thousands. Find the expected frequencies assuming the two characteristics are independent. Comment on the expected frequencies.

Table 11.33

Net worth	Married	Single	Widowed
100 to 249	225	50	65
250 to 499	60	20	25
500 to 999	15	4	3
1000 or more	10	2	3

Ans. The computations of the expected frequencies are shown in Table 11.34.

Table 11.34

Net worth	Married	Single	Widowed	Row total
100 – 249	$\frac{340\times310}{482}=218.67$	$\frac{340\times76}{482}=53.61$	$\frac{340\times96}{482}=67.72$	340
250 – 499	$\frac{105\times310}{482}=67.53$	$\frac{105\times76}{482}=16.56$	$\frac{105\times96}{482}=20.91$	105
500 – 999	$\frac{22\times310}{482}=14.15$	$\frac{22\times76}{482}=3.47$	$\frac{22\times96}{482}=4.38$	22
1000 or more	$\frac{15\times310}{482}=9.65$	$\frac{15\times76}{482}=2.37$	$\frac{15\times96}{482}=2.99$	15
Column total	310	76	96	482

Note that 4 of the 15, or 27%, of the expected frequencies are less than 5. The Chi-square test of independence is not appropriate.

SAMPLING DISTRIBUTION OF THE TEST STATISTIC FOR THE CHI-SQUARE INDEPENDENCE TEST

11.13 Use Minitab to test the hypothesis that opinions regarding crime is independent of the section of the country in problem 11.11 at level of significance $\alpha = .05$. Compare the expected frequencies in the Minitab output with those computed in problem 11.11.

Row	C1	C2	C3	C4
1	25	40	35	40
2	65	45	40	40
3	15	10	15	20

MTB > chisquare c1–c4

Expected counts are printed below observed counts.

	C1	C2	C3	C4	Total
1	25	40	35	40	140
	37.69	34.10	32.31	35.90	
2	65	45	40	40	190
	51.15	46.28	43.85	48.72	
3	15	10	15	20	60
	16.15	14.62	13.85	15.38	
Total	105	95	90	100	390

Chi-Sq = 4.274 + 1.020 + 0.224 + 0.469 + 3.748 + 0.036 + 0.337 + 1.560 + 0.082 + 1.457 + 0.096 + 1.385 = 14.688
DF = 6, P value = 0.023

Ans. Since the computed p value = 0.023 is less than $\alpha = .05$, the null hypothesis is rejected. The expected frequencies are the same as computed in problem 11.11.

11.14 Refer to problem 11.12. Combine the categories Single and Widowed into a new category called Nonmarried so that all cell expected frequencies exceed 5 and perform a test of independence.

Ans. Table 11.35 gives the new results after combining categories.

Table 11.35

Net worth	Married	Not Married
100 to 249	225	115
250 to 499	60	45
500 to 999	15	7
1000 or more	10	5

The Minitab analysis of the test of independence is as follows:

Data Display

Row	C1	C2
1	225	115
2	60	45
3	15	7
4	10	5

MTB > Chisquare c1 c2

Chi-Square Test
Expected counts are printed below observed counts

	C1	C2	Total
1	225	115	340
	218.67	121.33	
2	60	45	105
	67.53	37.47	
3	15	7	22
	14.15	7.85	
4	10	5	15
	9.65	5.35	
Total	310	172	482

Chi-Sq = $0.183 + 0.330 + 0.840 + 1.514 + 0.051 + 0.092 + 0.013 + 0.023 = 3.046$
DF = 3, P value = 0.385

SAMPLING DISTRIBUTION OF THE SAMPLE VARIANCE

11.15 A small population consists of the three values 10, 20, and 30. Determine the population variance and the sampling distribution of s^2 for samples of size 2.

Ans. The population mean is 20. The population variance is:

$$\sigma^2 = \frac{\Sigma(x-\mu)^2}{N} = \frac{(10-20)^2+(20-20)^2+(30-20)^2}{3} = 66.67$$

The three samples and their variances are: 10 and 20, $s^2 = 50$; 10 and 30, $s^2 = 200$; 20 and 30, $s^2 = 50$. The sampling distribution for s^2 is: $P(s^2 = 50) = \frac{2}{3}$, $P(s^2 = 200) = \frac{1}{3}$.

11.16 A sample of size 15 is taken from a normal population having $\sigma^2 = 14$. What is the probability that s^2 exceeds 21.064?

Ans. The statistic $\dfrac{(n-1)S^2}{\sigma^2} = \dfrac{14 s^2}{14} = s^2$ has a Chi-square distribution with 14 degrees of freedom. From the Chi-square table, the area to the right of 21.064 is 0.10 when df = 14. Therefore $P(s^2 > 21.064) = 0.10$.

INFERENCES CONCERNING THE POPULATION VARIANCE

11.17 The standard deviation for the annual medical costs of 30 heart disease patients was found to equal $1005. Find a 95% confidence interval for the standard deviation of the annual medical costs of all heart disease patients.

Ans. The general form of the confidence interval for the population variance is
$$\frac{(n-1)S^2}{\chi^2_{\alpha/2}} < \sigma^2 < \frac{(n-1)S^2}{\chi^2_{1-\alpha/2}}.$$

The level of confidence is $1 - \alpha = .95$. This implies that $\alpha = .05$, $\dfrac{\alpha}{2} = .025$, and $1 - \dfrac{\alpha}{2} = .975$.

The Chi-square table values are $\chi^2_{.025} = 45.722$ and $\chi^2_{.975} = 16.047$, for df = 29.

The lower limit for the 95% confidence interval for σ^2 is

$$\frac{(n-1)S^2}{\chi^2_{\alpha/2}} = \frac{29 \times 1,010,025}{45.722} = 640,626.5037.$$

The upper limit for the 95% confidence interval for σ^2 is

$$\frac{(n-1)S^2}{\chi^2_{1-\alpha/2}} = \frac{29 \times 1,010,025}{16.047} = 1,825,308.469.$$

The lower limit for the 95% confidence interval for σ is $\sqrt{640,626.5037} = \$800.39$.

The upper limit for the 95% confidence interval for σ is $\sqrt{1,825,308.469} = \$1,351.04$.

11.18 In manufacturing processes, it is desirable to control the variability of mass produced items. A machine fills containers with 5.68 liters of bleach. The variability in fills is acceptable provided the standard deviation is 0.15 liters or less. A sample of 10 containers is chosen to test the research hypothesis that the standard deviation exceeds 0.15 liters at level of significance $\alpha = .05$. What conclusion is made if the variance of the sample is found to equal 0.095 liters?

Ans. The null hypothesis is H_0: $\sigma = 0.15$ and the research hypothesis is H_a: $\sigma > 0.15$.

The degrees of freedom is df = $10 - 1 = 9$, and the critical value is $\chi^2_{.05} = 16.919$.

The computed value of the test statistic is $\chi^{2}* = \dfrac{(n-1)S^2}{\sigma_0^2} = \dfrac{9 \times .095}{.0225} = 38$.

Based on this sample, it is concluded that the variability is unacceptable and the filling machine needs to be checked.

Supplementary Problems

CHI-SQUARE DISTRIBUTION

11.19 Which one of the following terms best describes the Chi-square distribution curve: right-skewed, left skewed, symmetric, or uniform?

Ans. right-skewed

11.20 According to Chebyshev's theorem, at least 75% of any distribution will fall within 2 standard deviations of the mean. For a Chi-square distribution having 8 degrees of freedom, find two values a and b such that at least 75% of the area under the distribution curve will be between those values.

Ans. a = $8 - 2 \times 4 = 0$, b = $8 + 2 \times 4 = 16$

CHI-SQUARE TABLES

11.21 A Chi-square distribution has 11 degrees of freedom. Find the χ^2 values corresponding to the following right-hand tail areas.
(a) .005   (b) .975   (c) .990   (d) .025

Ans. (a) 26.757   (b) 3.816   (c) 3.053   (d) 21.920

11.22 A Chi-square distribution has 17 degrees of freedom. Find the left-hand tail area corresponding to the following values.
(a) 30.191   (b) 6.408   (c) 33.409   (d) 24.769

Ans. (a) .975   (b) .010   (c) .990   (d) .900

GOODNESS-OF-FIT TEST

11.23 A national survey found that among married Americans age 40 to 65 with household income above $50,000, 45% planned to work after retirement age, 45% planned not to work after retirement age, and 10% were not sure. A similar survey was conducted in Nebraska. Let p_1 represent the percent in Nebraska who plan to work after retirement age, p_2 represent the percent in Nebraska that plan not to work after retirement age, and p_3 represent the percent not sure. What null and research hypotheses would be tested to compare the multinomial probability distribution in Nebraska with the national distribution?

Ans. H_0: $p_1 = .45$, $p_2 = .45$, and $p_3 = .10$; H_a: The proportions are not as stated in the null hypothesis.

11.24 A sociological study concerning impaired coworkers was conducted. It was reported that 30% of American workers have worked with someone whose alcohol or drug use affected his/her job, 60% have not worked with such individuals, and 10% did not know. The Central Intelligence Agency (CIA) conducted a similar study of CIA employees. Let p_1 represent the proportion of CIA employees who have worked with someone whose alcohol or drug use affected his/her job, p_2 represent the proportion of CIA employees who have not worked with such coworkers, and p_3 represent the proportion who do not know. What null and research hypothesis would you test to determine if the CIA multinomial distribution was the same as that reported in the study?

Ans. H_0: $p_1 = .30$, $p_2 = .60$, and $p_3 = .10$; H_a: The proportions are not as stated in the null hypothesis.

OBSERVED AND EXPECTED FREQUENCIES

11.25 Refer to problem 11.23. The Nebraska survey found that 120 planned to work after retirement age, 160 did not plan to work after retirement age, and 35 were not sure. What are the expected frequencies?

Ans. $e_1 = 141.75$, $e_2 = 141.75$, and $e_3 = 31.5$

11.26 Refer to problem 11.24. The CIA study found that 37 had worked with someone whose alcohol or drug use had affected their job, 58 had not, and 17 did not know. What are the expected frequencies?

Ans. $e_1 = 33.6$, $e_2 = 67.2$, and $e_3 = 11.2$

SAMPLING DISTRIBUTION OF THE GOODNESS-OF-FIT TEST STATISTIC

11.27 Refer to problems 11.23 and 11.25. Perform the test at $\alpha = .01$.

Ans. The critical value is 9.21. The computed value of the test statistic is $\chi^2* = 6.076$. It cannot be concluded that the Nebraska distribution is different from the national distribution.

11.28 Refer to problems 11.24 and 11.26. Perform the test at $\alpha = .05$.

Ans. The critical value is 5.991. The computed value of the test statistic is $\chi^2* = 4.607$. It cannot be concluded that the CIA distribution is different from that reported in the sociological study.

CHI-SQUARE INDEPENDENCE TEST

11.29 Five hundred arrest records were randomly selected and the records were categorized according to age and type of violent crime. The results are shown in Table 11.36.

Table 11.36

Crime type	18–25	26–33	34–40	over 40
Murder	11	15	4	2
Rape	21	26	11	3
Robbery	120	85	3	2
Assault	147	42	5	3

Find the table of expected frequencies assuming independence, and comment on performing the Chi-square test of independence.

Ans. The expected frequencies are given in Table 11.37. The Chi-square test of independence is inappropriate because 37.5% of the expected frequencies are less than 5 and one is less than 1.

Table 11.37

Crime type	18–25	26–33	34–40	over 40
Murder	19.14	10.75	1.47	0.64
Rape	36.48	20.50	2.81	1.22
Robbery	125.58	70.56	9.66	4.20
Assault	117.81	66.19	9.06	3.94

11.30 Table 11.38 gives various cancer sites and smoking history for participants in a cancer study.

Table 11.38

Cancer site	Smoking history			
	Never	Light	Medium	Heavy
Lung	15	35	45	100
Oral-bladder	35	40	40	35
Other cancer	250	135	190	150
No cancer	2550	950	830	750

Find the expected frequencies, assuming that cancer site is independent of smoking history.

Ans. The expected frequencies are shown in Table 11.39.

Table 11.39

Cancer site	Smoking history			
	Never	Light	Medium	Heavy
Lung	90.37	36.78	35.04	32.82
Oral-bladder	69.51	28.29	26.95	25.24
Other cancer	335.98	136.75	130.26	122.01
No cancer	2354.15	958.18	912.75	854.93

SAMPLING DISTRIBUTION OF THE TEST STATISTIC FOR THE CHI-SQUARE INDEPENDENCE TEST

11.31 Combine the categories "34–40" and "Over 40" into a new category called "Over 33" and perform the Chi-square independence test for problem 11.29.

Ans. The new contingency table is given in Table 11.40. The expected frequencies are shown in parentheses.

Table 11.40

Crime type	18–25	26–33	Over 33
Murder	11 (19.14)	15 (10.75)	6 (2.11)
Rape	21 (36.48)	26 (20.50)	14 (4.03)
Robbery	120 (125.58)	85 (70.56)	5 (13.86)
Assault	147 (117.81)	42 (66.19)	8 (13.00)

Two of the cells have expected frequencies less than 5. Some statisticians would not perform the independence test because two of the expected frequencies are less than 5. The less restrictive rule states that the test is valid if all expected frequencies are at least one and at most 20% of the expected frequencies are less than one. The computed value of the test statistic is 71.918 and the p value is .000. The type of crime is most likely dependent on the age group.

11.32 Compute the value of the test statistic for the test of independence in problem 11.30 and test the hypothesis of independence at $\alpha = .01$.

Ans. The degrees of freedom is df = $(r - 1) \times (c - 1) = (4 - 1) \times (4 - 1) = 9$. The critical value is 21.666. The computed value of the test statistic is $\chi^2* = 327.960$. The cancer site is dependent upon the smoking history of an individual.

SAMPLING DISTRIBUTION OF THE SAMPLE VARIANCE

11.33 Fill in the following parentheses with s or σ.
(*a*) ( ) is a constant, but ( ) is a variable.
(*b*) ( ) has a distribution, but ( ) does not have a distribution.
(*c*) ( ) is a parameter and ( ) is a statistic.
(*d*) ( ) describes the variability of some characteristic for a sample, whereas ( ) describes the variability of a characteristic for a population.

Ans. (*a*) σ, s   (*b*) s, σ (*c*) σ, s   (*d*) s, σ

11.34 What distributional assumption is necessary in order that $\dfrac{(n-1)S^2}{\sigma^2}$ have a Chi-square distribution with $(n - 1)$ degrees of freedom?

Ans. The random sample from which S^2 is computed is assumed to be selected from a normal distribution.

INFERENCES CONCERNING THE POPULATION VARIANCE

11.35 Table 11.41 gives the monthly rent for 30 randomly selected 2-bedroom apartments in good condition from the state of New York. Find a 95% confidence interval for the standard deviation of all monthly rents in New York for 2-bedroom apartments in good condition.

Table 11.41

850	1100	975	675	990
750	1150	700	780	1100
1000	1025	1025	900	1250
1250	975	1150	1125	980
800	1050	1200	1500	1040
950	1350	800	1255	1350

Ans. The sample standard deviation is equal to $203.24.

The Chi-square table values are $\chi^2_{.025} = 45.722$ and $\chi^2_{.975} = 16.047$ for 29 degrees of freedom.

The 95% confidence interval extends from $161.86 to $273.22.

11.36 Table 11.42 gives the ages of 30 randomly selected airline pilots. Use the data to test the null hypothesis that the standard deviation of airline pilots is equal to 10 years vs. the alternative that the standard deviation is not equal to 10 years. Use level of significance $\alpha = .01$.

Ans. The standard deviation of the sample is 8.83 years. The critical values are 13.121 and 52.336. The computed value of the test statistic is 22.61. The sample data does not contradict the null hypothesis.

Table 11.42

45	55	58	38	44
50	54	57	44	40
45	29	47	37	38
28	44	46	35	56
30	42	49	57	34
32	41	52	52	47

Chapter 12

Analysis of Variance (ANOVA)

F DISTRIBUTION

Analysis of Variance (ANOVA) procedures are used to compare the means of several populations. ANOVA procedures might be used to answer the following questions: Do differences exist in the mean levels of repetitive stress injuries (RSI) for four different keyboard designs? Is there a difference in the mean amount awarded for age bias suits, race bias suits, and sex bias suits? Is there a difference in the mean amount spent on vacations for the three minority groups: Asians, Hispanics, and African Americans?

ANOVA procedures utilize a distribution called the *F distribution*. A given F distribution has two separate degrees of freedom, represented by df_1 and df_2. The first, df_1, is called the *degrees of freedom for the numerator* and the second, df_2, is called *degrees of freedom for the denominator*. For an F distribution with 5 degrees of freedom for the numerator and 10 degrees of freedom for the numerator, we write $df = (df_1, df_2) = (5, 10)$. Figure 12-1 shows one F distribution with $df = (10, 50)$ and another with $df = (10, 5)$.

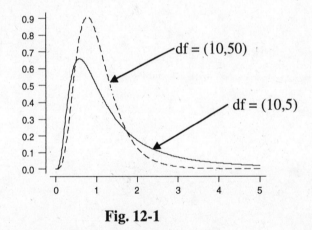

Fig. 12-1

Table 12.1 gives the basic properties of the F distribution.

Table 12.1

Properties of the F distribution
1. The total area under the F distribution curve is equal to one.
2. An F distribution curve starts at 0 on the horizontal axis and extends indefinitely to the right, approaching but never touching the horizontal axis.
3. An F distribution curve is always skewed to the right.
4. The mean of the F distribution is $\mu = \dfrac{df_2}{df_2 - 2}$ for $df_2 > 2$, where df_2 is the degrees of freedom for the denominator.

EXAMPLE 12.1 The F distribution with df = (10, 5) shown in Fig. 12-1 has mean equal to $\mu = \dfrac{df_2}{df_2 - 2} =$ $\dfrac{5}{5-2} = 1.67$, and the F distribution shown in the same figure with df = (10, 50) has mean $\mu = \dfrac{df_2}{df_2 - 2} = \dfrac{50}{50-2} =$ 1.04.

F TABLE

The F distribution table for 5% and 1% right-hand tail areas is given in Appendix 5. A technique for finding F values for the 5% and 1% left-hand tail areas will be given after illustrating how to read the table.

EXAMPLE 12.2 Consider the F distribution with $df_1 = 4$ and $df_2 = 6$. Table 12.2 shows a portion of the F distribution table, with right tail area equal to .05, given in Appendix 5.

Table 12.2

df_2	df_1					
	1	2	3	4	...	20
1						
2						
3						
4						
5						
6	5.99	5.14	4.76	**4.53**	...	3.87
...						
100						

Table 12.2 indicates that the area to the right of 4.53 is .05. This is shown in Fig. 12-2. The F distribution table with right tail area equal to .01 indicates that the area to the right of 9.15 is .01. This is shown in Fig. 12-3.

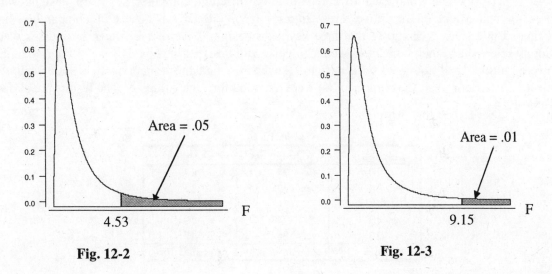

Fig. 12-2   Fig. 12-3

The following notation will be used with respect to the F distribution. The value, 4.53, shown in Fig. 12-2, is represented as $F_{.05}(4, 6) = 4.53$, and the value 9.15, as shown in Fig. 12-3 is represented as $F_{.01}(4, 6) = 9.15$.

Using the notation introduced in Example 12.2, the result shown in formula (*12.1*) holds for any F distribution.

$$F_{1-\alpha}(df_1, \ df_2) = \frac{1}{F_{\alpha}(df_2, \ df_1)} \tag{12.1}$$

EXAMPLE 12.3 Formula (*12.1*) may be used to find the F values for 1% and 5% left-hand tail areas for the F distribution discussed in Example 12.2. The symbol $F_{.95}(4, 6)$ represents the value for which the area to the right of $F_{.95}(4, 6)$ is .95 and therefore the area to the left of $F_{.95}(4, 6)$ is .05. Using formula (*12.1*), we find that

$$F_{.95}(4, \ 6) = \frac{1}{F_{.05}(6, \ 4)} = \frac{1}{6.16} = 0.1623$$

The value $F_{.05}(6, 4) = 6.16$ is shown in bold print in Table 12.3, which is taken from the F distribution table.

Table 12.3

df$_2$	\multicolumn{7}{c}{df$_1$}
	1	2	3	4	5	6	. . .	20
. . . 4	7.71	6.94	6.59	6.39	6.26	**6.16**		5.80

Similarly, the 1% left-hand tail area is

$$F_{.99}(4, \ 6) = \frac{1}{F_{.01}(6, \ 4)} = \frac{1}{15.21} = 0.0657$$

LOGIC BEHIND A ONE-WAY ANOVA

An experiment was conducted to compare three different computer keyboard designs with respect to their affect on repetitive stress injuries (rsi). Fifteen businesses of comparable size participated in a study to compare the three keyboard designs. Five of the fifteen businesses were randomly selected and their computers were equipped with design 1 keyboards. Five of the remaining ten were selected and equipped with design 2 keyboards, and the remaining five used design 3 keyboards. After one year, the number of rsi were recorded for each company. The results are shown in Table 12.4.

Table 12.4

Design 1	Design 2	Design 3
10	24	17
10	22	17
8	24	15
10	24	19
12	26	17
mean = 10	mean = 24	mean = 17

A Minitab dotplot of the data for the three designs is shown in Fig. 12-4.

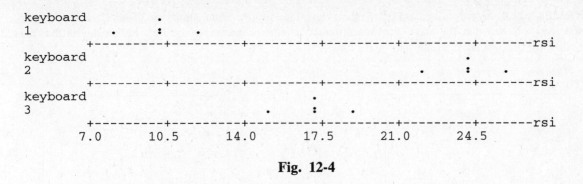

Fig. 12-4

If μ_1, μ_2, and μ_3 represent the mean number of rsi for design 1, design 2, and design 3, respectively, for the population of all such companies, the data indicate that the population means differ.

Suppose the data for the study were as shown in Table 12.5. A Minitab dotplot for these data is shown in Fig. 12-5.

Table 12.5

Design 1	Design 2	Design 3
10	34	29
12	14	17
5	24	5
1	19	10
22	29	24
mean = 10	mean = 24	mean = 17

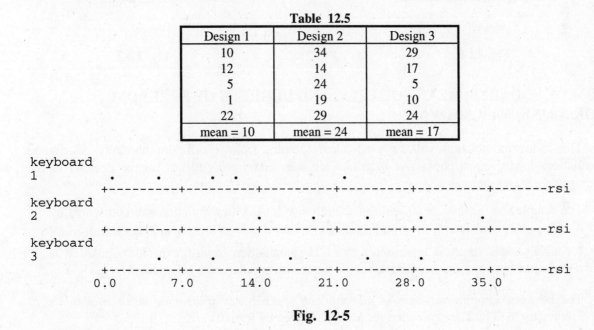

Fig. 12-5

Note that for the data shown in Fig. 12-5, it is not clear that the mean number of rsi differ for the three populations using the three different designs. The difference between the two sets of data can be explained by considering two different sources of variation. The *between samples variation* is measured by considering the variation between the sample means. In both cases, the sample means are $\bar{x}_1 = 10$, $\bar{x}_2 = 24$, and $\bar{x}_3 = 17$. The between samples variation is the same for the two sets of data. The *within samples variation* is measured by considering the variation within the samples. The dotplots indicate that the within samples variation is much greater for the data in Table 12.5 than for the data in Table 12.4. The measurement of these two sources of variation will be discussed in the next section. However, it is clear from the above discussion that a decision concerning the effectiveness of the three designs may be based on a consideration of the two sources of variation. In particular, it will be shown that the ratio of the between variation to the within variation may be used to decide whether μ_1, μ_2, and μ_3 are equal or not.

EXAMPLE 12.4 Figures 12-6 and 12-7 show boxplots for the data shown in Tables 12.4 and 12.5, respectively. Notice that Fig. 12-6 suggests different population means, while Fig. 12-7 suggests that it is possible that the population means may be the same.

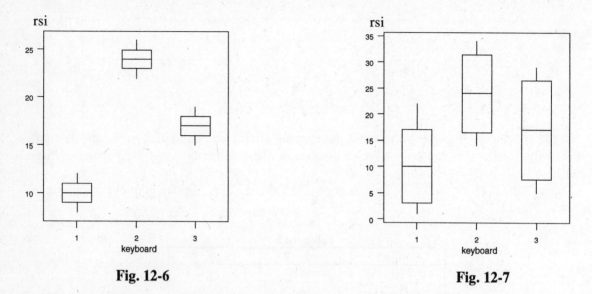

Fig. 12-6   Fig. 12-7

SUM OF SQUARES, MEAN SQUARES, AND DEGREES OF FREEDOM FOR A ONE-WAY ANOVA

The following notation will be used when samples from k different normally distributed populations having equal population variances are selected in order to test for the equality of the means of the k populations:

The sample size, sample mean, and sample variance for the ith population are represented by n_i, $\overline{x}_i$, and s_i^2, respectively. The total sample size is $n = n_1 + n_2 + \cdots + n_k$. The overall mean for all n sample values is represented by $\overline{x}$. The population mean for the ith population is represented by μ_i and the standard deviation for the ith population is represented by σ_i.

The between samples variation is measured by the *between treatments mean square* and is represented by MSTR. The expression for MSTR is given by formula (*12.2*):

$$MSTR = \frac{SSTR}{k-1} \qquad (12.2)$$

The numerator of formula (*12.2*), SSTR, is called the *treatment sum of squares*, and is computed by using formula (*12.3*):

$$SSTR = n_1(\overline{x}_1 - \overline{x})^2 + n_2(\overline{x}_2 - \overline{x})^2 + \cdots + n_k(\overline{x}_k - \overline{x})^2 \qquad (12.3)$$

The within samples variation is measured by the *error mean square* and is represented by MSE. The expression for MSE is given by

$$MSE = \frac{SSE}{n-k} \qquad (12.4)$$

The numerator of formula (*12.4*), SSE, is called the *error sum of squares*, and is computed by using formula (*12.5*), where s_1^2, s_2^2, . . . , s_k^2 are the sample variances.

$$SSE = (n_1 - 1)s_1^2 + (n_2 - 1)s_2^2 + \cdots + (n_k - 1)s_k^2 \qquad (12.5)$$

The denominator of formula (12.2), $k - 1$, is called the *degrees of freedom for treatments* and the denominator of formula (12.4), $n - k$, is called the *degrees of freedom for error*.

The sum of the treatment sum of squares and the error sum of squares is called the *total sum of squares*. The *total sum of squares* is represented by SST and is given by

$$SST = SSTR + SSE \qquad (12.6)$$

The total sum of squares may be computed directly by using formula (12.7), where the sum is over all n sample values. The *degrees of freedom for total* is equal to $n - 1$.

$$SST = \Sigma(x - \overline{x})^2 \qquad (12.7)$$

EXAMPLE 12.5 The data from Table 12.4 is listed below for convenient reference.

Design 1	Design 2	Design 3
10	24	17
10	22	17
8	24	15
10	24	19
12	26	17
mean = 10	mean = 24	mean = 17

Since the three sample sizes are equal, the overall mean is computed by finding the mean of the three treatment (design)means. The mean of 10, 24, and 17 is 17, and therefore $\overline{x} = 17$. The total sum of squares will be found first.

$$
\begin{aligned}
SST &= \Sigma(x - \overline{x})^2 \\
&= (10 - 17)^2 + (10 - 17)^2 + (8 - 17)^2 + (10 - 17)^2 + (12 - 17)^2 + (24 - 17)^2 + (22 - 17)^2 + (24 - 17)^2 \\
&\quad + (24 - 17)^2 + (26 - 17)^2 + (17 - 17)^2 + (17 - 17)^2 + (15 - 17)^2 + (19 - 17)^2 + (17 - 17)^2 \\
&= 49 + 49 + 81 + 49 + 25 + 49 + 25 + 49 + 49 + 81 + 0 + 0 + 4 + 4 + 0 \\
&= 514
\end{aligned}
$$

The treatment sum of squares is computed by using formula (12.3).

$$
\begin{aligned}
SSTR &= n_1(\overline{x}_1 - \overline{x})^2 + n_2(\overline{x}_2 - \overline{x})^2 + n_3(\overline{x}_3 - \overline{x})^2 \\
&= 5 \times (10 - 17)^2 + 5 \times (24 - 17)^2 + 5 \times (17 - 17)^2 \\
&= 5 \times 49 + 5 \times 49 + 5 \times 0 \\
&= 490
\end{aligned}
$$

We need to find the three sample variances before finding the error sum of squares. The formula for the sample variance is applied to each sample separately as follows:

$$S_1^2 = \frac{(10 - 10)^2 + (10 - 10)^2 + (8 - 10)^2 + (10 - 10)^2 + (12 - 10)^2}{5 - 1} = 2$$

$$S_2^2 = \frac{(24 - 24)^2 + (22 - 24)^2 + (24 - 24)^2 + (24 - 24)^2 + (26 - 24)^2}{5 - 1} = 2$$

$$S_3^2 = \frac{(17 - 17)^2 + (17 - 17)^2 + (15 - 17)^2 + (19 - 17)^2 + (17 - 17)^2}{5 - 1} = 2$$

The error sum of squares is computed using formula (12.5).

$$
\begin{aligned}
SSE &= (n_1 - 1)s_1^2 + (n_2 - 1)s_2^2 + (n_3 - 1)s_3^2 \\
&= 4 \times 2 + 4 \times 2 + 4 \times 2 = 24
\end{aligned}
$$

Note that SST = SSTR + SSE. It is clear from this example that the computation of the various sums of squares is rather time consuming and subject to computational errors. Computer software is usually employed to perform these computations. This will be discussed later. If it is necessary to perform the computations by hand, shortcut formulas are recommended. These shortcut formulas will now be discussed. The shortcut formula for computing the total sum of squares is given in (12.8):

$$SST = \Sigma x^2 - \frac{(\Sigma x)^2}{n} \tag{12.8}$$

The treatment sum of squares is computed by using formula (12.9), where T_i is the sum of the sample values for the ith treatment.

$$SSTR = \Sigma \frac{T_i^2}{n_i} - \frac{(\Sigma x)^2}{n} \tag{12.9}$$

After SST and SSTR are computed, SSE is found by using formula (12.10):

$$SSE = SST - SSTR \tag{12.10}$$

EXAMPLE 12.6 To find the sum of squares found in Example 12.5 using the shortcut formulas, refer to the following table.

Design 1	Design 2	Design 3
10	24	17
10	22	17
8	24	15
10	24	19
12	26	17
$T_1 = 50$	$T_2 = 120$	$T_3 = 85$

To find SST, it is necessary first to find Σx and Σx^2.

$$\Sigma x = 10 + 10 + 8 + 10 + 12 + 24 + 22 + 24 + 24 + 26 + 17 + 17 + 15 + 19 + 17 = 255$$

$$\Sigma x^2 = 100 + 100 + 64 + 100 + 144 + 576 + 484 + 576 + 576 + 676 + 289 + 289 + 225 + 361 + 289 = 4849$$

$$SST = \Sigma x^2 - \frac{(\Sigma x)^2}{n} = 4849 - 4335 = 514$$

The treatment sum of squares is found next.

$$SSTR = \Sigma \frac{T_i^2}{n_i} - \frac{(\Sigma x)^2}{n} = \frac{50^2}{5} + \frac{120^2}{5} + \frac{85^2}{5} - 4335 = 4825 - 4335 = 490$$

and, then the error sum of squares is found by subtraction:

$$SSE = SST - SSTR = 514 - 490 = 24$$

The degrees of freedom for total is $n - 1 = 14$, the degrees of freedom for treatments is $k - 1 = 2$, and the degrees of freedom for error is $n - k = 15 - 3 = 12$.

The between treatments mean square is MSTR $= \dfrac{\text{SSTR}}{k-1} = \dfrac{490}{2} = 245$, and the error mean square is MSE $=$

$\dfrac{\text{SSE}}{n-k} = \dfrac{24}{12} = 2$. This example illustrates that it is much easier to compute the sums of squares using the shortcut formulas than it is to use the defining formulas. Most researchers use statistical software to perform these computations.

SAMPLING DISTRIBUTION FOR THE ONE-WAY ANOVA TEST STATISTIC

The purpose behind a one-way ANOVA is to determine if the means of k populations are equal or not. In particular, we are interested in testing the null hypothesis H_0: $\mu_1 = \mu_2 = \cdots = \mu_k$ against the alternative hypothesis H_a: All k means are not equal. When the k samples are randomly and independently selected from normal populations having equal population standard deviations (or equivalently equal population variances), and the null hypothesis is true, the ratio of the treatment mean square to the error mean square has an F distribution with $df_1 = k - 1$, and $df_2 = n - k$. We express this as shown in formula (*12.11*). This ratio, sometimes referred to as the F ratio, is the test statistic used to test the above hypotheses.

$$F = \frac{\text{MSTR}}{\text{MSE}} \qquad\qquad (12.11)$$

EXAMPLE 12.7 In Examples 12.5 and 12.6, three different keyboard designs were compared with respect to the number of rsi found at 15 companies using the three designs. The test statistic given in formula (*12.11*) has an F distribution with $df_1 = k - 1 = 3 - 1 = 2$, and $df_2 = n - k = 15 - 3 = 12$. In Example 12.6, it was shown that MSTR = 245 and MSE = 2. The computed value of the test statistic is F* $= \frac{245}{2} = 122.5$. From the F distribution table in Appendix 5, the 5% and 1% right-hand tail critical values are as follows: $F_{.05}(2, 12) = 3.89$ and $F_{.01}(2, 12) = 6.93$. The extremely large value of the test statistic suggests that the population means are not equal and that the null hypothesis should be rejected. The differences in the sample means are significant, and indicate that design 1 may reduce the number of rsi.

BUILDING ONE-WAY ANOVA TABLES AND TESTING THE EQUALITY OF MEANS

The results of the computations in the proceeding sections are usually conveniently displayed in a *one-way ANOVA table*. The general structure of the one-way ANOVA table is given in Table 12.6.

Table 12.6

Source	df	SS	MS = SS/df	F Statistic
Treatment	k – 1	SSTR	MSTR	F
Error	n – k	SSE	MSE	
Total	n – 1	SST		

EXAMPLE 12.8 The computations in Examples 12.6 and 12.7 are summarized in the ANOVA Table 12.7.

Table 12.7

Source	df	SS	MS = SS/df	F Statistic
Treatment	2	490	245	122.5
Error	12	24	2	
Total	14	514		

The steps in testing the equality of means is summarized in Table 12.8.

Table 12.8

Steps for Testing the Equality of Means Using the One-Way ANOVA Procedure
Step 1: State the null and alternative hypothesis as follows: $H_0: \mu_1 = \mu_2 = \cdots = \mu_k$ H_a: All k means are not equal.
Step 2: Use the F distribution table and the level of significance, α, to determine the rejection region.
Step 3: Build the ANOVA Table, and from the table determine the computed value of the F ratio.
Step 4: State your conclusion. The null hypothesis is rejected if the computed value of the test statistic falls in the rejection region. Otherwise, the null hypothesis is not rejected.

Computer software is routinely used to perform the necessary computations to compute the test statistic used to test the equality of means. The software usually produces the ANOVA table and in addition provides a p value for the test. When the p value is given, the null hypothesis is rejected if the p value is less than the preset level of significance.

EXAMPLE 12.9 Two different Minitab commands will now be discussed which produce a one-way ANOVA table. The data for the number of rsi for the three keyboard designs is reproduced below.

Design 1	Design 2	Design 3
10	24	17
10	22	17
8	24	15
10	24	19
12	26	17

The Minitab command **aovoneway c1 – c3** requires that the data for the three designs be put into three separate columns. These data were put into columns c1, c2, and c3.

Row	design1	design2	design3
1	10	24	17
2	10	22	17
3	8	24	15
4	10	24	19
5	12	26	17

MTB > aovoneway c1 – c3
One-Way Analysis of Variance
Analysis of Variance

Source	DF	SS	MS	F	P
Factor	2	490.00	245.00	122.50	0.000
Error	12	24.00	2.00		
Total	14	514.00			

```
                                    Individual 95% CIs For Mean
                                    Based on Pooled St Dev
Level     N    Mean    St Dev    ---+---------+---------+---------+---
design1   5    10.00   1.414     (--*--)
design2   5    24.00   1.414                              (--*--)
design3   5    17.00   1.414                 (--*--)
                                    ---+---------+---------+---------+---
                                    10.0    15.0    20.0    25.0
```

The above ANOVA table is the same as the one given in Table 12.7. The Minitab output shows the p value associated with the test under the column labeled P. Recall from our previous discussion of p value, that the p value is the area to the right of $F^* = 122.50$ for an F distribution having $df_1 = 2$ and $df_2 = 12$. Or, stated in words, if the null hypothesis is true, i.e., if $\mu_1 = \mu_2 = \mu_3$, the probability of obtaining an F value this large or larger is .000. For this reason, the null hypothesis would be rejected.

The second Minitab command which produces a one-way ANOVA table is **oneway response in c2 treatment in c1.** The treatment groups are identified in one column and the responses are given in another column. The ouput is the same for both commands. The required data setup differs for the two commands.

Row	design	rsi
1	1	10
2	1	10
3	1	8
4	1	10
5	1	12
6	2	24
7	2	22
8	2	24
9	2	24
10	2	26
11	3	17
12	3	17
13	3	15
14	3	19
15	3	17

MTB > oneway response in c2 treatment in c1
One-Way Analysis of Variance
Analysis of Variance for rsi

Source	DF	SS	MS	F	P
Design	2	490.00	245.00	122.50	0.000
Error	12	24.00	2.00		
Total	14	514.00			

EXAMPLE 12.10 Before ending our discussion of the one-way ANOVA, it is instructive to compare the ANOVA tables for the two sets of data given in Tables 12.4 and 12.5 and illustrated in Figs. 12-4 and 12-5 as well as Figs. 12-6 and 12-7. The ANOVA for the data in Table 12.4 is given above in Example 12.9. The Minitab output for the ANOVA Table corresponding to the data in Table 12.5 is

One-Way Analysis of Variance

Analysis of Variance

Source	DF	SS	MS	F	P
Factor	2	490.0	245.0	3.30	0.072
Error	12	890.0	74.2		
Total	14	1380.0			

The ANOVA for the data in Table 12.5 does not indicate a difference in the three designs for $\alpha = .05$, since p = 0.072. Thus, the statistical test for the hypotheses concerning the means confirms what Figs. 12-4 and 12-5 as well as Figs. 12-6 and 12-7 suggested.

LOGIC BEHIND A TWO-WAY ANOVA

Suppose that rather than considering only the effect of the factor keyboard design, we would also like to consider the effect of the factor seating design on repetitive stress injuries (rsi). The one-way ANOVA allows us to analyze the effect of only one factor on the *response variable*, rsi. A two-way ANOVA allows one to analyze the effects of two factors on a response variable. Suppose there are three different keyboard designs and three different seating designs we are interested in testing. A 3 × 3 *factorial design* is one in which each keyboard design is matched with each seating design for a total of nine treatment combinations. Such a design allows one to test the *main effects* for keyboard designs and seating designs as well as test for *interaction* between the two factors. Suppose 18 businesses of comparable size are selected for the study. A randomization scheme is used and each of the nine treatments are used at two businesses. The number of rsi are recorded at each location and the results are shown in Table 12.9.

Table 12.9

Seating design	Keyboard design		
	1	2	3
1	10, 12	20, 22	16, 18
2	14, 16	25, 27	20, 22
3	8, 8	18, 16	14, 16

EXAMPLE 12.11 Table 12.10 is a table of means for the results shown in Table 12.9. The mean for each treatment (Keyboard–Seating design combination) is given as well as the marginal means. It is clear that keyboard design 2 with seating design 2 resulted in the largest mean number of rsi. Also, keyboard design 1 with seating design 3 resulted in the smallest mean number of rsi.

Table 12.10

Seating design	Keyboard design			Row mean
	1	2	3	
1	11	21	17	16.33
2	15	26	21	20.67
3	8	17	15	13.33
Column mean	11.33	21.33	17.67	

A graphical technique for analyzing the results of the experiment is a *main effects plot*. A Minitab main effects plot is shown in Fig. 12-8. The main effects plot for the factor, seating, is a plot of the row means given in Table 12.10. Each of these means is calculated for six businesses. The main effects plot for the factor, keyboard, is a plot of the column means given in Table 12.10. Each of these means is calculated for six businesses.

Main Effects Plot - Means for rsi

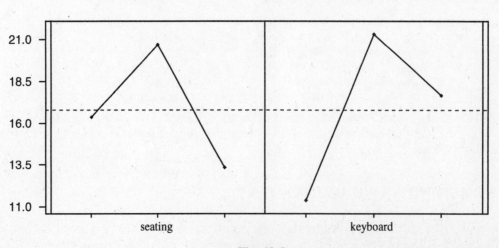

Fig. 12-8

Another plot used to help explain the experimental results is called the *interaction plot.* A Minitab interaction plot is shown in Fig. 12-9. The interaction plot indicates that regardless of the seating design used, keyboard design 1 results in the smallest number of rsi, keyboard design 2 results in the largest number of rsi, and the number of rsi for keyboard design 3 is between the other two. When the line segments are roughly parallel, as they are here, we say that there is no interaction between the factors.

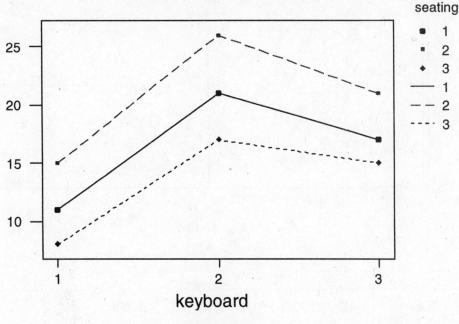

Fig. 12-9

EXAMPLE 12.12 Suppose the experiment described in Example 12.11 resulted in the data shown in Table 12.11.

Table 12.11

Seating design	Keyboard design		
	1	2	3
1	18, 20	20, 22	16, 18
2	14, 16	25, 27	20, 22
3	8, 8	18, 16	14, 16

The Minitab main effects plot is shown in Fig. 12-10 and the Minitab interaction plot is shown in Fig. 12-11. An understanding of interaction is accomplished by comparing Fig. 12-11 with Fig. 12-9. Recall that in Fig. 12-9, we were able to conclude that regardless of the seating design used, keyboard design 1 results in the smallest number of rsi, keyboard design 2 results in the largest number of rsi, and the number of rsi for keyboard design 3 is between the other two. Notice that this is not true in Fig. 12-11. In particular, when seating design 1 is used the smallest mean number of rsi occur for keyboard design 3, not keyboard design 1. In this study, the factors keyboard design and seating design are said to interact. Note that the line segments on the interaction plot are not parallel. The keyboard design which produces the minimum number of rsi depends on the seating design used. We cannot conclude that keyboard design 1 always produces the minimum number of rsi. *When interaction is present, the main effects must be interpreted carefully.*

The discussion and examples in this section are intended to provide some insight into the logic behind a two-factor factorial design. In the next section the ideas will be generalized and the analysis of variance associated with the two-factor design will be discussed. This section is intended to make the general discussion easier to understand.

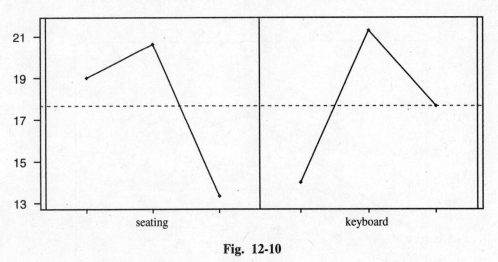

Fig. 12-10

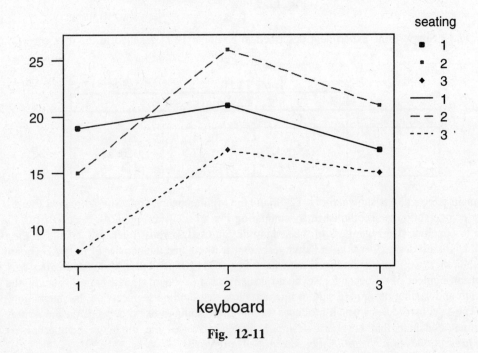

Fig. 12-11

SUM OF SQUARES, MEAN SQUARES, AND DEGREES OF FREEDOM FOR A TWO-WAY ANOVA

Let A and B be two factors whose influence on a response variable we wish to investigate. Let a be the number of *levels* of factor A and let b be the number of *levels* of factor B. Each level of factor A is combined with each level of factor B to form a total of ab *treatments*. Equal size samples are taken from each of the ab treatments. If each sample is of size m, the total number of sample values is n = mab.

EXAMPLE 12.13 Table 12.12 gives the compensatory awards (in thousands)for wrongful firing based on age, race, or disability for males and females. Either bias type or gender may be called factor A. Suppose we let factor A be bias type. Then A has three levels and B has two levels. There are six populations or treatments. The terms population, treatment, or treatment combination are used for the six groups. We shall use the term treatment. The six treatments are: (age bias and male), (race bias and male), (disability bias and male), (age bias and female), (race bias and female), and (disability bias and female). The values for a, b, m, and n are 3, 2, 3, and 18, respectively.

Table 12.12

Gender	Bias type		
	Age	Race	Disability
Male	200, 175, 215	150, 125, 135	100, 95, 115
Female	185, 210, 225	130, 145, 115	90, 80, 110

Initially, a two factor factorial design may be analyzed using a one-way ANOVA. The number of treatments is k = ab, and the total sample size is n. The one-way ANOVA for the two factor design is given in Table 12.13.

Table 12.13

Source	df	SS	MS = SS/df	F Statistic
Treatment	ab − 1	SSTR	MSTR	F
Error	n − ab	SSE	MSE	
Total	n − 1	SST		

The Treatment sum of squares, SSTR, is now partitioned in three parts as shown in formula (*12.12*), where SSA is called *factor A sum of squares*, SSB is called *factor B sum of squares*, and SSAB is called *interaction sum of squares*.

$$SSTR = SSA + SSB + SSAB \qquad (12.12)$$

The treatment degrees of freedom, ab − 1, is also partitioned into three parts as follows:

$$\text{df for A} = a - 1, \ \ \text{df for B} = b - 1, \ \ \text{df for AB} = (a-1)(b-1)$$

The degrees of freedom for A, B, and AB add up to the degrees of freedom for treatment as given by formula (*12.13*):

$$ab - 1 = (a-1) + (b-1) + (a-1)(b-1) \qquad (12.13)$$

Mean squares are also defined for A, B, and AB. The factor A mean square, MSA, is given by formula (*12.14*):

$$MSA = \frac{SSA}{a-1} \qquad (12.14)$$

The factor B mean square, MSB, is given by

$$MSB = \frac{SSB}{b-1} \tag{12.15}$$

The interaction mean square, MSAB, is given by

$$MSAB = \frac{SSAB}{(a-1)(b-1)} \tag{12.16}$$

The formulas for the sum of squares for A, B, and AB will not be discussed since the analysis of factorial designs are almost always performed by computer software.

EXAMPLE 12.14 In Example 12.13, the total degrees of freedom is $n-1 = 18-1 = 17$, the treatment degrees of freedom is $ab-1 = 6-1 = 5$, and the error degrees of freedom is $n-ab = 18-6 = 12$. Furthermore, the 5 degrees of freedom for treatments is partitioned into $a-1 = 3-1 = 2$ for factor A, $b-1 = 2-1 = 1$ for factor B, and $(a-1)(b-1) = 2 \times 1 = 2$ for interaction.

BUILDING TWO-WAY ANOVA TABLES

The sum of squares, mean squares, degrees of freedom, and test statistics for a factorial design are summarized in a two-way ANOVA table. The structure for a two-way ANOVA table is given in Table 12.14.

Table 12.14

Source	df	SS	MS = SS/df	F Statistic
Factor A	$a-1$	SSA	MSA	$F_A = MSA/MSE$
Factor B	$b-1$	SSB	MSB	$F_B = MSB/MSE$
Interaction	$(a-1)(b-1)$	SSAB	MSAB	$F_{AB} = MSAB/MSE$
Error	$n-ab$	SSE	MSE	
Total	$n-1$	SST		

EXAMPLE 12.15 A two-way ANOVA table for the data given in Table 12.12 is given in Table 12.15.

Table 12.15

Source	df	SS	MS	F Statistic
Bias type	2	SSA	MSA	F_A
Sex	1	SSB	MSB	F_B
Interaction	2	SSAB	MSAB	F_{AB}
Error	12	SSE	MSE	
Total	17	SST		

SAMPLING DISTRIBUTIONS FOR THE TWO-WAY ANOVA

When there is no interaction between factors A and B, the test statistic F_{AB} has an F distribution with $df_1 = (a-1)(b-1)$ and $df_2 = n-ab$. When there are no differences among the means for the main effect A, the test statistic F_A has an F distribution with $df_1 = a-1$ and $df_2 = n-ab$. When there are no differences among the means for the main effect B, the test statistic F_B has an F distribution with $df_1 = b-1$ and $df_2 = n-ab$.

When there are no differences among the means for the factor A, we say there is no main effect due to factor A. Otherwise, we say there is a main effect due to factor A. Similar terminology is used for factor B.

EXAMPLE 12.16 In Example 12.15, the test statistic F_A has an F distribution with $df_1 = 2$ and $df_2 = 12$. The test statistic F_B has an F distribution with $df_1 = 1$ and $df_2 = 12$. The test statistic F_{AB} has an F distribution with $df_1 = 2$ and $df_2 = 12$. These distributions may be used to test for significant interaction and main effects and is discussed in the next section when the computer analysis of the data is discussed.

TESTING HYPOTHESIS CONCERNING MAIN EFFECTS AND INTERACTION

The steps for testing interaction and main effects in a two-factor factorial design is summarized in Table 12.16.

Table 12.16

Steps for Testing for Interaction and Main Effects Using the Two-way ANOVA Procedure
Step 1: State the null and alternative hypothesis for interaction and main effects as follows: H_0: There is no interaction between factors A and B H_a: Factors A and B interact H_0: There are no differences among the means for main effect A H_a: At least two of the main effect A means differ H_0: There are no differences among the means for main effect B H_a: At least two of the main effect B means differ
Step 2: Use the F distribution table and the level of significance α to determine the rejection regions.
Step 3: Build the ANOVA table and from the table determine the computed values of the test statistics.
Step 4: State your conclusions. The null hypothesis is rejected if the computed value of the test statistic falls in the rejection region. Otherwise, the null hypothesis is not rejected.

EXAMPLE 12.17 Table 12.9 gave the number of repetitive stress injuries (rsi)for a 3 by 3 factorial design and the table is reproduced below. The Minitab analysis for this experiment is also shown.

Seating design	Keyboard design		
	1	2	3
1	10, 12	20, 22	16, 18
2	14, 16	25, 27	20, 22
3	8, 8	18, 16	14, 16

Two different Minitab commands will be discussed to analyze the data. The first to be discussed is the command **Twoway.** First of all note the form of the data as given below. The data in the above table must be put in the form shown before performing the analysis. Each data line identifies the row, the column, and the data value in that row and column.

```
Row     seating    keyboard    rsi
 1         1           1        10
 2         1           1        12
 3         1           2        20
 4         1           2        22
 5         1           3        16
 6         1           3        18
 7         2           1        14
 8         2           1        16
 9         2           2        25
10         2           2        27
11         2           3        20
12         2           3        22
13         3           1         8
14         3           1         8
15         3           2        18
16         3           2        16
17         3           3        14
18         3           3        16
```

The command **Twoway 'rsi' 'seating' 'keyboard';** starts with the keyword Twoway, followed by the response and then the row and column names. The subcommand **Means 'seating' 'keyboard'.** gives row and column means and confidence intervals.

```
MTB > Twoway 'rsi' 'seating' 'keyboard';
SUB C >    Means 'seating' 'keyboard'.
```

Two-Way Analysis of Variance
```
Analysis of Variance for rsi
Source        DF      SS        MS
seating        2    163.11    81.56
keyboard       2    307.11   153.56
Interaction    4      4.89     1.22
Error          9     16.00     1.78
Total         17    491.11
```

```
                      Individual 95% CI
seating     Mean    --+---------+---------+---------+---------
   1       16.33                  (----*----)
   2       20.67                               (----*----)
   3       13.33    (----*----)
                    --+---------+---------+---------+---------
                  12.50     15.00     17.50     20.00
```

```
                      Individual 95% CI
keyboard    Mean    -------+---------+---------+---------+----
   1       11.33    (---*---)
   2       21.33                                (---*---)
   3       17.67                      (---*---)
                    -------+---------+---------+---------+----
                     12.00     15.00     18.00     21.00
```

The test statistic value for the null hypothesis H_0: No interaction between factors A and B is easily computed, since MSAB = 1.22 and MSE = 1.78. The computed value for F_{AB}^* is 1.22/1.78 = 0.69. The critical value for $\alpha = .05$ is $F_{.05}(4, 9) = 3.63$. Since F_{AB}^* does not exceed 3.63, the null hypothesis is not rejected. Therefore, interaction is not significant. This confirms what the interaction plot in Fig. 12-9 indicates.

Suppose we call keyboard design factor A. The test statistic value for the null hypothesis H_0: There are no differences among the means for main effect A is easily computed, since MSA = 153.56 and MSE = 1.78. The computed value for F_A^* is 153.56/1.78 = 86.27. The critical value for $\alpha = .05$ is $F_{.05}(2, 9) = 4.26$. Since F_A^* far exceeds 4.26, we reject the null hypothesis and conclude that the means for the three keyboard designs differ.

The seating design is factor B. The test statistic value for the null hypothesis H_0: There are no differences among the means for main effect B is easily computed, since MSB = 81.56 and MSE = 1.78. The computed value for F_B^* is 81.56/1.78 = 45.82. The critical value for $\alpha = .05$ is $F_{.05}(2, 9) = 4.26$. Since F_A^* far exceeds

4.26, we reject the null hypothesis and conclude that the means for the three seating designs differ. The main effects plot in Fig. 12-8 indicates that seating design 3 and keyboard design 1 is a good combination for reducing repetitive stress injuries.

A second Minitab command for doing the analysis is the command **ANOVA 'rsi' = 'seating' 'keyboard' 'seating'*'keyboard'**. The output below is produced by this command. This output may be preferable, since it provides p values. By considering the p values, we see immediately that interaction is not significant, but that both main effects are significant.

```
MTB > ANOVA 'rsi' = 'seating' 'keyboard' 'seating'*'keyboard'
```

Analysis of Variance (Balanced Designs)

```
Factor        Type Levels Values
seating       fixed    3     1     2     3
keyboard      fixed    3     1     2     3

Analysis of Variance for rsi

Source             DF        SS        MS       F       P
seating             2   163.111    81.556   45.87   0.000
keyboard            2   307.111   153.556   86.37   0.000
seating*keyboard    4     4.889     1.222    0.69   0.619
Error               9    16.000     1.778
Total              17   491.111
```

EXAMPLE 12.18 The data from Table 12.11 are reproduced below.

	Keyboard design		
Seating design	1	2	3
1	18, 20	20, 22	16, 18
2	14, 16	25, 27	20, 22
3	8, 8	18, 16	14, 16

The Minitab analysis for these data is shown below.

```
MTB > ANOVA 'rsi' = 'seating' 'keyboard' 'seating'*'keyboard'
```

Analysis of Variance (Balanced Designs)

```
Factor        Type  Levels  Values
seating       fixed    3      1     2     3
keyboard      fixed    3      1     2     3

Analysis of Variance for rsi
Source             DF        SS        MS       F       P
seating             2   177.333    88.667   49.88   0.000
keyboard            2   161.333    80.667   45.38   0.000
seating*keyboard    4    65.333    16.333    9.19   0.003
Error               9    16.000     1.778
Total              17   420.000
```

From this output, we immediately see that there is significant interaction. This confirms what the interaction plot shows in Fig 12-11. It would be a good idea for you to review the discussion given in Example 12.12 concerning the nature of the interaction. Remember, when interaction is present, the main effects must be interpreted carefully.

EXAMPLE 12.19 The 3 by 2 factorial data for the factors bias type and gender given in Table 12.12 are reproduced on the next page.

Gender	Bias type		
	Age	Race	Disability
Male	200, 175, 215	150, 125, 135	100, 95, 115
Female	185, 210, 225	130, 145, 115	90, 80, 110

The Minitab output for the Table 12.12 data is given below.

```
MTB > ANOVA 'award' =  'gender' 'biastype' 'gender'*'biastype'
```

Analysis of Variance (Balanced Designs)

```
Factor      Type   Levels Values
gender      fixed     2.   1       2
biastype    fixed     3    1       2       3

Analysis of Variance for award
Source            DF        SS          MS        F       P
gender             1        22.2        22.2      0.09    0.774
biastype           2     33144.4     16572.2     64.50    0.000
gender*biastype    2       344.4       172.2      0.67    0.530
Error             12      3083.3       256.9
Total             17     36594.4
```

The p values indicate that there is no significant interaction. There is no significant effect due to gender. However, the mean awards differ according to the type of bias.

Solved Problems

F DISTRIBUTION

12.1 What does it mean to say that the F distribution curve is asymptotic to the horizontal axis of the rectangular coordinate system?

Ans. This means that as we move to the right from the origin along the horizontal axis, the F distribution curve approaches the horizontal axis, but never touches it.

F TABLE

12.2 Find the following using the F distribution in Appendix 5.
(a) $F_{.01}(7, 3)$ (b) $F_{.05}(4, 6)$

Ans. (a) 27.67   (b) 4.53

12.3 Find the following using the F distribution in Appendix 5.
(a) $F_{.99}(7, 3)$ (b) $F_{.95}(4, 6)$

Ans. (a) $F_{.99}(7, 3) = \dfrac{1}{F_{.01}(3, 7)} = \dfrac{1}{8.45} = 0.1183$ (b) $F_{.95}(4, 6) = \dfrac{1}{F_{.05}(6, 4)} = \dfrac{1}{6.16} = 0.1623$

12.4 Find the critical value of F for the following.
(a) df = (4, 8) and area in the right tail equal to .01.
(b) df = (6, 6) and area in the right tail equal to .05.

Ans. (a) 7.01   (b) 4.28

LOGIC BEHIND A ONE-WAY ANOVA

12.5 Eighteen individuals, diagnosed as having mild high blood pressure, were randomly divided into three groups of six each and were assigned to one of three treatment groups. The treatment 1 group served as a control and made no changes in their diet. The individuals in the treatment 2 group replaced 50% of their diet with fruits and vegetables. The individuals in treatment group 3 replaced 50% of their diet with fruits and vegetables and reduced fat calories to 15% of their daily total. After six months, the change in diastolic blood pressure was measured and the results are given in Table 12.17. The change in blood pressure was determined by subtracting the reading after the six-month study from the initial reading.

Table 12.17

Treatment 1	Treatment 2	Treatment 3
4	6	11
3	8	6
−2	8	16
2	10	11
0	10	14
−1	12	8
mean = 1	mean = 9	mean = 11

A Minitab dotplot of the data given in Table 12.17 is shown in Fig 12-12. Describe in words what the dotplot suggests concerning the three treatments.

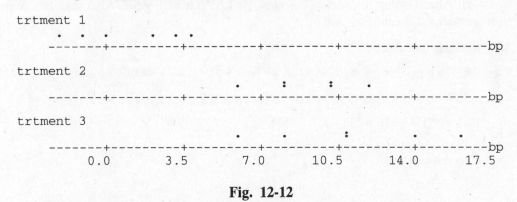

Fig. 12-12

Ans.   The plot suggests that the means for treatments 2 and 3 are greater than the mean for treatment 1. It is not clear whether the means for treatments 2 and 3 are different.

12.6 Table 12.18 is another set of data obtained for the same experiment described in Problem 12.5. A Minitab dotplot of the data is shown in Fig. 12-13. Describe in words what the dotplot suggests concerning the three treatments.

Table 12.18

Treatment 1	Treatment 2	Treatment 3
−10	0	0
10	16	22
4	8	11
−4	9	5
0	19	16
6	2	12
mean = 1	mean = 9	mean = 11

A Minitab dotplot of the data is shown in Fig. 12-13. Describe in words what the dotplot suggests concerning the three treatments.

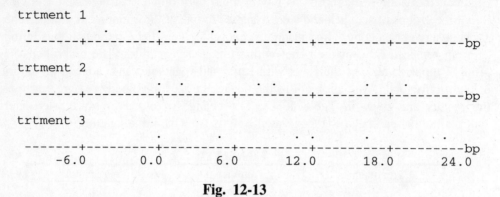

Fig. 12-13

Ans.   It is not clear whether the treatment means differ or not since the data points overlap for the three treatments.

SUM OF SQUARES, MEAN SQUARES, AND DEGREES OF FREEDOM FOR A ONE-WAY ANOVA

12.7 For the data in Table 12.17, find SST, SSTR, and SSE using both the defining and the shortcut formulas. After finding the sum of squares, find MSTR and MSE. Also give the degrees of freedom for total, treatments, and error.

Ans.   Defining formulas:

$$SSTR = n_1(\bar{x}_1 - \bar{x})^2 + n_2(\bar{x}_2 - \bar{x})^2 + n_3(\bar{x}_k - \bar{x})^2 = 6(1-7)^2 + 6(9-7)^2 + 6(11-7)^2 = 336$$

$$SSE = (n_1 - 1)s_1^2 + (n_2 - 1)s_2^2 + (n_3 - 1)s_3^2 = 5 \times 2.366^2 + 5 \times 2.098^2 + 5 \times 3.688^2 = 118$$

$$SST = SSTR + SSE = 336 + 118 = 454$$

Shortcut formulas:

$$SST = \Sigma x^2 - \frac{(\Sigma x)^2}{n} = 1336 - 882 = 454$$

$$SSTR = \Sigma \frac{T_i^2}{n_i} - \frac{(\Sigma x)^2}{n} = \frac{6^2}{6} + \frac{54^2}{6} + \frac{66^2}{6} - 882 = 336$$

$$SSE = SST - SSTR = 454 - 336 = 118$$

Degrees of freedom: for total = n – 1 = 17, for treatments = k – 1 = 2, for error = n – k = 15

$$MSTR = \frac{SSTR}{k-1} = 168$$

$$MSE = \frac{SSE}{n-k} = 7.87$$

12.8 For the data in Table 12.18, find SST, SSTR, and SSE using both the defining and the shortcut formulas. After finding the sum of squares, find MSTR and MSE. Also give the degrees of freedom for total, treatments, and error.

Ans.   Defining formulas:

$$SSTR = n_1(\overline{x}_1 - \overline{x})^2 + n_2(\overline{x}_2 - \overline{x})^2 + n_3(\overline{x}_k - \overline{x})^2 = 6(1-7)^2 + 6(9-7)^2 + 6(11-7)^2 = 336$$

$$SSE = (n_1 - 1)s_1^2 + (n_2 - 1)s_2^2 + (n_3 - 1)s_3^2 = 5 \times 7.239^2 + 5 \times 7.483^2 + 5 \times 7.797^2 = 846$$

$$SST = SSTR + SSE = 336 + 846 = 1182$$

Shortcut formulas:

$$SST = \Sigma x^2 - \frac{(\Sigma x)^2}{n} = 2064 - 882 = 1182$$

$$SSTR = \Sigma \frac{T_i^2}{n_i} - \frac{(\Sigma x)^2}{n} = \frac{6^2}{6} + \frac{54^2}{6} + \frac{66^2}{6} - 882 = 336$$

$$SSE = SST - SSTR = 1182 - 336 = 846$$

Degrees of freedom: for total $= n - 1 = 17$, for treatments $= k - 1 = 2$, for error $= n - k = 15$

$$MSTR = \frac{SSTR}{k-1} = 168$$

$$MSE = \frac{SSE}{n-k} = 56.4$$

SAMPLING DISTRIBUTION FOR THE ONE-WAY ANOVA TEST STATISTIC

12.9 (*a*) For the experiment described in problem 12.5, what conditions are necessary in order that $F = \dfrac{MSTR}{MSE}$ have an F distribution with $df_1 = 2$ and $df_2 = 15$?

(*b*) What is the computed value of F, F*, for the data in problem 12.5?

(*c*) What is the critical value for testing equal means at $\alpha = .05$? Would you reject the null hypothesis at this significance level?

Ans.   (*a*) The populations represented by the three samples are normally distributed. The three populations have equal variances. The null hypothesis is assumed to be true.

(*b*) The computed value of the test statistic is $F^* = \dfrac{168}{7.87} = 21.35$.

(*c*) $F_{.05}(2, 15) = 3.68$. Reject the null hypothesis because $F^* > 3.68$.

12.10 (*a*) For the experiment described in problem 12.6, what conditions are necessary in order that $F = \dfrac{MSTR}{MSE}$ have an F distribution with $df_1 = 2$ and $df_2 = 15$?

(*b*) What is the computed value of F, F*, for the data in problem 12.6?

(*c*) What is the critical value for testing equal means at $\alpha = .05$? Would you reject the null hypothesis at this significance level?

Ans. (a) The populations represented by the three samples are normally distributed. The three populations have equal variances. The null hypothesis is assumed to be true.

(b) The computed value of the test statistic is $F^* = \dfrac{168}{56.4} = 2.98$.

(c) $F_{.05}(2, 15) = 3.68$. Do not reject the null hypothesis, since F^* does not exceed 3.68.

BUILDING ONE-WAY ANOVA TABLES AND TESTING THE EQUALITY OF MEANS

12.11 Refer to Problems 12.5, 12.7, and 12.9. Use the results in these three problems to build the one-way ANOVA table for this experiment.

Ans. The one-way ANOVA table is shown below. The ANOVA table is used to systematically summarize the lengthy computations for the procedure and to give the computed value of the test statistic. By referring to the F distribution table, we see that the null hypothesis of equal means for the three treatments would be rejected at both the .05 and the .01 levels. Replacing 50% of their diet with fruits and vegetables appears to reduce the blood pressure for individuals with mild high blood pressure based on the results of this study.

ANOVA Table for the Data in Table 12.17

Source	Df	SS	MS = SS/df	F Statistic
Treatment	2	336	168	21.35
Error	15	118	7.87	
Total	17			

12.12 Refer to problems 12.6, 12.8, and 12.10. Use the results in these three problems to build the one-way ANOVA table for this experiment.

Ans. The one-way ANOVA table is shown below. The ANOVA table is used to systematically summarize the lengthy computations for the procedure and to give the computed value of the test statistic. By referring to the F distribution table, we see that the null hypothesis of equal means for the three treatments would not be rejected at either the .05 or the .01 levels. Based on the results of this study, we cannot claim that replacing 50% of their diets by fruits and vegetables will reduce the blood pressures for individuals with mild high blood pressure.

ANOVA Table for the Data in Table 12.18

Source	Df	SS	MS = SS/df	F Statistic
Treatment	2	336	168	2.98
Error	15	846	56.4	
Total	17			

LOGIC BEHIND A TWO-WAY ANOVA

12.13 A medical study utilized a 3 by 2 factorial design. One factor was the type of surgical procedure and the other factor was the temperature of the patients' environment during surgery. Some patients were kept warm with blankets and intravenous fluids and others were kept cool. Four patients were available for each surgical procedure–temperature combination. For each patient, the length of hospital stay was recorded and the results of the study are given in Table 12.19.

Table 12.19

Temperature	Surgical Procedure		
	1	2	3
Warm	2, 3, 3, 4	5, 6, 8, 5	9, 9, 10, 12
Cool	4, 5, 5, 6	9, 10, 11, 10	13, 14, 14, 15

The mean for each treatment (Surgical Procedure–Temperature combination), as well as marginal means, is given in Table 12.20.

Table 12.20

Temperature	Surgical Procedure			Row mean
	1	2	3	
Warm	3	6	10	6.3
Cool	5	10	14	9.7
Column mean	4	8	12	

A Minitab main effects plot is shown in Fig. 12-14 and an interaction plot is shown in Fig. 12-15.

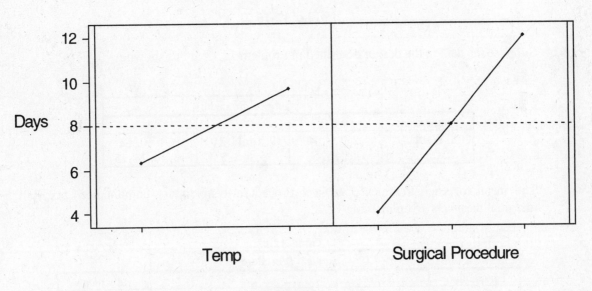

Fig. 12-14

Explain in words the interaction plot and the main effects plot shown in Figures 12-14 and 12-15.

Ans. The solid line segment in Fig. 12-15 corresponds to patients who were kept warm during surgery and the dashed line corresponds to patients who were kept cool during surgery. The interaction plot indicates that regardless of surgical procedure, the mean length of hospital stay is less when the patient is kept warm during surgery since the solid line is always below the dashed line. This indicates that there is no interaction between temperature and the type of surgical procedure. The main effects plot for temperature shows that the mean length of stay for patients kept warm during surgery is less than the mean length of stay for those kept cool during surgery. The main effects plot for surgical procedure contrasts the mean lengths of stay for the three surgical procedures.

Interaction Plot - Means for days

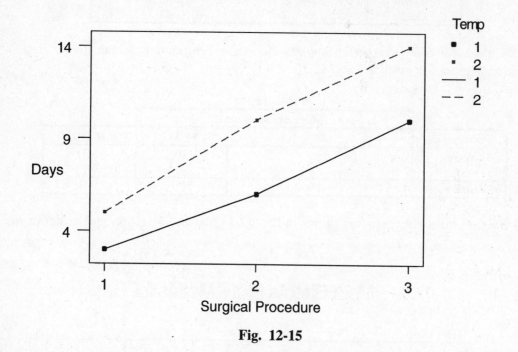

Fig. 12-15

12.14 Suppose the data in the design described in Problem 12.13 are as shown in Table 12.21.

Table 12.21

Temperature	Surgical Procedure		
	1	2	3
Warm	2, 3, 3, 4	9, 10, 11, 10	9, 9, 10, 12
Cool	4, 5, 5, 6	5, 6, 8, 5	13, 14, 14, 15

The mean for each treatment (Surgical Procedure–Temperature combination), as well as marginal means, is given in Table 12.22.

Table 12.22

Temperature	Surgical Procedure			
	1	2	3	Row mean
Warm	3	10	10	7.67
Cool	5	6	14	8.33
Column mean	4	8	12	

Explain in words the interaction plot and the main effects plot shown in Figs. 12-16 and 12-17.

Main Effects Plot - Means for Days

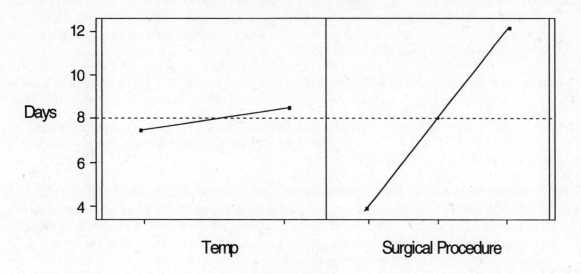

Fig. 12-16

Interaction Plot – Means for days

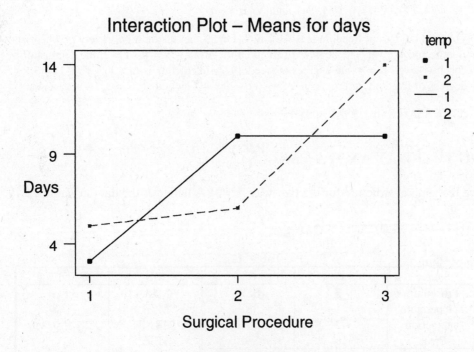

Fig. 12-17

Ans. The interaction plot in Fig. 12-17 illustrates a totally different situation than the interaction plot in Fig. 12-15. The interaction plot in Fig. 12-17 indicates that keeping the patient warm during surgery shortens the length of hospital stay for patients undergoing surgical procedures 1 and 3, but lengthens the stay for those undergoing procedure 2. The response lines are not parallel and we say there is interaction present. This was not true in Fig. 12-15. When interaction is present, the main effects are more difficult to explain.

SUM OF SQUARES, MEAN SQUARES, AND DEGREES OF FREEDOM FOR A TWO-WAY ANOVA

12.15 Refer to problem 12.13. By initially considering the two-way design as a one-way design, give the degrees of freedom for total, treatments, and error. Then, break the degrees of freedom for treatments down into degrees of freedom for temperature, surgical procedure, and temperature-surgical procedure interaction.

Ans. As a one-way design, the total degrees of freedom is $n - 1 = 24 - 1 = 23$. The degrees of freedom for treatments is $ab - 1 = 2 \times 3 - 1 = 5$. The degrees of freedom for error is $n - ab = 24 - 6 = 18$.

The 5 degrees of freedom for treatments is partitioned into $a - 1 = 2 - 1 = 1$ degree of freedom for temperature, $b - 1 = 3 - 1 = 2$ degrees of freedom for surgical procedures, and $(a - 1)(b - 1) = 1 \times 2 = 2$ degrees of freedom for interaction.

12.16 Sixty plots were used in an agricultural experiment. Four varieties of wheat and five different levels of fertilizer were used in a 4×5 factorial experiment. Each variety of wheat and each fertilizer level combination was used on three randomly chosen plots. List the 20 treatments, give the complete breakdown for the total degrees of freedom, and express the total sum of squares as a sum of four separate sums of squares.

Ans. One treatment would be the first variety of wheat combined with the first level of fertilizer, represented as (V1, F1). The other 19 treatments are: (V1, F2), (V1, F3), (V1, F4), (V1, F5), (V2, F1), (V2, F2), (V2, F3), (V2, F4), (V2, F5), (V3, F1), (V3, F2), (V3, F3), (V3, F4), (V3, F5), (V4, F1), (V4, F2), (V4, F3), (V4, F4), and (V4, F5).

The total degrees of freedom is $n - 1 = 60 - 1 = 59$. Let factor A be variety of wheat and let factor B be fertilizer level. The degrees of freedom for A is $a - 1 = 4 - 1 = 3$, the degrees of freedom for B is $b - 1 = 5 - 1 = 4$, the degrees of freedom for interaction is $(a - 1) \times (b - 1) = 3 \times 4 = 12$. The degrees of freedom for error is $n - ab = 60 - 20 = 40$. Note that $59 = 3 + 4 + 12 + 40$.

$$SST = SSA + SSB + SSAB + SSE.$$

BUILDING TWO-WAY ANOVA TABLES

12.17 Give the general structure for the two-way ANOVA table for the data in Table 12.19.

Ans. The results are given in Table 12.23.

Table 12.23

Source	df	SS	MS	F Statistic
Procedure	2	SSA	MSA	F_A
Temperature	1	SSB	MSB	F_B
Interaction	2	SSAB	MSAB	F_{AB}
Error	18	SSE	MSE	
Total	23	SST		

12.18 Give the general structure for the two-way ANOVA table for the experiment described in problem 12.16.

Ans. The results are given in Table 12.24.

CHAP. 12] ANALYSIS OF VARIANCE (ANOVA) 299

Table 12.24

Source	Df	SS	MS	F Statistic
Variety	3	SSA	MSA	F_A
Fertilizer level	4	SSB	MSB	F_B
Interaction	12	SSAB	MSAB	F_{AB}
Error	40	SSE	MSE	
Total	59	SST		

SAMPLING DISTRIBUTIONS FOR THE TWO-WAY ANOVA

12.19 Find the critical values for F_A, F_B, and F_{AB} in problem 12.17. Use $\alpha = .05$.

Ans. Factor A is the surgical procedure and factor B is the temperature of the patients' environment. The test statistic F_A has an F distribution with $df_1 = 2$ and $df_2 = 18$. The critical value for F_A is 3.55. The test statistic F_B has an F distribution with $df_1 = 1$ and $df_2 = 18$. The critical value for F_B is 4.41. The test statistic F_{AB} has an F distribution with $df_1 = 2$ and $df_2 = 18$ and the critical value for F_{AB} is 3.55.

12.20 Find the critical values for F_A, F_B, and F_{AB} in problem 12.18. Use $\alpha = .01$.

Ans. Factor A is the variety of wheat and factor B is the level of fertilizer. The test statistic F_A has an F distribution with $df_1 = 3$ and $df_2 = 40$. The critical value for F_A is 4.31. The test statistic F_B has an F distribution with $df_1 = 4$ and $df_2 = 40$. The critical value for F_B is 3.83. The test statistic F_{AB} has an F distribution with $df_1 = 12$ and $df_2 = 40$ and the critical value for F_{AB} is 2.66

TESTING HYPOTHESIS CONCERNING MAIN EFFECTS AND INTERACTION

12.21 The Minitab output for the data given in Table 12.19 is shown below. After reviewing Problems 12.13, 12.15, 12.17, and 12.19, as well as the Minitab output, test for main effects as well as interaction at $\alpha = .05$.

```
Analysis of Variance for days

Source          DF        SS         MS        F       P
Temp             1    66.667     66.667    60.00   0.000
Surgproc         2   256.000    128.000   115.20   0.000
Temp*Surgproc    2     5.333      2.667     2.40   0.119
Error           18    20.000      1.111
Total           23   348.000
```

Ans. The Minitab output confirms the results discussed in problems 12.13, 12.15, 12.17, and 12.19. The p value for interaction, 0.119, is greater than α and interaction is not significant. This makes the interpretation of main effects easier. The p values for temperature and surgical procedure are both 0.000. The mean length of hospital stay in days differs for the three surgical procedures and for the two operation temperatures at which the patients are kept.

12.22 The Minitab output for the data given in Table 12.21 is shown below. After reviewing Problem 12.14, as well as the Minitab output, test for main effects as well as interaction at $\alpha = .05$.

Analysis of Variance for days

Source	DF	SS	MS	F	P
Temp	1	2.667	2.667	2.40	0.139
Surgproc	2	256.000	128.000	115.20	0.000
Temp*Surgproc	2	69.333	34.667	31.20	0.000
Error	18	20.000	1.111		
Total	23	348.000			

Ans. Since the interaction is significant, i. e., the p value $= 0.000 < \alpha$, we explain the nature of the interaction rather than make broad generalizations about the main effects. In Fig. 12-17, the solid line segment corresponds to the patients who were kept warm during surgery and the line segment made up of dashes corresponds to patients who were kept cool during surgery. The interaction plot suggests that the hospital stay is shorter if the patient is kept warm during surgery for procedures 1 and 3. However, it appears that for procedure 2, keeping the patient warm during surgery increases the hospital stay.

Supplementary Problems

F DISTRIBUTION

12.23 Fill in the following blanks with the appropriate distribution. Choose from the words standard normal, student t, Chi-square, and F.

(a) The _____ distribution is symmetrical about zero and has standard deviation equal to one.

(b) The _____ distribution is skewed to the right. The shape of the distribution curve is determined by the number of degrees of freedom.

(c) The _____ distribution is symmetrical about zero. The shape of the distribution curve is determined by the number of degrees of freedom.

(d) The shape of the _____ distribution is determined by two separate degrees of freedom.

Ans. (a) standard normal   (b) Chi-square   (c) student t   (d) F

F TABLE

12.24 Find the following using the F distribution in Appendix 5.
(a) $F_{.01}(2, 8)$ (b) $F_{.05}(2, 8)$

Ans. (a) 8.65   (b) 4.46

12.25 Find the following using the F distribution in Appendix 5.
(a) $F_{.99}(8, 2)$ (b) $F_{.95}(8, 2)$

Ans. (a) 0.1156   (b) 0.2242

12.26 The random variable F_{ratio} has an F distribution with $df_1 = 5$ and $df_2 = 5$. Find two values a and b such that $P(a < F_{ratio} < b) = .90$.

Ans. a = 0.1980 and b = 5.05

LOGIC BEHIND A ONE-WAY ANOVA

12.27 Thirty individuals were randomly divided into 3 groups of 10 each. Each member of one group completed a questionnaire concerning the Internal Revenue Service (IRS). The score on the questionnaire is called the Customer Satisfaction Index (CSI). The higher the score, the greater the satisfaction. A second set of CSI scores were obtained for garbage collection from the second group, and a third set of scores were obtained for long distance telephone service. The scores are given in Table 12.25. A Minitab boxplot for the data in Table 12.25 is shown in Fig. 12-18. Describe what the boxplot suggests concerning the mean CSI scores for the IRS, garbage collection, and long distance phone service.

Table 12.25

IRS	Garbage collection	Long distance service
40	55	60
45	60	65
55	65	70
50	50	70
45	55	70
45	55	75
50	60	75
50	70	70
40	65	80
60	60	80
mean = 48.0	mean = 59.5	mean = 71.5

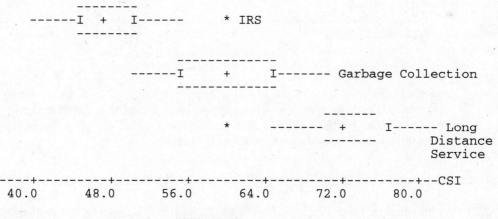

Fig. 12-18

Ans. Even though the whiskers overlap for the three different groups, the boxes that contain the middle 50% of the data, do not overlap for the three groups. Note that the plus signs are above the median values and are fairly spread apart. The data suggest that the mean CSI scores for the three populations are probably different. An analysis of variance may be performed to test the null hypothesis of equal population means.

12.28 Suppose the survey described in problem 12.27 resulted in the data given in Table 12.26. Describe what the boxplot in Fig. 12-19 suggests concerning the mean CSI scores for the IRS, garbage collection, and long distance phone service.

Table 12.26

IRS	Garbage collection	Long distance service
15	20	35
45	50	65
30	65	70
70	50	80
45	25	60
45	85	75
30	60	75
70	90	80
40	65	80
90	85	95
mean = 48	mean = 59.5	mean = 71.5

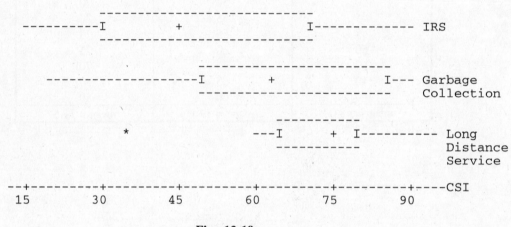

Fig. 12-19

Ans. The boxplots for the three sets of CSI scores shown in Fig. 12-19 exhibit a considerable amount of overlap. This is the type of results we might expect if there is no difference in the three population means. The boxplots indicate that the assumption that the three populations have normal distributions with equal variances should be checked.

SUM OF SQUARES, MEAN SQUARES, AND DEGREES OF FREEDOM FOR A ONE-WAY ANOVA

12.29 For the data in Table 12.25, find SST, SSTR, and SSE using both the defining and the shortcut formulas. After finding the sum of squares, find MSTR and MSE. Also give the degrees of freedom for total, treatments, and error.

> *Ans.* SSTR = 2761.7   SSE = 1035.0   SST = 3796.7   MSTR = 1380.8   MSE = 38.3
> Degrees of freedom: for total = 29, for treatments = 2, and for error = 27.

12.30 For the data in Table 12.26, find SST, SSTR, and SSE using both the defining and the shortcut formulas. After finding the sum of squares, find MSTR and MSE. Also give the degrees of freedom for total, treatments, and error.

> *Ans.* SSTR = 2761.7   SSE =12085   SST = 14847   MSTR = 1380.8   MSE = 448
> Degrees of freedom: for total = 29, for treatments = 2, and for error = 27.

SAMPLING DISTRIBUTION FOR THE ONE-WAY ANOVA TEST STATISTIC

12.31 Refer to problems 12.27 and 12.29. If the populations of CSI responses concerning the IRS, garbage collection, and long distance phone service are normally distributed, have equal variances and equal means, what is the distribution of $F = \dfrac{MSTR}{MSE}$? What is the $\alpha = .05$ critical value for testing the null hypothesis that the population means are equal? Give the computed value of the test statistic and your conclusion.

   Ans. The test statistic $F = \dfrac{MSTR}{MSE}$ has an F distribution with $df_1 = 2$ and $df_2 = 27$. The critical value, $F_{.05}(2, 27)$, is between 3.39 and 3.32. The computed value of the test statistic is $F^* = 36.05$. The null hypothesis is rejected.

12.32 Refer to problems 12.28 and 12.30. If the populations of CSI responses concerning the IRS, garbage collection, and long distance phone service are normally distributed, have equal variances and equal means, what is the distribution of $F = \dfrac{MSTR}{MSE}$? What is the $\alpha = .05$ critical value for testing the null hypothesis that the population means are equal? Give the computed value of the test statistic and your conclusion.

   Ans. The test statistic $F = \dfrac{MSTR}{MSE}$ has an F distribution with $df_1 = 2$ and $df_2 = 27$. The critical value, $F_{.05}(2, 27)$, is between 3.39 and 3.32. The computed value of the test statistic is $F^* = 3.08$. The null hypothesis is not rejected.

BUILDING ONE-WAY ANOVA TABLES AND TESTING THE EQUALITY OF MEANS

12.33 Table 12.27 gives the cost in thousands of dollars for randomly selected weddings with approximately 100 guests for three geographical regions in the U.S. Build a one-way ANOVA and test the null hypothesis that the mean costs for such weddings do not differ for the three regions. Test at a 5% level of significance.

Table 12.27

South	North	West
15.5	17.8	15.5
13.0	20.0	17.0
19.5	21.3	22.0
18.8	24.3	18.4
17.5	18.8	19.9
19.0	19.0	21.5
16.6	19.5	
14.9	20.5	
	22.0	
	23.4	
mean = 16.85	mean = 20.66	mean = 19.05

   Ans. The Minitab output for the above data is as follows:

```
Analysis of Variance
Source      DF       SS        MS        F        P
Factor       2      64.55     32.27     6.26     0.007
Error       21     108.20      5.15
Total       23     172.75
```

The p value is less than the preset alpha. The mean costs differ for the three regions. Such weddings appear to cost less in the South than in the North or West on the average.

12.34 A sociological study compared the time spent watching TV per week for children ages 2 to 11 for the years 1994, 1995, 1996, and 1997. The times in hours per week are shown in Table 12.28. Is there a difference in the means for the four years? Test at a 5% level of significance.

Table 12.28

1994	1995	1996	1997
22	21	20	25
24	19	24	20
25	25	21	18
18	20	18	20
19	18	15	15
25	15	19	21
17	19	24	20
24	21	24	15
15	19	18	23
25	20	17	21
25	16	23	20
25	25	23	15
25	22	20	19
16	24	16	25
20	20	20	25
20	20	16	23
17	18	15	16
22	18	19	22
25	18	24	21
16	20	18	15
mean = 21.3	mean = 19.9	mean = 19.7	mean = 20.0

Ans. The Minitab output for the one-way ANOVA is shown below.

```
Analysis of Variance
Source     DF        SS        MS        F        P
Factor      3       30.1      10.0      0.95     0.421
Error      76      802.7      10.6
Total      79      832.8
```

The means seem to have decreased since 1994. However, the p value = .421 indicates that the means are not different.

LOGIC BEHIND A TWO-WAY ANOVA

12.35 Table 12.29 gives the salaries in thousands of dollars for 20 individuals. Half are Nurse Practitioners and half are Physician Assistants. In addition, they are classified as practicing in a rural or an urban setting.

Table 12.29

	Physician Assistant	Nurse Practitioner
Urban	45, 51, 52, 48, 54	42, 44, 47, 49, 50
Rural	46, 50, 50, 50, 52	45, 48, 49, 47, 48

The interaction plots for the data given in Table 12.29 are given in Figs. 12-20 and 12-21. Explain the plots in words.

Interaction Plot - Means for salary

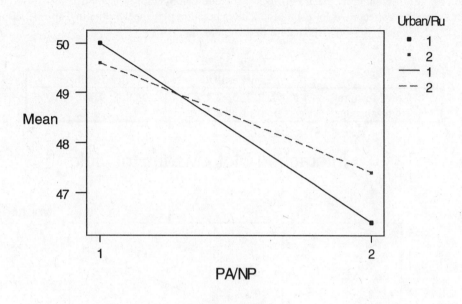

Fig. 12-20

Interaction Plot - Means for salary

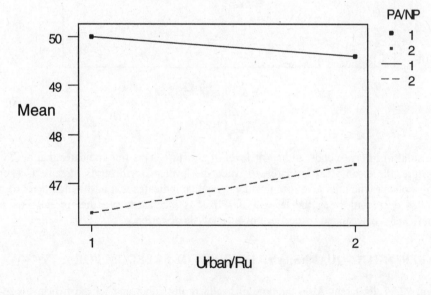

Fig. 12-21

Ans. The solid line in Fig. 12-20 corresponds to urban health care workers. The solid line shows that urban Physician Assistants in the study had a greater mean salary than urban Nurse Practitioners. The dashed line corresponds to rural health care workers. This line shows that the mean salary for rural Physician Assistants in the study was also greater than the mean for Nurse Practitioners. The solid line in Fig. 12-21 corresponds to Physician Assistants and the dashed line corresponds to Nurse Practitioners. This Figure shows that the mean for urban Physician Assistants exceeds the mean for rural Physician Assistants. The dashed line shows the opposite to be true for Nurse Practitioners.

12.36 Table 12.30 gives the results for a 2 × 2 factorial design. The yield in kilograms of okra per plot is given for 28 different plots. Each Fertilizer–Moisture combination was applied to 7 different plots and the total yield was recorded for each plot. The interaction plot for the data in Table 12.30 is shown in Fig. 12-22. Explain, in words, the nature of the interaction.

Table 12.30

	Low Fertilizer	High Fertilizer
Low Moisture	2, 3, 4, 5, 6, 4, 4	8, 8, 8, 9, 10, 6, 10
High Moisture	9, 9, 11, 8, 8, 9, 8	1, 1, 2, 2, 2, 3, 3

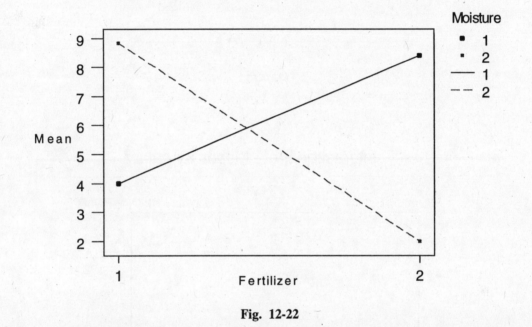

Fig. 12-22

Ans. The solid line corresponds to the low level of moisture. This line indicates that at the low level of moisture, the mean yield is increased when the level of fertilizer is increased. The dashed line corresponds to the high level of moisture. This line indicates that at the high level of moisture, the yield is decreased when the level of fertilizer is increased. Whether or not this interaction is significant is determined by performing an analysis of variance.

SUM OF SQUARES, MEAN SQUARES, AND DEGREES OF FREEDOM FOR A TWO-WAY ANOVA

12.37 In problem 12.35, let factor A be the type of health professional and let the two levels of factor A be Physician Assistant and Nurse Practitioner. Let factor B be the population setting and let the two levels of factor B be urban and rural. Give the degrees of freedom for the following sources of variation: total, A, B, AB, and error.

Ans. The degrees of freedom for the sources are: total df = 19, A df = 1, B df = 1, AB df = 1, error df = 16.

12.38 In problem 12.36, let factor A be fertilizer level and let the two levels of factor A be low and high. Let factor B be the moisture level and let the two levels of factor B be low and high. Give the degrees of freedom for the following sources of variation: total, A, B, AB, and error.

Ans. The degrees of freedom for the sources are: total df = 27, A df = 1, B df = 1, AB df = 1, error df = 24.

BUILDING TWO-WAY ANOVA TABLES

12.39 Give the general structure for the two-way ANOVA table for the data in Table 12.29. Name the factors and levels as given in problem 12.37.

Ans. The results are given in Table 12.31.

Table 12.31

Source	df	SS	MS	F Statistic
A	1	SSA	MSA	F_A
B	1	SSB	MSB	F_B
AB	1	SSAB	MSAB	F_{AB}
Error	16	SSE	MSE	
Total	19	SST		

12.40 Give the general structure for the two-way ANOVA table for the data in Table 12.30. Name the factors and levels as given in problem 12.38.

Ans. The results are given in Table 12.32.

Table 12.32

Source	Df	SS	MS	F Statistic
A	1	SSA	MSA	F_A
B	1	SSB	MSB	F_B
AB	1	SSAB	MSAB	F_{AB}
Error	24	SSE	MSE	
Total	27	SST		

SAMPLING DISTRIBUTIONS FOR THE TWO-WAY ANOVA

12.41 Give the critical values for the test statistics F_A, F_B, and F_{AB} in problem 12.39. Use $\alpha = .05$.

Ans. The critical value for all three is 4.49.

12.42 Give the critical values for the test statistics F_A, F_B, and F_{AB} in problem 12.40. Use $\alpha = .05$.

Ans. The critical value for all three is 4.26.

TESTING HYPOTHESIS CONCERNING MAIN EFFECTS AND INTERACTION

12.43 The Minitab output for the data given in Table 12.29 is shown below. After reviewing problems 12.35, 12.37, 12.39, and 12.41, as well as the below Minitab output, test for main effects as well as interaction at $\alpha = .05$.

```
Analysis of Variance for salary
Source           DF       SS       MS       F       P
Urban/ru          1    0.450    0.450    0.06   0.812
PA/NP             1   42.050   42.050    5.44   0.033
Urban/ru*PA/NP    1    2.450    2.450    0.32   0.581
Error            16  123.600    7.725
Total            19  168.550
```

Ans. The p values indicate that at the $\alpha = .05$ level of significance, the interaction is not significant. There is no significant difference in the mean salaries between urban and rural settings. However, the mean salary for Physician Assistants exceeds the mean salary for Nurse Practitioners.

12.44 The Minitab output for the data given in Table 12.30 is shown below. After reviewing Problems 12.36, 12.38, 12.40, and 12.42, as well as the below Minitab output, test for main effects as well as interaction at $\alpha = .05$.

```
Analysis of Variance for yield
Source              DF       SS        MS         F      P
Moisture             1     4.321     4.321      3.18   0.087
Fertlzer             1    10.321    10.321      7.61   0.011
Moisture*Fertlzer    1   222.893   222.893   164.24   0.000
Error               24    32.571     1.357
Total               27   270.107
```

Ans. Because of the significant moisture-fertilizer interaction, the main effects must be interpreted in the presence of this interaction.

Chapter 13

Regression and Correlation

STRAIGHT LINES

Suppose eight individuals, employed at large companies were interviewed, and the number of years of service, x, and the number of annual paid days off, y, were determined for each. The results are shown in Table 13.1.

Table 13.1

x	y
2	12
4	13
8	15
10	16
16	19
20	21
24	23
30	26

A *scatter plot* for these data is shown in Fig. 13-1.

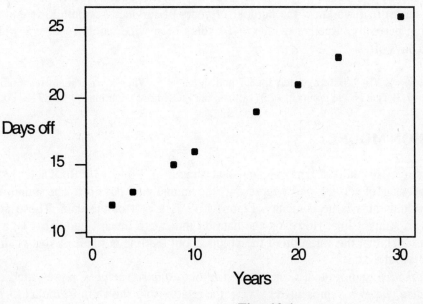

Fig. 13-1

The data plotted in Fig. 13-1 have a very special property. The points fall perfectly on a straight line. The equation of the line is $y = 11.0 + 0.5x$. The number 11.0 is called the y intercept. This is the value of y when $x = 0$. The number 0.5 is called the slope of the line. The equation may also be expressed as $y = 0.5x + 11.0$. When straight lines are studied in algebra, the slope-intercept form of a line is given as $y = mx + b$. When lines are discussed in statistics, the equation is often written as

$y = \beta_0 + \beta_1 x$. In this form, β_0 is the y intercept and β_1 is the slope. Figure 13-2 shows the line $y = 11.0 + 0.5x$ and the points from Table 13.1 on the same graph.

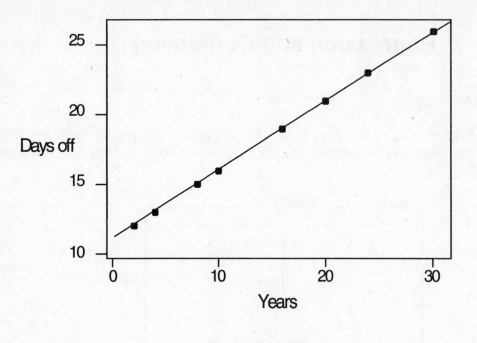

Fig. 13-2

If the relationship between years of service and annual paid days off for all employees of large companies were described by the equation $y = 11.0 + 0.5x$, then we could predict perfectly the number of annual paid days off if we knew the number of years of service. Assuming the equation described the relationship perfectly, a person with $x = 14$ years of service would receive $y = 11 + .5 \times 14 = 18$ annual paid days off.

EXAMPLE 13.1 The line $y = 1.7 - 4.5x$ has y intercept 1.7 and slope -4.5. When $x = 2$, the corresponding y value is $y = 1.7 - 4.5(2) = -7.3$. That is, the point $(2, -7.3)$ falls on the line whose equation is $y = 1.7 - 4.5x$.

LINEAR REGRESSION MODEL

Table 13.2 contains a more realistic data set than that shown in Table 13.1. In Table 13.2, x represents the number of years of service, and y represents the annual paid days off. These data were obtained from 20 individuals at a large company. Figure 13-3 is a plot of the data. These points clearly do not fall along a straight line. However, the data do indicate a linear trend. That is, a line could be fit to the points such that the variation of the points about the line is small. A line is shown which provides a good fit to the data.

A *linear regression model* assumes that some *dependent* or *response variable*, represented by y, is related to an *independent variable,* represented by x, by the relationship shown in formula (*13.1*).

$$y = \beta_0 + \beta_1 x + e \qquad\qquad (13.1)$$

The *error term*, e, is a normally distributed random variable with mean equal to 0 and standard deviation equal to σ. Each value of x determines a population of y values.

Table 13.2

x	y	x	y
2	10	16	20
2	14	20	20
2	12	20	22
4	11	20	21
4	15	24	23
8	14	24	24
8	16	24	22
10	14	30	27
10	18	30	25
16	18	30	26

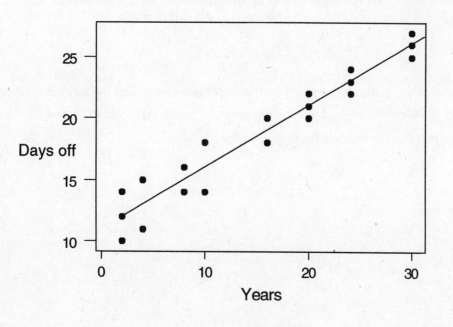

Fig. 13-3

The linear regression model, given in formula (*13.1*), describes the relationship between y and x for some population. The intercept, β_0, and the slope, β_1, are unknown parameters and must be estimated from sample data. The next section discusses the technique used to estimate β_0 and β_1. Table 13.3 contains a summary of the basic properties of the linear regression model.

Table 13.3

Basic Properties of the Linear Regression Model
1. The regression model is $y = \beta_0 + \beta_1 x + e$, where β_0 is the y intercept of the line given by $\beta_0 + \beta_1 x$, β_1 is the slope of the line given by $\beta_0 + \beta_1 x$, and e is the error or deviation of the actual y value from the line given by $\beta_0 + \beta_1 x$.
2. The error term e is a random variable with a mean of 0, i.e., $E(e) = 0$.
3. The expected value of y is $E(y) = \beta_0 + \beta_1 x$.
4. The variance of y equals σ^2 and is the same for all values of x.
5. The values of e are independent.
6. The error term e is normally distributed.

EXAMPLE 13.2 Consider the linear regression model $y = 2.5 + 3x + e$, where the normal random variable e has mean equal to 0 and standard deviation equal to 1.2. The mean response for y when $x = 2$ is $E(y) = 2.5 + 3(2) = 8.5$ and the mean response for y when $x = 4$ is $E(y) = 2.5 + 3(4) = 14.5$. In this example we are assuming that we know the values for β_0 and β_1. However, this is not usually the case. We usually have to estimate these two parameters from sample data taken from the population.

LEAST SQUARES LINE

The line shown in Fig. 13-3 is called the *least-squares line*. The least-squares line is determined such that the sum of the squares of the deviations of the data points from the line is a minimum. For point (x_i, y_i), the deviation from the fitted line, whose equation is represented as $\hat{y} = b_0 + b_1 x$, is given by $d_i = y_i - (b_0 + b_1 x_i)$. The estimates b_0 and b_1 are determined so that the sum of squares of deviations about the line is minimized. The sum of squares of deviations is given by

$$D = \sum d_i^2 = \sum [y_i - (b_0 + b_1 x_i)]^2 \tag{13.2}$$

The estimated y intercept, b_0, is also represented by the symbols a and $\hat{\beta}_0$ and the estimated slope is represented by b and $\hat{\beta}_1$. We shall use b_0 and b_1 as the symbols for the estimates of β_0 and β_1. Using calculus, it may be shown that the estimate for β_1 is given by

$$b_1 = \frac{S_{xy}}{S_{xx}} \tag{13.3}$$

where S_{xx} is given by

$$S_{xx} = \sum x^2 - \frac{(\sum x)^2}{n} \tag{13.4}$$

and S_{xy} is given by

$$S_{xy} = \sum xy - \frac{(\sum x)(\sum y)}{n} \tag{13.5}$$

The estimate for β_0 is given by

$$b_0 = \bar{y} - b_1 \bar{x} \tag{13.6}$$

EXAMPLE 13.3 The data from Table 13.2 is reproduced below. Recall that x represents the number of years of service and y represents the annual paid days off.

x	y	x	y
2	10	16	20
2	14	20	20
2	12	20	22
4	11	20	21
4	15	24	23
8	14	24	24
8	16	24	22
10	14	30	27
10	18	30	25
16	18	30	26

For these data, $\Sigma xy = 2 \times 10 + 2 \times 14 + 2 \times 12 + \cdots + 30 \times 27 + 30 \times 25 + 30 \times 26 = 6600$.
$\Sigma x = 2 + 2 + 2 + \cdots + 30 + 30 + 30 = 304$, and $\overline{x} = \frac{304}{20} = 15.2$.

$\Sigma y = 10 + 14 + 12 + \cdots + 27 + 25 + 26 = 372$, and $\overline{y} = \frac{372}{20} = 18.6$.

$\Sigma x^2 = 4 + 4 + 4 + \cdots + 900 + 900 + 900 = 6512$, $\Sigma y^2 = 100 + 196 + 144 + \cdots + 729 + 625 + 676 = 7426$.

$S_{xx} = \Sigma x^2 - \dfrac{(\Sigma x)^2}{n} = 6512 - 4620.8 = 1891.2$, $S_{xy} = \Sigma xy - \dfrac{(\Sigma x)(\Sigma y)}{n} = 6600 - 5654.4 = 945.6$.

$b_1 = \dfrac{S_{xy}}{S_{xx}} = \dfrac{945.6}{1891.2} = 0.5$ and $b_0 = \overline{y} - b_1 \overline{x} = 18.6 - 0.5 \times 15.2 = 11.0$.

The equation of the line is $\hat{y} = b_0 + b_1 x = 11.0 + 0.5x$. The fitted line is referred to by several different names. The fitted line is called *the line of best fit, the least-squares line, the estimated regression line, and the prediction line*. The computations for b_0 and b_1 are rarely performed by hand in practice. Computer software is normally used to find the equation. This is illustrated in Example 13.4.

EXAMPLE 13.4 The Minitab procedure for computing the equation of the regression line is shown in Fig. 13-4. The command **Regress 'Daysoff' 1 'Years'** requires that the column containing the dependent variable values be named after the word Regress, followed by the number of independent variables, followed by the column containing the independent variable values. The regression equation is printed out as the first line of output in Fig. 13-4. The remaining output will be discussed in later sections.

```
MTB > Regress 'Daysoff' 1 'Years';
SUBC>    Constant.

Regression Analysis

The regression equation is
Daysoff = 11.0 + 0.500 Years

Predictor        Coef        St Dev          T          P
Constant      11.0000        0.5703      19.29      0.000
Years          0.5000        0.0316      15.82      0.000

S = 1.374       R-Sq = 93.3%      R-Sq(adj) = 92.9%

Analysis of Variance

Source          DF          SS          MS          F          P
Regression       1      472.80      472.80     250.31      0.000
Error           18       34.00        1.89
Total           19      506.80
```

Fig. 13-4

The amount of computation saved by this procedure is hard to appreciate. Before the availability of computer software, these computations were done by use of a hand held calculator .

ERROR SUM OF SQUARES

The *error sum of squares*, denoted by SSE, is given by formula (*13.7*):

$$SSE = \Sigma(y_i - \hat{y}_i)^2$$

(*13.7*)

The differences, $(y_i - \hat{y}_i)$, are called *residuals*. A residual measures the deviation of an observed data point from the estimated regression line. If the estimated regression line fits the data points perfectly, as is the case for the data in Table 13.1, then SSE = 0. The more the variability of the data points away from the line, the larger the value for SSE. The computation of SSE is illustrated in Example 13.5.

EXAMPLE 13.5 The data in Table 13.2 gives 20 observations for the number of years of service, x, and the number of annual paid days off, y. The equation of the estimated regression line is found to be $\hat{y} = b_0 + b_1 x = 11.0 + 0.5x$ in Example 13.3. The computation of the residuals and SSE is shown in Table 13.4. Note that the sum of the residuals, $\Sigma(y_i - \hat{y}_i)$, is equal to zero. The error sum of squares, SSE, is equal to 34. The error sum of squares is shown in Fig. 13-4 as part of the Minitab output. It is shown under the Analysis of Variance portion of the output and is located at the intersection of the Error row and the SS column.

Table 13.4

Employee	x_i	y_i	$\hat{y}_i = 11 + 0.5x_i$	$(y_i - \hat{y}_i)$	$(y_i - \hat{y}_i)^2$
1	2	10	12	−2	4
2	2	14	12	2	4
3	2	12	12	0	0
4	4	11	13	−2	4
5	4	15	13	2	4
6	8	14	15	−1	1
7	8	16	15	1	1
8	10	14	16	−2	4
9	10	18	16	2	4
10	16	18	19	−1	1
11	16	20	19	1	1
12	20	20	21	−1	1
13	20	22	21	1	1
14	20	21	21	0	0
15	24	23	23	0	0
16	24	24	23	1	1
17	24	22	23	−1	1
18	30	27	26	1	1
19	30	25	26	−1	1
20	30	26	26	0	0
				$\Sigma(y_i - \hat{y}_i) = 0$	$\Sigma(y_i - \hat{y}_i)^2 = 34$

A convenient formula for computing SSE that is equivalent to formula (*13.7*) is given in formula (*13.8*). The computation of S_{yy} is the same as that for S_{xx} using the y values instead of the x values.

$$SSE = S_{yy} - \frac{S_{xy}^2}{S_{xx}} \qquad (13.8)$$

EXAMPLE 13.6 To illustrate the computation of SSE in Example 13.5 using the computation formula, recall that in Example 13.3, we found that $S_{xx} = 1891.2$ and $S_{xy} = 945.6$. The computation of S_{yy} is now illustrated:

$$S_{yy} = \Sigma y^2 - \frac{(\Sigma y)^2}{n} = 7426 - \frac{372^2}{20} = 506.8 \quad \text{and} \quad SSE = S_{yy} - \frac{S_{xy}^2}{S_{xx}} = 506.8 - \frac{945.6^2}{1891.2} = 34$$

STANDARD DEVIATION OF ERRORS

The standard deviation of the error term in the model $y = \beta_0 + \beta_1 x + e$ is represented by σ and the variance of the error term is σ^2. The variance σ^2 is estimated by S^2, where S^2 is given by formula (*13.9*):

$$S^2 = \frac{SSE}{n-2} \qquad\qquad (13.9)$$

The square root of S^2 is called the *standard deviation of errors*, and is given by formula (*13.10*):

$$S = \sqrt{\frac{SSE}{n-2}} \qquad\qquad (13.10)$$

EXAMPLE 13.7 The standard deviation of errors for the data given in Table 13.2 is found by recalling that $n = 20$ and in Example 13.6, we found that SSE = 34. The standard deviation of errors is:

$$S = \sqrt{\frac{SSE}{n-2}} = \sqrt{\frac{34}{20-2}} = 1.374$$

The value for S is shown in the Minitab output given in Fig. 13-4.

TOTAL SUM OF SQUARES

Suppose we were requested to estimate the mean number of annual paid days off given by large companies, but that we did not know the years of service, x, for each sampled employee. Our best estimate would be the mean of the 20 values for y given in Table 13.2. The mean of these 20 values is $\bar{y} = 18.6$ days per year. The accuracy of the estimate is related to the variation of the individual y values about the mean. The sum of squares about the mean is called the *total sum of squares*, and is given by

$$SST = \Sigma(y_i - \bar{y})^2 = \Sigma y^2 - \frac{(\Sigma y)^2}{n} \qquad\qquad (13.11)$$

EXAMPLE 13.8 By referring to Example 13.6, it is seen that the total sum of squares is equal to S_{yy}. Therefore, from Example 13.6, we see that SST = 506.8. The total sum of squares is given in the Minitab output shown in Fig. 13-4.

REGRESSION SUM OF SQUARES

When the regression line is not used in estimating the mean annual paid days off, the total sum of squares measures the variation about the mean, 18.6, as discussed in the previous section. When the regression line is used, there is still some unexplained variation about the regression line. This *unexplained variation* is given by SSE. This implies that the regression line explains an amount of variation equal to SST − SSE. This *explained variation* is called the *regression sum of squares*. The regression sum of squares is given by formula (*13.12*):

$$SSR = SST - SSE \qquad\qquad (13.12)$$

EXAMPLE 13.9 When the mean, $\bar{y} = 18.6$ days per year, is used to estimate the mean number of days off per year, the variation of the y values about this mean is SST = 506.8. When the number of years of service is taken into account, the estimated regression line was found to be $\hat{y} = 11.0 + 0.5x$. There is variation about this line that is not explained by the estimated regression line. This unexplained variation is given by SSE = 34. This implies that the regression line explained SST – SSE = 506.8 – 34 = 472.8 or 93.3% of the variation of the values about the mean. The regression sum of squares is SSR = 472.8.

The regression sum of squares is directly computable by the formula

$$SSR = \Sigma(\hat{y}_i - \overline{y})^2 \qquad (13.13)$$

EXAMPLE 13.10 Table 13.5 illustrates the computation of the regression sum of squares using formula (*13.13*). Note that SSR = 472.80. In Examples 13.5 and 13.8, we found that SSE = 34.0 and SST = 506.8. We see that SST = SSR + SSE. These three sum of squares is shown under the SS column of the analysis of variance portion of Fig. 13-4.

Table 13.5

Employee	x_i	y_i	$\hat{y}_i = 11 + 0.5x_i$	$\hat{y}_i - \overline{y}$	$(\hat{y}_i - \overline{y})^2$
1	2	10	12	–6.6	43.56
2	2	14	12	–6.6	43.56
3	2	12	12	–6.6	43.56
4	4	11	13	–5.6	31.36
5	4	15	13	–5.6	31.36
6	8	14	15	–3.6	12.96
7	8	16	15	–3.6	12.96
8	10	14	16	–2.6	6.76
9	10	18	16	–2.6	6.76
10	16	18	19	0.4	0.16
11	16	20	19	0.4	0.16
12	20	20	21	2.4	5.76
13	20	22	21	2.4	5.76
14	20	21	21	2.4	5.76
15	24	23	23	4.4	19.36
16	24	24	23	4.4	19.36
17	24	22	23	4.4	19.36
18	30	27	26	7.4	54.76
19	30	25	26	7.4	54.76
20	30	26	26	7.4	54.76
				sum = 0	sum = 472.80

COEFFICIENT OF DETERMINATION

The *coefficient of determination* is defined by

$$r^2 = \frac{SSR}{SST} \qquad (13.14)$$

When the data points fall perfectly on a straight line as in Fig. 13-1, SSE = 0 and therefore SST = SSR. In this case, the coefficient of determination is equal to 1. When the estimated regression line explains none of the variation in y, SSR = 0, and the coefficient of determination is equal to 0. When the coefficient of determination is expressed as a percentage, it can be thought of as the percentage of the total sum of squares that can be explained using the estimated regression equation.

EXAMPLE 13.11 For the data in Table 13.2 and discussed in the previous examples, the coefficient of determination is $r^2 = \dfrac{SSR}{SST} \times 100\% = \dfrac{472.8}{506.8} \times 100 = 93.3\%$.

The coefficient of determination may be determined by the use of formula (*13.15*):

$$r^2 = \frac{S_{xy}^2}{S_{xx} S_{yy}} \tag{13.15}$$

EXAMPLE 13.12 For the data in Table 13.2, it was found in Example 13.3 that $S_{xx} = 1891.2$ and $S_{xy} = 945.6$. Also in Example 13.6 we found that $S_{yy} = 506.8$. Therefore, the coefficient of determination is found as follows:

$$r^2 = \frac{S_{xy}^2}{S_{xx} S_{yy}} = \frac{945.6^2}{1891.2 \times 506.8} = .933 \text{ or } 93.3\%$$

The value for r^2 is also given in Fig. 13-4.

MEAN, STANDARD DEVIATION, AND SAMPLING DISTRIBUTION OF THE SLOPE OF THE ESTIMATED REGRESSION EQUATION

The slope of the estimated regression line, b_1, is a statistic and has a sampling distribution. That is, if several different surveys concerning the number of years of service, x, and the number of annual paid days off, y, were conducted, and the estimated regression line found in each case, the values for the slope and the y intercept would not all be equal. The estimated slope, b_1, has a sampling distribution which is normally distributed. The mean of b_1 is β_1 and the standard deviation of b_1 is given in the following:

$$\sigma_{b_1} = \frac{\sigma}{\sqrt{S_{xx}}} \tag{13.16}$$

Since σ is unknown, the standard deviation of errors is substituted for σ to obtain the standard error for b_1. The standard error for b_1 is given in formula (*13.17*):

$$S_{b_1} = \frac{S}{\sqrt{S_{xx}}} \tag{13.17}$$

EXAMPLE 13.13 For the data given in Table 13.2, the value for S_{xx} was found to equal 1891.2 in Example 13.3 and in Example 13.7, the value for S was found to equal 1.374. The standard error for b_1 is therefore equal to $S_{b_1} = \frac{S}{\sqrt{S_{xx}}} = \frac{1.374}{\sqrt{1891.2}} = 0.0316$. This value is given in the computer output in Fig. 13-4 at the intersection of the Years row and the St Dev column. The standard error for b_1 is needed to set confidence intervals on and test hypothesis concerning the slope of the population regression line.

INFERENCES CONCERNING THE SLOPE OF THE POPULATION REGRESSION LINE

A $(1 - \alpha) \times 100\%$ confidence interval for β_1 is obtained by using the distribution theory given in the previous section. The confidence interval is given in formula (*13.18*), where $t_{\alpha/2}$ is obtained from the Student t distribution using $n - 2$ degrees of freedom.

$$b_1 \pm t_{\alpha/2} \, S_{b_1} \tag{13.18}$$

EXAMPLE 13.14 Consider the continuing Example concerning the number of annual paid days off as a function of the number of years of service. Suppose we wish to determine a 95% confidence interval for β_1. The values for b_1 and S_{b_1} were obtained in Examples 13.3 and 13.13. They may also be obtained from the computer printout in Fig. 13-4. The values are $b_1 = 0.5$ and $S_{b_1} = 0.0316$. The value for $t_{.025}$ is obtained from the Student t table using df $= 20 - 2 = 18$. The value is $t_{.025} = 2.101$. The 95% margin of error is $\pm t_{\alpha/2} S_{b_1} = \pm 2.101 \times 0.0316 = \pm 0.066$. The confidence interval extends from $0.5 - 0.066 = 0.434$ to $0.5 + 0.066 = 0.566$.

The steps for testing an hypothesis about the value of the slope of the population regression line are given in Table 13.6.

Table 13.6

Steps for Testing the Value of the Slope of the Population Regression Line
Step 1: The null hypothesis is H_0: $\beta_1 = \beta_{10}$ and the alternative hypothesis is either lower, upper, or two-tailed.
Step 2: Use the Student t distribution table and the level of significance α to determine the rejection regions. The degrees of freedom is $n - 2$.
Step 3: The test statistic is computed as $t^* = \dfrac{b_1 - \beta_{10}}{S_{b1}}$.
Step 4: State your conclusions. The null hypothesis is rejected if the computed value of the test statistic falls in the rejection region. Otherwise, the null hypothesis is not rejected.

EXAMPLE 13.15 One of the most often tested hypothesis concerning β_1 is that it equals zero. This is equivalent to testing the hypothesis that x does not determine y and that there is no significant relationship between x and y. Suppose we wish to test H_0: $\beta_1 = 0$ vs. H_a: $\beta_1 \neq 0$ at significance level $\alpha = .05$ using the data in Table 13.2. Recall from Example 13.3 that $b_1 = 0.5$ and from Example 13.13 that $S_{b_1} = 0.0316$. The value for β_{10} in Table 13.6 is 0. The critical values are determined by finding the t value with .025 area in the right tail and df $= 18$. The values are ± 2.101. The computed value of the test statistic is found as follows:

$$t^* = \frac{b_1 - \beta_{10}}{S_{b1}} = \frac{0.5 - 0}{0.0316} = 15.82$$

Since this value falls far beyond the value 2.101, the null hypothesis is rejected. Note that the value 15.82 is found in the printout in Fig. 13-4 as the t value corresponding to the predictor years and the corresponding p value is given as 0.000. Most researchers would use the information given in the printout to test this hypothesis.

ESTIMATION AND PREDICTION IN LINEAR REGRESSION

Suppose we are interested in estimating the mean number of annual paid days off for all employees of large companies who have 20 years of service at such companies. Recalling basic property 3 in Table 13.3, we have $E(y) = \beta_0 + \beta_1 x$. The mean number of annual paid days off would be equal to $\beta_0 + 20\beta_1$. Since we do not know β_0 and β_1, we would use b_0 and b_1 to estimate β_0 and β_1. The point estimate for the mean would be $b_0 + 20b_1$. The standard error of this estimate is needed to set a confidence interval on the mean or to test an hypothesis about this mean. A general expression for a confidence interval when predicting the mean response will now be given.

A $100(1 - \alpha)$ % confidence interval for the mean response $E(y) = \beta_0 + \beta_1 x_0$ is given by formula (*13.19*):

$$b_0 + b_1 x_0 \pm t_{\alpha/2} S \sqrt{\frac{1}{n} + \frac{(x_0 - \overline{x})^2}{S_{xx}}} \qquad (13.19)$$

EXAMPLE 13.16 The data in Table 13.2 will be used to obtain a 95% confidence interval for the mean number of annual paid days off for employees having 20 years of service. In Example 13.3, the estimated regression equation was found to be $\hat{y} = b_0 + b_1x = 11.0 + 0.5x$. The point estimate of the mean is $\hat{y} = 11.0 + 0.5(20) = 21$ days. The standard error of this point estimated is found by recalling from Example 13.3 that $\bar{x} = 15.2$, $S_{xx} = 1891.2$, and $n = 20$. In Example 13.7, we found that $S = 1.374$. Also, $x_0 = 20$. The standard error is equal to

$$S\sqrt{\frac{1}{n} + \frac{(x_0 - \bar{x})^2}{S_{xx}}} = 1.374\sqrt{\frac{1}{20} + \frac{(20 - 15.2)^2}{1891.2}} = 0.343$$

For $df = n - 2 = 18$ degrees of freedom, the student t value is $t_{.025} = 2.101$. The 95% margin of error is

$$t_{\alpha/2}S\sqrt{\frac{1}{n} + \frac{(x_0 - \bar{x})^2}{S_{xx}}} = 2.101 \times 0.343 = 0.72$$

The 95% confidence interval for the mean number of annual days off for employees having 20 years of service extends from $21 - 0.72 = 20.28$ days to $21 + 0.72 = 21.72$ days.

Suppose we wished to predict the number of annual days off for a single individual having 20 years of service rather than the mean number for all employees having 20 years of service. The prediction is still determined by using the estimated regression line as in the case of estimating the mean. However, the standard error is larger because a single observation is more uncertain than the mean of the population distribution.

A $100(1 - \alpha)$ % prediction interval when predicting a single observation y at $x = x_0$ given by formula (*13.20*):

$$b_0 + b_1x_0 \pm t_{\alpha/2}S\sqrt{1 + \frac{1}{n} + \frac{(x_0 - \bar{x})^2}{S_{xx}}} \qquad (13.20)$$

EXAMPLE 13.17 A 95% prediction interval for the number of annual days off for an individual having 20 years of service is found as follows. The point estimate is the same as that found when estimating the mean in Example 13.16, 21 years. The standard error associated with the prediction is found as follows:

$$t_{\alpha/2}S\sqrt{1 + \frac{1}{n} + \frac{(x_0 - \bar{x})^2}{S_{xx}}} = 1.374\sqrt{1 + \frac{1}{20} + \frac{(20 - 15.2)^2}{1891.2}} = 1.42 \text{ days}$$

The margin of error associated with this prediction is $2.101 \times 1.42 = 2.98$. The prediction interval extends from $21 - 2.98 = 18.02$ days to $21 + 2.98 = 23.98$ days.

EXAMPLE 13.18 In the Minitab output shown in Fig. 13-4, if the subcommand **Predict 20;** is added to the commands, the following additional output is obtained.

```
   Fit      StDev Fit        95.0% CI            95.0% PI
 21.000       0.343      (20.280, 21.720)     (18.023, 23.977)
```

The fit value is the value obtained by substituting 20 into the estimated regression equation. The StDev Fit value is the standard error associated with estimating the mean. In addition, the 95% confidence interval and the 95% prediction interval are also given.

LINEAR CORRELATION COEFFICIENT

The *linear correlation coefficient*, also known as simply the *correlation coefficient*, is a measure of the strength of the linear association between two variables. The sample correlation coefficient is given by formula (*13.21*). The correlation coefficient is also sometimes referred to as the *Pearson correlation coefficient*.

$$r = \frac{S_{xy}}{\sqrt{S_{xx}S_{yy}}} \qquad (13.21)$$

EXAMPLE 13.19 The correlation coefficient for the data in Table 13.2 will be determined. In Example 13.3 we found that $S_{xy} = 945.6$ and $S_{xx} = 1891.2$. In Example 13.6, we found that $S_{yy} = 506.8$. The correlation coefficient is therefore found as follows.

$$r = \frac{S_{xy}}{\sqrt{S_{xx}S_{yy}}} = \frac{945.6}{\sqrt{1891.2 \times 506.8}} = 0.966.$$

The Minitab computation for r is as shown below. The x values are put into column 1 and the y values are put into column 2 and the command **corr c1 c2** is used to obtain the correlation.

```
MTB > corr c1 c2
```

Correlations (Pearson)

```
Correlation of Years and Daysoff = 0.966
```

The basic properties of the sample correlation coefficient are given in Table 13.7.

Table 13.7

Basic Properties of r
1. The value of r is always between –1 and +1, i.e., $-1 \le r \le +1$.
2. The magnitude of r indicates the strength of the linear relationship, and the sign of r indicates whether the relation is direct or inverse. If r is positive, then y tends to increase linearly as x increases, with the tendency being greater the closer that r is to 1. If r is negative, then y tends to decrease linearly as x increases, with the tendency being greater the closer that r is to –1. If all the points on the scatter diagram lie perfectly on a straight line with a positive slope, the r = +1. If all the points on the scatter diagram lie perfectly on a straight line with a negative slope, then r = –1.
3. A value of r close to –1 or +1 represents a strong linear relationship.
4. A value of r close to 0 represents a weak linear relationship.

If the formula for r is applied to every pair of (x, y) values in the population, we obtain the *population correlation coefficient*. The Greek letter ρ represents the population correlation coefficient.

EXAMPLE 13.20 The data given in Table 13.1 and the corresponding scatter plot shown in Fig. 13-1 are reproduced below and on the next page.

x	y
2	12
4	13
8	15
10	16
16	19
20	21
24	23
30	26

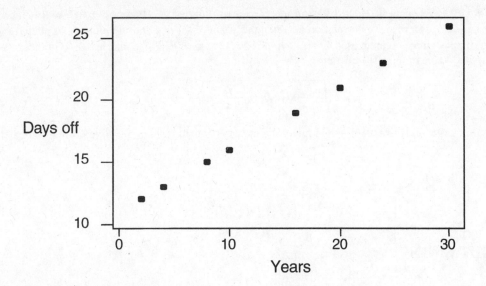

For the data shown in the table, we have:

$n = 8, \Sigma x = 114, \Sigma y = 145, \Sigma x^2 = 2316, \Sigma y^2 = 2801,$ and $\Sigma xy = 2412.$

$S_{xx} = \Sigma x^2 - \dfrac{(\Sigma x)^2}{n} = 2316 - 1624.5 = 691.5, \quad S_{yy} = \Sigma y^2 - \dfrac{(\Sigma y)^2}{n} = 2801 - 2628.125 = 172.875,$ and

$S_{xy} = \Sigma xy - \dfrac{(\Sigma x)(\Sigma y)}{n} = 2412 - 2066.25 = 345.75.$ The correlation coefficient is then found as follows:

$$r = \frac{S_{xy}}{\sqrt{S_{xx}S_{yy}}} = \frac{345.75}{\sqrt{691.5 \times 172.875}} = 1.0$$

Since the points fall exactly on a straight line with a positive slope, the correlation coefficient is equal to 1.

INFERENCE CONCERNING THE POPULATION CORRELATION COEFFICIENT

One of the most important uses of the sample correlation coefficient is the determination of whether or not a correlation exists between two population variables. Recall that r measures the correlation between x and y in the sample and ρ is the corresponding population correlation. In particular, we are often interested in testing the null hypothesis that $\rho = 0$ vs. one of three alternatives. The alternative $\rho \neq 0$ states that a correlation is present. The alternative $\rho < 0$ states that an inverse correlation exists, and the alternative $\rho > 0$ states that a direct correlation exists. Table 13.8 gives the steps for testing a hypothesis concerning ρ.

Table 13.8

Steps for Testing for No Population Correlation
Step 1: The null hypothesis is H_0: $\rho = 0$ and the alternative hypothesis is either lower, upper, or two tailed.
Step 2: Use the Student t distribution table and the level of significance α to determine the rejection regions. The degrees of freedom is $n - 2$.
Step 3: The test statistic is computed as $t^* = r\sqrt{\dfrac{n-2}{1-r^2}}$.
Step 4: State your conclusions. The null hypothesis is rejected if the computed value of the test statistic falls in the rejection region. Otherwise, the null hypothesis is not rejected.

EXAMPLE 13.21 The data in Table 13.2 and the computed value of r in Example 13.19 will be used to test whether or not there is a positive correlation between the years of service and the number of annual paid days off for employees of large companies. The null hypothesis is $H_0: \rho = 0$ and the alternative hypothesis is $H_a: \rho > 0$. The computed value of the test statistic is:

$$t^* = r\sqrt{\frac{n-2}{1-r^2}} = .966\sqrt{\frac{20-2}{1-.966^2}} = 15.85$$

Because of this large value of t*, we would reject the null and conclude that a positive correlation exists in the population.

Solved Problems

STRAIGHT LINES

13.1 Find the slope and y intercept of the following straight lines.

(a) y = 1.5x – 2.5 (b) y = 3.0 – 2.5x (c) 2y = 4x –6 (d) 16x – 32y + 8 = 0

Ans. In order to find the slope and y intercept, each equation will be put into the form y = mx + b. The number m is the slope and the number b is the y intercept. (a) The slope is m = 1.5 and the y intercept is b = –2.5. (b) The slope is m = –2.5 and the y intercept is b = 3.0. (c) The given line is equivalent to y = 2x – 3 and slope is m = 2 and the y intercept is b = –3. (d) The given line is equivalent to y = .5x + .25 and the slope is m = .5 and the y intercept is b = .25.

13.2 Determine which of the following points fall on the line y = –2x + 4.

(a) (2, 0) (b) (0, 2) (c) (50, 96) (d) (–10, 24) (e) (14.23, –24.46)

Ans. The points given in (a), (d), and (e) fall on the line since they satisfy the equation.

LINEAR REGRESSION MODEL

13.3 Identify the values of β_0, β_1, and σ in the linear regression model y = 12.1 + 3.5x + e, where e is a normal random with mean 0 and standard 1.2.

Ans. $\beta_0 = 12.1$, $\beta_1 = 3.5$ and $\sigma = 1.2$.

13.4 Consider the linear regression model y = –1.5x + 5.6 + e, where e is a normal random variable with mean 0 and variance $\sigma^2 = 4$. Determine the mean and standard deviation of y when x = 2 and when x = 4.5.

Ans. When x = 2, y = –1.5 (2) + 5.6 + e = 2.6 + e. E(y) = E(2.6 + e) = 2.6. Since 2.6 is a constant, the variance of y is the same as the variance of e or 4. Therefore the standard deviation of y is 2 when x = 2.

When x = 4.5, y = –1.5(4.5) + 5.6 + e = –1.15 + e. E(y) = E(–1.15 + e) = –1.15. Since –1.15 is a constant, the variance of y is the same as the variance of e or 4. Therefore the standard deviation of y is 2 when x = 4.5. Notice that the standard deviation of y is 2, regardless of the value of x.

LEAST SQUARES LINE

13.5 The number of hours spent per week viewing TV, y, and the number of years of education, x, were recorded for 10 randomly selected individuals. The results are given in Table 13.9. In addition, this table gives computations needed in finding b_0 and b_1. Find the least-squares line for these data.

Ans. $S_{xx} = \Sigma x^2 - \dfrac{(\Sigma x)^2}{n} = 2085 - 1988.1 = 96.9 \quad S_{xy} = \Sigma xy - \dfrac{(\Sigma x)(\Sigma y)}{n} = 1351 - 1494.6 = -143.6$

$\bar{x} = 14.1 \quad \bar{y} = 10.6 \quad b_1 = \dfrac{S_{xy}}{S_{xx}} = \dfrac{-143.6}{96.9} = -1.4819 \quad b_0 = \bar{y} - b_1\bar{x} = 10.6 - (-1.4819)(14.1)$

$b_0 = 10.6 + 20.8948 = 31.4948$

The equation of the least-squares line is $\hat{y} = b_0 + b_1x = 31.495 - 1.482x$.

Table 13.9

x	y	x^2	y^2	xy
12	10	144	100	120
14	9	196	81	126
11	15	121	225	165
16	8	256	64	128
16	5	256	25	80
18	4	324	16	72
12	20	144	400	240
20	4	400	16	80
10	16	100	256	160
12	15	144	225	180
$\Sigma x = 141$	$\Sigma y = 106$	$\Sigma x^2 = 2085$	$\Sigma y^2 = 1408$	$\Sigma xy = 1351$

13.6 The Minitab output for the data in problem 13.5 is shown in Fig. 13-5. Give the estimated regression line.

```
MTB > Regress 'TVhours' 1 'Yearsedu';
SUBC> Constant;
SUBC> Predict 15.

Regression Analysis
The regression equation is
TVhours = 31.5 - 1.48 Yearsedu

Predictor       Coef       StDev          T         P
Constant       31.495      4.388       7.18     0.000
Yearsedu      -1.4819      0.3039      -4.88     0.000

S = 2.992      R-Sq = 74.8%     R-Sq(adj) = 71.7%

Analysis of Variance
Source        DF        SS          MS         F         P
Regression     1      212.81      212.81     23.78     0.000
Error          8       71.59        8.95
Total          9      284.40

Fit    St Dev Fit      95.0% CI           95.0% PI
9.266      0.985    (6.995, 11.538)   (2.002, 16.531)
```

Fig. 13-5

Ans. From Fig. 13-5, we see that the estimated regression equation is given in the following portion of the output in Fig. 13-5. Note that this is the same equation as the least-squares line found in problem 13.5.

```
The regression equation is TVhours = 31.5 − 1.48 Yearsedu

Predictor        Coef       StDev            T          P
Constant        31.495      4.388         7.18      0.000
Yearsedu        -1.4819     0.3039        -4.88     0.000
```

ERROR SUM OF SQUARES

13.7 Compute the error sum of squares for the data in Table 13.9 using both the defining formula and the computational formula.

Ans. Table 13.10 gives the details of the computation of SSE $= \Sigma(y_i - \hat{y}_i)^2$. The columns from left to right give the following information: Individual or subject, Number of years of education, Number of hours spent per week watching TV, The fitted or estimated value using the least squares line, The residual, and The squares of the residuals.

Table 13.10

Subject	x_i	y_i	$\hat{y}_i = 31.495 - 1.482x$	$(y_i - \hat{y}_i)$	$(y_i - \hat{y}_i)^2$
1	12	10	13.7121	−3.71207	13.7795
2	14	9	10.7482	−1.74819	3.0562
3	11	15	15.1940	−0.19401	0.0376
4	16	8	7.7843	0.21569	0.0465
5	16	5	7.7843	−2.78431	7.7524
6	18	4	4.8204	−0.82043	0.6731
7	12	20	13.7121	6.28793	39.5380
8	20	4	1.8566	2.14345	4.5944
9	10	16	16.6760	−0.67595	0.4569
10	12	15	13.7121	1.28793	1.6588
				$\Sigma(y_i - \hat{y}_i) = 0$	$\Sigma(y_i - \hat{y}_i)^2 = 71.593$

The error sum of squares may also be computed using SSE $= S_{yy} - \dfrac{S_{xy}^2}{S_{xx}}$. From problem 13.5, we have

$S_{xx} = 96.9$ and $S_{xy} = -143.6$. In addition, $S_{yy} = \Sigma y^2 - \dfrac{(\Sigma y)^2}{n} = 1408 - 1123.6 = 284.4$. Using these values,

we find SSE $= S_{yy} - \dfrac{S_{xy}^2}{S_{xx}} = 284.4 - \dfrac{(-143.6)^2}{96.9} = 71.593$.

13.8 Using the Minitab output shown in Fig. 13-5, locate SSE and compare it with the result found in problem 13.7.

Ans. The error sum of squares is shown in bold in the following portion of Fig. 13-5.

```
Analysis of Variance

Source        DF       SS          MS           F          P
Regression    1        212.81      212.81     23.78      0.000
Error         8        71.59       8.95
Total         9        284.40
```

STANDARD DEVIATION OF ERRORS

13.9 Use the computed value for SSE in problems 13.7 and 13.8 to find the standard deviation of errors, S.

Ans. The standard deviation of errors is given by the formula $S = \sqrt{\dfrac{SSE}{n-2}} = \sqrt{\dfrac{71.593}{8}} = 2.9915$.

13.10 Locate the standard deviation of errors in Fig. 13-5 and confirm that it is the same as that computed in problem 13.9.

Ans. The following row taken from Fig. 13-5 gives the value for S.

$$\textbf{S = 2.992} \qquad \text{R-Sq} = 74.8\% \qquad \text{R-Sq(adj)} = 71.7\%$$

TOTAL SUM OF SQUARES

13.11 Find the total sum of squares for the data given in Table 13.9.

Ans. The total sum of squares is given by $SST = \Sigma(y_i - \bar{y})^2 = \Sigma y^2 - \dfrac{(\Sigma y)^2}{n} = 1408 - \dfrac{106^2}{10} = 1408 -$ $1123.6 = 284.4$. The values for Σy^2 and Σy are found at the bottom of Table 13.9.

13.12 Locate the Total sum of squares in Fig. 13-5 and confirm that it is the same as that found in problem 13.11.

Ans. The following portion of Fig. 13-5 gives the total sum of squares. It is the same as that found in problem 13.11.

Source	DF	SS	MS	F	P
Regression	1	212.81	212.81	23.78	0.000
Error	8	71.59	8.95		
Total	9	**284.40**			

REGRESSION SUM OF SQUARES

13.13 Find the regression sum of squares for the data given in Table 13.9 by subtraction as well as by direct computation.

Ans. The regression sum of squares is given by SSR = SST − SSE = 284.4 − 71.593 = 212.807, where SST is computed in problems 13.11 and 13.12 and SSE is computed in problems 13.7 and 13.8. The direct computation is illustrated in Table 13.11. The mean of the y values is 10.6.

Table 13.11

Subject	x_i	y_i	$\hat{y}_i = 31.495 - 1.482x$	$(y_i - \bar{y})$	$(y_i - \bar{y})^2$
1	12	10	13.7121	3.11207	9.6850
2	14	9	10.7482	0.14819	0.0220
3	11	15	15.1940	4.59401	21.1050
4	16	8	7.7843	−2.81569	7.9281
5	16	5	7.7843	−2.81569	7.9281
6	18	4	4.8204	−5.77957	33.4034
7	12	20	13.7121	3.11207	9.6850
8	20	4	1.8566	−8.74345	76.4479
9	10	16	16.6760	6.07595	36.9172
10	12	15	13.7121	3.11207	9.6850
				Sum = 0	Sum = 212.81

13.14 Locate the regression sum of squares in Fig. 13-5 and confirm that you get the same value as that computed in problem 13.13.

Ans. The regression sum of squares is shown in the following portion of output selected from Fig. 13-5.

```
Source        DF       SS        MS        F       P
Regression    1      212.81    212.81    23.78    0.000
Error         8       71.59      8.95
Total         9      284.40
```

COEFFICIENT OF DETERMINATION

13.15 Calculate the coefficient of determination for the results given in Table 13.9 and explain its meaning.

Ans. The coefficient of determination is given by $r^2 = \dfrac{\text{SSR}}{\text{SST}} = \dfrac{212.81}{284.4} = 0.748$. The estimated regression equation accounts for about 75% of the variation in TV viewing time.

13.16 Locate the coefficient of determination in Fig. 13-5 and confirm that you get the same value as that computed in problem 13.15.

Ans. The r^2 value shown in the following row, taken from Fig. 13-5, is the same as that in problem 13.15.

```
S = 2.992        R-Sq = 74.8%        R-Sq(adj) = 71.7%
```

MEAN, STANDARD DEVIATION, AND SAMPLING DISTRIBUTION OF THE SLOPE OF THE ESTIMATED REGRESSION EQUATION

13.17 Find the standard error for b_1 using the data in Table 13.9.

Ans. The standard error for b_1 is given by $S_{b_1} = \dfrac{S}{\sqrt{S_{XX}}} = \dfrac{2.992}{\sqrt{96.9}} = 0.304$. The value for S is found in problem 13.9 and the value for S_{xx} is given in problem 13.5.

13.18 Locate the value for the standard error for b_1 in Fig. 13-5 and compare it with the value computed in problem 13.17.

Ans. The standard error of b_1 is shown in bold print in the following selection taken from Fig. 13-5. It is the same as that computed in problem 13.17.

```
Predictor     Coef      St Dev       T        P
Constant     31.495     4.388      7.18     0.000
Yearsedu     -1.4819    0.3039    -4.88     0.000
```

INFERENCES CONCERNING THE SLOPE OF THE POPULATION REGRESSION LINE

13.19 Use the data in Table 13.9 to test H_0: $\beta_1 = 0$ vs. H_a: $\beta_1 < 0$ at significance level $\alpha = .05$.

Ans. First, let us determine the critical value. The degrees of freedom is df = n − 2 = 8. The t value for a right tail area equal to .05 is $t_{.05} = 1.860$. Since this is a lower-tail test, the critical value is

−1.860. The computed value of the test statistic is $t^* = \dfrac{b_1 - \beta_{10}}{S_{b_1}}$. From problem 13.5, we know that $b_1 = -1.4819$, and from problem 13.17, $S_{b_1} = 0.304$. The hypothesized value for the slope is $\beta_{10} = 0$. The computed value of the test statistic is therefore $t^* = \dfrac{-1.4819 - 0}{.304} = -4.87$. Since this value is smaller than −1.860, we reject the null hypothesis and conclude that the slope is less than zero. This indicates that the number of hours spent viewing TV and the number of years of education are inversely related.

13.20 Locate the computed value of the test statistic in problem 13.19 in the Minitab output given in Fig. 13-5.

Ans. The following selection from Fig. 13-5 gives the computed test statistic as −4.88 with a corresponding p value equal to 0.000.

```
Predictor       Coef        StDev         T          P
Constant       31.495       4.388       7.18      0.000
Yearsedu       -1.4819      0.3039      -4.88      0.000
```

ESTIMATION AND PREDICTION IN LINEAR REGRESSION

13.21 Find a 95% confidence interval for the mean number of hours spent per week watching TV for all individuals with 15 years of education and a 95% prediction interval for an individual with 15 years of education using the data in Table 13.9.

Ans. The 95% confidence interval for the mean is given by $b_0 + b_1 x_0 \pm t_{\alpha/2} S \sqrt{\dfrac{1}{n} + \dfrac{(x_0 - \bar{x})^2}{S_{xx}}}$. The point estimate for the mean is $\hat{y} = b_0 + b_1(15) = 31.495 - 1.482(15) = 9.265$. The margin of error associated with this estimate is $t_{\alpha/2} S \sqrt{\dfrac{1}{n} + \dfrac{(x_0 - \bar{x})^2}{S_{xx}}}$. The value for $t_{.025}$ is found by using 8 degrees of freedom to be 2.306. The following values have been found in previous problems: $S = 2.992$, $n = 10$, $\bar{x} = 14.1$, $S_{xx} = 96.9$, and $x_0 = 15$. Therefore, the margin of error is: $2.306 \times 2.992 \times \sqrt{\dfrac{1}{10} + \dfrac{(15 - 14.1)^2}{96.9}} = 2.27$. The 95% confidence interval for the mean number of hours watching TV for all individuals having 15 years of education extends from $9.265 - 2.27 = 6.995$ to $9.265 + 2.27 = 11.535$. The 95% prediction interval for an individual is given by $b_0 + b_1 x_0 \pm t_{\alpha/2} S \sqrt{1 + \dfrac{1}{n} + \dfrac{(x_0 - \bar{x})^2}{S_{xx}}}$.

The point estimate is the same as given above when setting a confidence interval on the mean and the t value is the same as above. The margin of error is computed as follows: $2.306 \times 2.992 \times \sqrt{1 + \dfrac{1}{10} + \dfrac{(15 - 14.1)^2}{96.9}} = 7.264$. The 95% prediction interval extends from $9.265 - 7.264 = 2.001$ to $9.265 + 7.264 = 16.529$. Note that the prediction interval is much wider than the confidence interval for the mean.

13.22 Locate the 95% confidence interval and the 95% prediction interval in Fig. 13-5 and compare with the results in problem 13.21.

Ans. The following portion of the output given in Fig. 13-5 gives the 95% confidence interval and prediction interval. The results are the same except for a small amount of round off error.

```
Fit          St Dev Fit      95.0% CI            95.0% PI
9.266          0.985      (6.995, 11.538)      (2.002, 16.531)
```

LINEAR CORRELATION COEFFICIENT

13.23 Compute the linear correlation coefficient for the data in Table 13.9.

Ans. The correlation coefficient is given by the formula $r = \dfrac{S_{xy}}{\sqrt{S_{xx} S_{yy}}}$. In problem 13.5, we found that $S_{xx} = 96.9$ and $S_{xy} = -143.6$. In problem 13.7, we found that $S_{yy} = 284.4$. The correlation coefficient is equal to $\dfrac{-143.6}{\sqrt{96.9 \times 284.4}} = -.865.$

13.24 Determine the correlation coefficient using the computer output in Fig. 13-5.

Ans. In problem 13.16, we found that $r^2 = .748$. Therefore $r = \pm\sqrt{.748} = \pm .865$. The negative sign is taken since we know that the variables are inversely related. We know they are inversely related since the slope of the estimated regression line is negative. Therefore $r = -.865$.

INFERENCE CONCERNING THE POPULATION CORRELATION COEFFICIENT

13.25 Use the data in Table 13.9 to test H_0: $\rho = 0$ vs. H_a: $\rho < 0$ at $\alpha = .01$.

Ans. The critical value is determined by noting that $df = n - 2 = 10 - 2 = 8$ and that $t_{01} = 2.896$. Since this is a lower-tail test, the critical value is -2.896. The test statistic is computed as $t^* = r\sqrt{\dfrac{n-2}{1-r^2}} = -.865\sqrt{\dfrac{8}{1-.748225}} = -4.876$. Since this value is less than the critical value, we conclude that the two variables are negatively correlated.

13.26 A sample of 100 federal taxpayers were interviewed and the annual income and cost to prepare their return was determined for each. The correlation coefficient between the two variables was found to be 0.37. Do these results indicate a positive correlation between these two variables in the population? Use $\alpha = .05$.

Ans. Because of the large sample size, we may use the standard normal distribution table to find the critical value. The critical value is 1.645. The computed value of the test statistic is $t^* = r\sqrt{\dfrac{n-2}{1-r^2}} = .37\sqrt{\dfrac{98}{.8631}} = 3.94$. Since the computed value of the test statistic exceeds the critical value, we conclude that there is a positive population correlation.

Supplementary Problems

STRAIGHT LINES

13.27 Figure 13-6 shows the line whose equation is $y = 4 + 2x$ and 8 points, labeled A through H. Points A, C, F, and G have the following coordinates: A:(2, 6), C:(2, 10), G:(5, 16), and F(5, 12). Find the coordinates for the points B, D, E, and H. Also find the distance between the following pairs of points: A and B, B and C, H and G, and H and F. Find the sum of the squares of the distances between the four pairs of points.

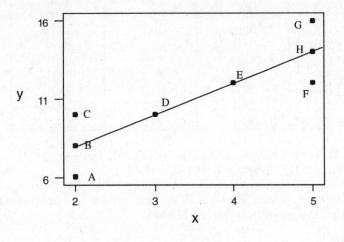

Fig. 13-6

Ans.   B: (2, 8), D: (3, 10), E: (4, 12), and H: (5, 14)
         All four of the distances are equal to 2 units. The sum of the squares of the distances is equal to 16.

13.28   Use the coordinates of the points B and H found in problem 13.27 to find the slope of the straight line passing through these points and compare this with the slope of the line $y = 4 + 2x$.

Ans.   The slope of the line is $m = \dfrac{y_2 - y_1}{x_2 - x_1} = \dfrac{14 - 8}{5 - 2} = 2$. The slope of the line $y = 4 + 2x$ is $m = 2$.

LINEAR REGRESSION MODEL

13.29   A linear regression model has a slope of 12.5 and a y intercept equal to 19.2. The error term has a standard deviation equal to 2.5. Give the equation of the linear regression model.

Ans. $y = 19.2 + 12.5x + e$

13.30   For the linear regression model described in problem 13.29, find the mean values for y when x equals 2 and 5.

Ans.   44.2 and 81.7

LEAST-SQUARES LINE

13.31   Table 13.12 gives the household income in thousands of dollars, x, and cost of filing federal taxes in dollars, y, for 20 randomly selected federal taxpayers. Find the following: Σx, Σx^2, Σy, Σy^2, Σxy, S_{xx}, S_{yy}, S_{xy}, b_1, and b_0.

Table 13.12

x	y	x	y
17.5	35.0	28.0	75.0
37.5	60.5	22.5	70.0
47.5	88.5	25.0	60.0
25.0	70.5	29.5	65.0
55.5	125.0	65.0	150.0
35.0	63.0	51.0	100.0
15.5	30.0	39.3	75.0
12.0	30.0	33.0	40.0
32.0	65.0	45.0	75.0
42.3	80.0	75.0	200.0

Ans. $\Sigma x = 733.10$, $\Sigma x^2 = 31,992$, $\Sigma y = 1557.50$, $\Sigma y^2 = 153,407$, $\Sigma xy = 68,864$, $S_{xx} = 5120.2195$, $S_{yy} = 32116.6875$, $S_{xy} = 11773.8375$, $b_1 = 2.2995$, $b_0 = -6.4132$.

13.32 The Minitab output for the regression of y on x for the data in Table 13.12 is shown in Fig. 13-7. Give the portion of the output that shows the values for b_0 and b_1.

```
MTB > Regress 'Cost' 1 'Income';
SUBC>    Predict 40.
Regression Analysis


The regression equation is
Cost = - 6.42 + 2.30 Income

Predictor      Coef       St Dev        T          P
Constant      -6.420 .     9.353      -0.69      0.501
Income         2.2997      0.2339      9.83      0.000

S = 16.73      R-Sq = 84.3%     R-Sq(adj) = 83.4%

                   Analysis of Variance

Source        DF        SS          MS         F          P
Regression    1        27076       27076      96.70      0.000
Error         18        5040         280
Total         19       32116

     Fit    St Dev Fit      95.0% CI          95.0% PI
    85.57      3.82      (77.53, 93.60)    (49.50, 121.64)
```

Fig. 13-7

Ans. The estimate regression equation is shown in bold print. The differences between these values and those given in problem 13.31 are due to round off error.

ERROR SUM OF SQUARES

13.33 Use the results given in problem 13.31 to find SSE for the regression line fit to the data in Table 13.12.

Ans. $\text{SSE} = S_{yy} - \dfrac{S_{xy}^2}{S_{xx}} = 32116.6875 - 27073.6927 = 5042.9948$

13.34 The Minitab output for the regression of y on x for the data in Table 13.12 is shown in Fig. 13-7. Give the portion of the output that shows the value for SSE.

Ans.   Error   18   **5040**   280
The difference in the values for SSE in problems 13.33 and 13.34 are due to round-off error.

STANDARD DEVIATION OF ERRORS

13.35   Use the result found in problem 13.33 to find the standard deviation of errors for the regression line fit to the data in Table 13.12.

Ans. $S = \sqrt{\dfrac{SSE}{n-2}} = \sqrt{\dfrac{5042.9948}{18}} = 16.7382$

13.36   The Minitab output for the regression of y on x for the data in Table 13.12 is shown in Fig. 13-7. Give the portion of the output that shows the value for the standard deviation of errors.

Ans.   **S = 16.73**   R-Sq = 84.3%   R-Sq(adj) = 83.4%

TOTAL SUM OF SQUARES

13.37   Use the results given in problem 13.31 to find the total sum of squares for the data given in Table 13.12.

Ans. $SST = S_{yy} = 32116.6875$

13.38   The Minitab output for the regression of y on x for the data in Table 13.12 is shown in Fig. 13-7. Give the portion of the output that shows the value for the total sum of squares.

Ans.   Total   19   **32116**

REGRESSION SUM OF SQUARES

13.39   Use the results of problems 13.33 and 13.37 to find the regression sum of squares for the data in Table 13.12.

Ans. $SSR = SST - SSE = 32116.6875 - 5042.9948 = 27073.6927$

13.40   The Minitab output for the regression of y on x for the data in Table 13.12 is shown in Fig. 13-7. Give the portion of the output that shows the value for the regression sum of squares.

Ans.   Regression   1   **27076**   27076   96.70   0.000
The difference in the answers given in problems 13.39 and 13.40 is due to round-off error.

COEFFICIENT OF DETERMINATION

13.41   Use the total sum of squares found in problem 13.37 and the regression sum of squares found in problem 13.39 to find the coefficient of determination for the linear regression model applied to the data in Table 13.12.

Ans. $r^2 = \dfrac{SSR}{SST} = \dfrac{27073.6927}{32116.6875} = 0.843$

13.42   The Minitab output for the regression of y on x for the data in Table 13.12 is shown in Fig. 13-7. Give the portion of the output that shows the value of the coefficient of determination.

Ans.   S = 16.73   **R-Sq = 84.3%**   R-Sq(adj) = 83.4%

MEAN, STANDARD DEVIATION, AND SAMPLING DISTRIBUTION OF THE SLOPE OF THE ESTIMATED REGRESSION EQUATION

13.43 Find the standard error for b_1, the slope of the estimated regression equation found in problem 13.31.

Ans. $S_{b_1} = \dfrac{S}{\sqrt{S_{xx}}} = \dfrac{16.7382}{\sqrt{5120.2195}} = 0.2339$

13.44 The Minitab output for the regression of y on x for the data in Table 13.12 is shown in Fig. 13-7. Give the portion of the output that shows the value standard error for b_1.

Ans. Income   2.2997   **0.2339**   9.83   0.000

INFERENCES CONCERNING THE SLOPE OF THE POPULATION REGRESSION LINE

13.45 Using the data in Table 13.12, test the hypothesis that the slope of the line in the population regression model is equal to 0. Use level of significance $\alpha = .05$.

Ans. The critical values are ± 2.101. The computed value of the test statistic is $t^* = \dfrac{b_1 - \beta_{10}}{S_{b_1}} =$

$\dfrac{2.2995 - 0}{.2339} = 9.8311$. The slope of the regression line is different from 0.

13.46 Give the portion of Fig. 13-7 that is used to test the null hypothesis that the slope of the regression line is equal to 0. The regression model is $y = \beta_0 + \beta_1 x + e$, where x = household income in thousands, and y = cost of filing federal taxes. Assume $\alpha = .01$.

Ans.

Predictor	Coef	St Dev	T	P
Constant	-6.420	9.353	-0.69	0.501
Income	2.2997	0.2339	**9.83**	**0.000**

The computed test statistic is seen to be the same as in problem 13.45. The p value, 0.000, indicates that the null hypothesis should be rejected.

ESTIMATION AND PREDICTION IN LINEAR REGRESSION

13.47 Find a 95% confidence interval for the mean cost of filing federal income taxes for all those individuals with household income equal to $40,000 and a 95% prediction interval for an individual with household income equal to $40,000 using the data in Table 13.12.

Ans. The 95% confidence interval for the mean is given by $b_0 + b_1 x_0 \pm t_{\alpha/2} S \sqrt{1 + \dfrac{1}{n} + \dfrac{(x_0 - \overline{x})^2}{S_{xx}}}$. The

point estimate of the mean is $-6.4132 + 2.2995 \times 40 = 85.5668$. The margin of error is $\pm 2.101 \times$

$16.7382 \times \sqrt{\dfrac{1}{20} + \dfrac{(40 - 36.655)^2}{5120.2195}}$ or ± 8.0336. The 95% confidence interval extends from 85.5668

$- 8.0336 = \$77.53$ to $85.5668 + 8.0336 = \$93.60$. The 95% prediction interval for an individual is

given by $b_0 + b_1 x_0 \pm t_{\alpha/2} S \sqrt{1 + \dfrac{1}{n} + \dfrac{(x_0 - \overline{x})^2}{S_{xx}}}$. The point estimate for the individual is 85.5668.

The margin of error is $\pm 2.101 \times 16.7382 \times \sqrt{1 + \dfrac{1}{20} + \dfrac{(40 - 36.655)^2}{5120.2195}}$ or ± 36.0729. The 95%

prediction interval extends from $85.5668 - 36.0729 = \$49.49$ to $85.5668 + 36.0729 = \$121.64$.

13.48 Locate the 95% confidence interval and the 95% prediction interval in Fig. 13-7 and compare with the results in problem 13.47.

Ans.
```
Fit        St Dev Fit     95.0% CI          95.0% PI
85.57         3.82      (77.53, 93.60)    (49.50, 121.64)
```
The 95% confidence interval and the 95% prediction intervals are seen to be the same as those found in problem 13.47.

LINEAR CORRELATION COEFFICIENT

13.49 Calculate the correlation coefficient between the household income and the cost of filing federal taxes using the data in Table 13.12.

Ans. The correlation coefficient is given by $r = \dfrac{S_{xy}}{\sqrt{S_{xx}S_{yy}}} = \dfrac{11773.8375}{\sqrt{5120.2195 \times 32116.6875}} = 0.918$.

13.50 Using the output shown in Fig. 13-7, determine the correlation coefficient for the data given in Table 13.12.

Ans. $S = 16.73$ **R-Sq = 84.3%**   R-Sq(adj) = 83.4%

Since $r^2 = .843$, $r = \sqrt{.843} = 0.918$.

INFERENCE CONCERNING THE POPULATION CORRELATION COEFFICIENT

13.51 Use the computed value for r in problem 13.49 to test the hypothesis H_0: $\rho = 0$ vs. H_a: $\rho \neq 0$ at level of significance $\alpha = .05$, where ρ represents the correlation between household income and the cost of filing federal taxes in the population.

Ans. The test statistic is computed as $t^* = r\sqrt{\dfrac{n-2}{1-r^2}} = .918\sqrt{\dfrac{18}{1-.842724}} = 9.82$. The critical values are ± 2.101. The null hypothesis is rejected at $\alpha = .05$.

13.52 How can the Minitab output in Fig. 13-7 be used to test the hypothesis H_0: $\rho = 0$ vs. H_a: $\rho \neq 0$ at level of significance $\alpha = .05$, where ρ represents the correlation between household income and the cost of filing federal taxes in the population.

Ans. The hypothesis H_0: $\rho = 0$ vs. H_a: $\rho \neq 0$ is equivalent to the hypothesis H_0: $\beta_1 = 0$ vs. H_a: $\beta_1 \neq 0$. The following portion of the output is used to test H_0: $\beta_1 = 0$ vs. H_a: $\beta_1 \neq 0$.

```
Predictor     Coef      St Dev        T         P
Constant    -6.420      9.353     -0.69     0.501
Income      2.2997      0.2339      9.83     0.000
```
The p value = 0.000 indicates that the null hypothesis is rejected.

Chapter 14

Nonparametric Statistics

NONPARAMETRIC METHODS

Many of the hypotheses tests discussed in previous chapters required various assumptions such as normality of the characteristic being measured or equality of population variances in two sample tests for example. What can we do if the assumptions are not satisfied? In addition, many research studies involve low-level data such as nominal or ordinal data to which previously discussed procedures do not apply. In a taste test, the response may be Pepsi or Coke when asked which of two colas are preferred. In situations such as these, *nonparametric* statistical methods are often used.

Nonparametric tests often replace the raw data with *ranks*, *signs*, or both ranks and signs. The nonparametric tests then analyze the resulting ranks and/or signs. Since the original data are not actually used, one criticism of these procedures is that they are wasteful of information.

EXAMPLE 14.1 Table 14.1 gives a set of data and the corresponding ranks associated with the data values. The smallest number receives a rank of 1, the next a rank of 2, and so on. If two or more values are tied, they receive the average of the ranks that would normally occur. Note that 13.5 occurred 3 times and the ranks would have been 3, 4, and 5. The average of 3, 4, and 5 is 4 and so 4 was assigned to each occurrence of the value 13.5. The sum of the ranks 1 through n is always equal to $\frac{n(n+1)}{2}$. If n = 10, then the sum of the ranks is $\frac{10 \times 11}{2} = 55$. This sum is obtained even if ties occur in the data. If the ranks are added in Table 14.1, the sum 55 is obtained. The replacement of data with ranks will be utilized in many of the following sections.

Table 14.1

Value	17.2	13.5	15.5	12.5	13.5	16.0	11.5	13.5	14.3	18.0
Rank	9	4	7	2	4	8	1	4	6	10

EXAMPLE 14.2 In a taste test, each individual was asked to taste a pizza with a thick crust and a pizza with a thin crust, and to state which one they preferred. A coin was flipped to determine which one they tasted first in order to eliminate the order of tasting as a factor. Table 14.2 gives the response for each individual and each response is coded as + if the response was thick and – if the response was thin. The results seem to indicate that the thin crust is preferred over the thick crust. But are these sample results significant? That is, what do these results tell us about the set of preferences in the population? Seventy percent of the sample preferred thin over thick crust. What are the chances of these results if in fact there is no difference in preference in the population? The sign test in the next section will help us answer this question.

Table 14.2

Value	thin	thick	thin	thick	thin	thin	thin	thick	thin	thin
Sign	–	+	–	+	–	–	–	+	–	–

SIGN TEST

The *sign test* is one of the simplest and easiest to understand of the nonparametric tests. In Example 14.2, the purpose of the taste test is to decide if there is a difference in taste preference between thin and thick pizza crust. The null hypothesis is that there is no difference in preference for the two types of crusts. The research hypothesis might be one-tail or two-tail. Suppose it is a two-tail research hypothesis. As usual, we proceed under the assumption that the null hypothesis is true and only reject that assumption if our sample results lead us to reject it. If there is no difference in preference, then the probability of a + on any trial is .5 and the probability of a − is also .5. The number of + signs in the 10 trials, x, is a binomial random variable. The p value associated with the outcome x = 3 is computed by finding $P(x \leq 3)$ and then doubling the result since this is a two-tail test. Using the binomial distribution tables in Appendix 1 with n = 10 and p = .5, we find that $P(x \leq 3)$ is equal to .0010 + .0098 + .0439 + .1172 = .1719 and the p value = 2 × .1719 = 0.3438. At the conventional level of significance $\alpha = .05$ we cannot reject the null hypothesis since the p value is not less than the preselected level of significance. We see from this discussion that the sign test uses the binomial distribution to perform the sign test. The p value may be computed by using Minitab. The computation using Minitab is as follows:

```
MTB > cdf 3;
SUBC> binomial n = 10 p = .5.
```

Cumulative Distribution Function

Binomial with n = 10 and p = 0.500000

```
     x        P(x <= x)
   3.00         0.1719
```

The p value is then found by doubling 0.1719 to get 0.3438 as found above.

EXAMPLE 14.3 In order to test the hypothesis that the median price for a home in a city is $105,000, a sample of 10 recently sold homes is obtained. If the selling price exceeds $105,000, a plus is recorded. If the price is less than $105,000, a negative sign is recorded. If the price is equal to $105,000, the selling price is not used in the analysis and the sample size is reduced. Table 14.3 gives the selling prices in thousands and the signs are recorded as described.

Table 14.3

Price	110	130	102	125	140	112	117	130	200	119
Sign	+	+	−	+	+	+	+	+	+	+

Suppose the alternative hypothesis is that the median selling price exceeds $105,000. If x represents the number of + signs, then the p value = $P(x \geq 9)$. The computation of the p value is as follows.

```
MTB > cdf 8;
SUBC> binomial n = 10 p = .5.
```

Cumulative Distribution Function

Binomial with n = 10 and p = 0.500000

```
     x        P(x <= x)
   8.00         0.9893
```

Since $P(x \geq 9) = 1 - P(x \leq 8) = 1 - .9893 = 0.0107$, the p value is 0.0107. The null hypothesis is rejected and we conclude that the median price exceeds $105,000 at level of significance $\alpha = .05$. The normal approximation to the binomial distribution may be used to determine the p value also. It is recommended that the student review this approximation which is discussed in Chapter 6. The approximation is valid provided that $np \geq 5$ and $nq \geq 5$. np and nq both equal 5 in this example. The mean is computed as $\mu = np = 10 \times .5 = 5$ and the standard deviation is computed as $\sigma = \sqrt{npq} = \sqrt{10 \times .5 \times .5} = 1.581$. The z value is computed as $z = \frac{8.5 - 5}{1.581} = 2.21$.

The p value is the area to the right of 2.21 under the standard normal curve. $P(z > 2.21) = .5 - .4864 = .0136$. The normal approximation to the binomial distribution provides a reasonably good result if n is 10 or greater.

Many books recommend that n be 20 or 25, but if the continuity correction is used the result will be fairly good for n equal to 10 or more.

EXAMPLE 14.4 The sign test described in Example 14.3 may be performed by using the sign test procedure in Minitab. The prices given in Table 14.3 are entered in column c1 and the following commands are used. The subcommand **Alternative 1.** indicates that the alternative is upper tailed.

```
MTB > STest 105 C1;
SUBC> Alternative 1.
```

Sign Test for Median

```
Sign test of median = 105.0 vs.  >  105.0

          N   Below  Equal  Above     P      Median
C1        10     1      0      9    0.0107    122.0
```

Note that the same p value is given here as was found in Example 14.3.

EXAMPLE 14.5 The sign test is also used to test for differences in matched paired experiments. Table 14.4 gives the blood pressure readings before and six weeks after starting hypertensive medication as well as the differences in the readings.

Table 14.4

Patient	1	2	3	4	5	6	7	8	9	10	11	12	13	14	15
Before reading	95	90	98	99	87	95	95	97	90	90	99	95	95	92	96
After reading	85	80	91	88	90	90	88	84	93	91	90	90	97	94	86
Difference	10	10	7	11	-3	5	7	13	-3	-1	9	5	-2	-2	10

The research hypothesis is that the medication will reduce the blood pressure. The difference is found by subtracting the after reading from the before reading. If the medication is effective, the majority of the differences will be positive. If x represents the number of positive signs in the 15 differences and if the null hypothesis that the medication has no effect is true, then x will have a binomial distribution with n = 15 and p = .5. The null hypothesis should be rejected for large values of x. That is, as this study is described, this is an upper-tail test. The Minitab solution is as follows. If the medication is not effective, the median difference should be zero. The differences are entered into column 1.

```
10    10     7    11    -3     5     7    13    -3    -1     9     5
-2    -2    10
```

```
MTB > STest median = 0.0 data in C1;
SUBC> Alternative 1.
```

Sign Test for Median

```
Sign test of median = 0.00000 vs.  >  0.00000

          N   Below  Equal  Above     P      Median
C1        15     5      0     10    0.1509    7.000
```

The p value is seen to equal 0.1509. This indicates that the null hypothesis would not be rejected at level of significance $\alpha = .05$. A dotplot of the differences is shown in Fig. 14-1. The dotplot illustrates a weakness of the sign test. Note that the 5 minus values are much smaller in absolute value than are the absolute values of the 10 positive differences. The sign test does not utilize this additional information. The signed-rank test discussed in the next section uses this additional information.

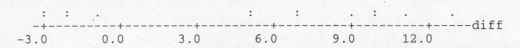

Fig. 14-1

The sign test is a nonparametric alternative to the one sample t test. The one sample t test assumes that the characteristic being measured has a normal distribution. The sign test does not require this assumption. If the selling prices of the homes discussed in Example 14.3 are normally distributed, then the one sample t test could be used to test the null hypothesis that the mean is $105,000.

WILCOXON SIGNED-RANK TEST FOR TWO DEPENDENT SAMPLES

In Example 14.5, it was noted that the sign test did not make use of the information that the 10 positive differences were all larger in absolute value than the absolute value of any of the 5 negative differences. The sign test used only the information that there were 10 pluses and 5 negatives. The data in Table 14.4 will be used to describe the *Wilcoxon signed-rank test*. Table 14.5 gives the computations needed for the Wilcoxon signed-rank test.

Table 14.5

Before reading	After reading	Difference, D	$|D|$	Rank of $|D|$	Signed rank
95	85	10	10	12	12
90	80	10	10	12	12
98	91	7	7	8.5	8.5
99	88	11	11	14	14
87	90	−3	3	4.5	−4.5
95	90	5	5	6.5	6.5
95	88	7	7	8.5	8.5
97	84	13	13	15	15
90	93	−3	3	4.5	−4.5
90	91	−1	1	1	−1
99	90	9	9	10	10
95	90	5	5	6.5	6.5
95	97	−2	2	2.5	−2.5
92	94	−2	2	2.5	−2.5
96	86	10	10	12	12

The first column gives the before blood pressure readings, the second column gives the after readings, the third column gives the difference D = Before − After, the fourth column gives the absolute values of the differences, the fifth column gives the ranks of the absolute differences, and the sixth column gives the *signed-rank* which restores the sign of the difference to the rank of the absolute difference. Two sum of ranks are defined as follows: W^+ = sum of positive ranks and W^- = absolute value of sum of negative ranks. For the data in Table 14.5, we have the following:

$$W^+ = 12 + 12 + 8.5 + 14 + 6.5 + 8.5 + 15 + 10 + 6.5 + 12 = 105$$

$$W^- = 4.5 + 4.5 + 1 + 2.5 + 2.5 = 15$$

To help in understanding the logic behind the signed-rank test, it is helpful to note some interesting facts concerning the above data. The sum of the ranks of the absolute differences must equal the sum of the first 15 positive integers. Using the formula given in Example 14.1, the sum of the ranks 1 through 15 is $\frac{15 \times 16}{2}$ = 120. Note that $W^+ + W^-$ = 120. If the null hypothesis is true and the medication has no effect, we would expect a 60, 60 split in the two sum of ranks. That is if H_0 is true, we would expect to find W^+ = 60 and W^- = 60. If the medication is effective for every patient, we would expect W^+ = 120 and W^- = 0. Either of the rank sums, W^+ or W^-, may be used to perform the hypothesis test. Tables for the Wilcoxon test statistic are available for determining critical values for the Wilcoxon signed-rank test. However, statistical software would most often be used to perform the test in real world applications. The Minitab solution is given in Example 14.6.

EXAMPLE 14.6 The differences in column 3 of Table 14.5 are entered into column c1. The command **WTest 0.0 C1;** indicates that the Wilcoxon signed-rank test is to be used to test that the median of the differences is 0 and that the differences are in column c1. The subcommand **Alternative 1.**indicates that the test is upper-tailed. The output gives the Wilcoxon statistic, W^+, computed in the above discussion and most importantly it gives the p value for the test. The probability of obtaining a value of W^+ equal to 105 or larger is 0.006. This p value is much less than .05 and indicates that the medication is effective. Note that the p value obtained when using the sign test in Example 14.5 is 0.1509. The Wilcoxon signed-rank test is more powerful than the sign test as indicated by this example.

```
Diff
     10        10         7        11        -3         5         7        13        -3        -1
      9         5        -2        -2        10

MTB > WTest 0.0 C1;
SUBC> Alternative 1.
```

Wilcoxon Signed Rank Test

```
Test of median = 0.000000 vs. median > 0.000000

                 N for    Wilcoxon              Estimated
          N      Test     Statistic      P      Median
Diff      15     15        105.0       0.006     5.000
```

A normal approximation procedure is also used for the Wilcoxon signed-rank test procedure when the sample size is 15 or more. The test statistic is given by formula (*14.1*), where W is either W^+ or W^- depending on the nature of the alternative hypothesis, and n is the number of nonzero differences.

$$Z = \frac{W - n(n+1)/4}{\sqrt{n(n+1)(2n+1)/24}} \qquad (14.1)$$

EXAMPLE 14.7 The normal approximation will be applied to the blood pressure data in Table 14.5 and the results compared to the results in Example 14.6. The computed test statistic is as follows:

$$Z = \frac{105 - 15(16)/4}{\sqrt{15(16)(31)/24}} = 2.56$$

The p value is $P(z > 2.56) = .5 - .4948 = 0.0052$. The p value in Example 14.6 is equal to 0.006. Note that the two values are very close. The difference may be due to round-off error or it may be due the Minitab software using the exact distribution for the test statistic rather than the normal approximation.

Like the sign test, the Wilcoxon signed rank test is a nonparametric alternative to the one sample t test. However, the Wilcoxon signed-rank test is usually more powerful than the sign test. There are situations such as taste preference tests where the sign test is applicable but the Wilcoxon signed-rank test is not.

WILCOXON RANK-SUM TEST FOR TWO INDEPENDENT SAMPLES

Consider an experiment designed to compare the lifetimes of rats restricted to two different diets. Eight rats were randomly divided into two groups of four each. The rats in the control group were allowed to consume all the food they desired and the rats in the experimental group were allowed only 80% of the food that they normally consumed. The experiment was continued until all rats expired and their lifetimes are given in Table 14.6.

Table 14.6

Control	3.4	3.6	4.0	2.5
Experimental	3.7	4.2	4.2	4.5

A nonparametric procedure for comparing the two groups was developed by Wilcoxon. Mann and Whitney developed an equivalent but slightly different nonparametric procedure. Some books discuss the *Wilcoxon rank-sum test* and some books discuss the *Mann-Whitney U test*. Both procedures make use of a ranking procedure we will now describe. Imagine that the data values are combined together and ranked. Table 14.7 gives the same data as shown in Table 14.6 along with the combined rankings shown in parentheses.

Table 14.7

Control	3.4 (2)	3.6 (3)	4.0 (5)	2.5 (1)
Experimental	3.7 (4)	4.2 (6.5)	4.2 (6.5)	4.5 (8)

We let R_1 be the sum of ranks for group 1, the control group, and R_2 be the sum of ranks for group 2. For the data in Table 14.7, we have $R_1 = 11$ and $R_2 = 25$. The sum of all eight ranks must equal $\dfrac{8 \times 9}{2} = 36$. Note that $R_1 + R_2 = 36$.

Now, consider some hypothetical situations. Suppose there is no difference in the lifetimes of the control and experimental diets. We would expect $R_1 = 18$ and $R_2 = 18$. Suppose the experimental diet is highly superior to the control diet, then we might expect all rats in the experimental group to be alive after all the rats in the control diet have died, and as a result obtain $R_1 = 10$ and $R_2 = 26$. Critical values for the testing the hypothesis of no difference in the two groups may be obtained from a table of the Mann-Whitney statistic or a table of the Wilcoxon rank-sum statistic. These tables are available in most elementary statistics texts. Most users of Statistics will use a computer software package to evaluate their results. Example 14.8 illustrates the use of Minitab to evaluate the results in the experiment described above.

EXAMPLE 14.8 The Minitab analysis for the data in Table 14.6 is shown below. Note that the data are stored in separate columns. The command **Mann-Whitney 95.0 C1 C2;** will perform a Mann-Whitney test on the data in columns c1 and c2. The subcommand **Alternative -1.** will test the alternative that the median lifetime for the control group is less than the median lifetime for the experimental group. The output line **W = 11** is the value for R_1 computed in the above discussion of this experiment. The output line **Test of ETA1 = ETA2 vs. ETA1 < ETA2 is significant at 0.0303** gives the p value as 0.0303 for the lower-tail alternative hypothesis. At level of significance $\alpha = .05$, the null hypothesis would be rejected since the p value $< \alpha$. This rather small study seems to indicate that restricting the diet prolongs the lifetime of such rats.

```
Row    Control  Expermtl

 1       3.4      3.7
 2       3.6      4.2
 3       4.0      4.2
 4       2.5      4.5

MTB > Mann-Whitney 95.0 C1 C2;
SUBC> Alternative -1.
```

Mann-Whitney Confidence Interval and Test

```
Control    N = 4       Median =  3.500
Expermtl   N = 4       Median =  4.200
Point estimate for ETA1-ETA2 is -0.700
97.0 Percent CI for ETA1-ETA2 is (-2.000,0.300)
W = 11.0
Test of ETA1 = ETA2 vs. ETA1 < ETA2 is significant at 0.0303
The test is significant at 0.0295 (adjusted for ties)
```

A normal approximation procedure is also used for the Wilcoxon rank-sum test procedure when the two sample sizes are large. If both sample sizes, n_1 and n_2, are 10 or more, the approximation is usually good. The test statistic is given by formula (14.2) where R_1 is the rank-sum associated with sample size n_1.

$$Z = \frac{R_1 - n_1(n_1 + n_2 + 1)/2}{\sqrt{n_1 n_2 (n_1 + n_2 + 1)/12}} \qquad (14.2)$$

EXAMPLE 14.9 A test instrument which measures the level of type A personality behavior was administered to a group of lawyers and a group of doctors. The results and the combined rankings are given in Table 14.8. The higher the score on the instrument, the higher the level of type A behavior. The null hypothesis is that the medians are equal for the two groups, and the alternative hypothesis is that the medians are unequal.

Table 14.8

Lawyers	Doctors
28 (5)	20 (1)
51 (20)	26 (4)
54 (21)	32 (8)
34 (10)	36 (11)
33 (9)	40 (13)
38 (12)	23 (2)
44 (15)	24 (3)
45 (16)	30 (6.5)
47 (17)	30 (6.5)
48 (18)	42 (14)
56 (22)	
50 (19)	
$R_1 = 184$	$R_2 = 69$

We note that $n_1 = 12$, $n_2 = 10$, and $R_1 = 184$. The computed value of the test statistic is found as follows;

$$Z^* = \frac{184 - 12(12 + 10 + 1)/2}{\sqrt{12 \times 10 \times (12 + 10 + 1)/12}} = 3.03$$

The null hypothesis is rejected for level of significance $\alpha = .05$, since the critical values are ± 1.96 and the computed value of the test statistic exceeds 1.96. The median score for lawyers exceeds the median score for doctors.

In closing this section, we note that the parametric equivalent test for the Wilcoxon rank-sum test and the Mann-Whitney U test is the two-sample t test discussed in Chapter 10.

KRUSKAL-WALLIS TEST

In Chapter 12, the one-way ANOVA was discussed as a statistical procedure for comparing several populations in terms of means. The procedure assumed that the samples were obtained from normally distributed populations with equal variances. The *Kruskal-Wallis test* is a nonparametric statistical method for comparing several populations. Consider a study designed to compare the net worth of three different groups of senior citizens (ages 51– 61). The net worth (in thousands) for each senior citizen in the study is shown in Table 14.9. The null hypothesis is that the distributions of net worths are the same for the three groups and the alternative hypothesis is that the distributions are different.

Table 14.9

Whites	African-Americans	Hispanics
250	80	70
175	150	100
130	70	130
205	90	75
225	110	95

The data are combined into one group of 15 numbers and ranked. The rankings are shown in Table 14.10. The sums of the ranks for each of the three samples are shown as R_1, R_2, and R_3.

Table 14.10

Whites	African-Americans	Hispanics
15	4	1.5
12	11	7
9.5	1.5	9.5
13	5	3
14	8	6
$R_1 = 63.5$	$R_2 = 29.5$	$R_3 = 27$

The sum of the ranks from 1 to 15 is equal to 120. If the null hypothesis were true, we would expect the sum of ranks for each of the three groups to equal 40. That is, if H_0 is true we would expect that $R_1 = R_2 = R_3 = 40$. The Kruskal-Wallis test statistic measures the extent to which the rank sums for the samples differs from what would be expected if the null were true. The Kruskal-Wallis test statistic is given by formula (*14.3*), where k = the number of populations, n_i = the number of items in sample i, R_i = sum of ranks for sample i, and n = the total number of items in all samples.

$$W = \frac{12}{n(n+1)}\left[\sum_1^k \frac{R_i^2}{n_i}\right] - 3(n+1) \qquad (14.3)$$

W has an approximate Chi-square distribution with df = k – 1 when all sample sizes are 5 or more.

EXAMPLE 14.10 The computed value for W for the data in Table 14.9 is determined as follows:

$$W* = \frac{12}{15 \times 16}\left[\frac{63.5^2}{5} + \frac{29.5^2}{5} + \frac{27^2}{5}\right] - 3 \times 16 = 8.315$$

The critical value from the Chi-square table with df = 2 and α = .05 is 5.991. The population distributions are not the same, since W* exceeds the critical value. Suppose the sample rank sums were each the same. That is, suppose $R_1 = R_2 = R_3 = 40$. The computed value for W would then equal the following:

$$W* = \frac{12}{15 \times 16}\left[\frac{40^2}{5} + \frac{40^2}{5} + \frac{40^2}{5}\right] - 3 \times 16 = 0$$

It is clear that this is always a one-tailed test. That is, the null is rejected only for large values of the computed test statistic.

EXAMPLE 14.11 The Minitab analysis for the data in Table 14.9 is given below. The value for H is the same as that for W* found in Example 14.10. The p value is 0.016, indicating that the null hypothesis would be rejected for $\alpha = .05$, the same conclusion reached in Example 14.10.

```
Row     Group   Networth
 1        1        250
 2        1        175
 3        1        130
 4        1        205
 5        1        225
 6        2         80
 7        2        150
 8        2         70
 9        2         90
10        2        110
11        3         70
12        3        100
13        3        130
14        3         75
15        3         95
```

```
MTB > Kruskal-Wallis 'Networth' 'Group'.
```

Kruskal–Wallis Test

```
Kruskal-Wallis Test on Networth

Group      N      Median     Ave Rank      Z
  1        5      205.00       12.7       2.88
  2        5       90.00        5.9      -1.29
  3        5       95.00        5.4      -1.59
Overall   15                    8.0

H = 8.31   DF = 2   P = 0.016
```

RANK CORRELATION

The Pearson correlation coefficient was introduced in chapter 13 as a measure of the strength of the linear relationship of two variables. The Pearson correlation coefficient is defined by formula (*14.4*). The *Spearman rank correlation coefficient* is computed by replacing the observations for x and y by their ranks and then applying formula (*14.4*) to the ranks. The Spearman rank correlation coefficient is represented by r_s.

$$r = \frac{S_{xy}}{\sqrt{S_{xx}S_{yy}}} \qquad (14.4)$$

Table 14.11

Basic Properties of the Spearman Rank Correlation Coefficient
1. The value of r_s is always between −1 and +1, i.e., $-1 \le r_s \le +1$
2. A value of r_s near +1 indicates a strong positive association between the rankings. A value of r_s near −1 indicates a strong negative association between the rankings.
3. The Spearman rank correlation coefficient may be applied to data at the ordinal level of measurement or above.

EXAMPLE 14.12 Table 14.12 gives the number of tornadoes and number of deaths due to tornadoes per month for a sample of 10 months. In addition, the ranks are shown for the values for each variable.

Table 14.12

Number of tornadoes, x	Rank of x	Number of deaths, y	Rank of y
35	1	0	2
85	4	0	2
90	6	2	5
94	7	1	4
50	2	0	2
76	3	3	6
98	8	4	7.5
104	9	4	7.5
88	5	5	9
128	10	7	10

The formula for the Pearson correlation coefficient given in formula (*14.4*) is applied to the ranks of x and the ranks of y.

$\Sigma x = 1 + 4 + 6 + \cdots + 10 = 55$, $\Sigma x^2 = 1 + 16 + 36 + \cdots + 100 = 385$, $\Sigma y = 55$, $\Sigma y^2 = 4 + 4 + 25 + \cdots + 100 = 382.5$, $\Sigma xy = 1 \times 2 + 4 \times 2 + 6 \times 5 + \cdots + 10 \times 10 = 363.5$.

$S_{xx} = \Sigma x^2 - \dfrac{(\Sigma x)^2}{n} = 385 - 302.5 = 82.5$, $S_{yy} = \Sigma y^2 - \dfrac{(\Sigma y)^2}{n} = 382.5 - 302.5 = 80$, and $S_{xy} = \Sigma xy - \dfrac{(\Sigma x)(\Sigma y)}{n} = 363.5 - 302.5 = 61$.

$$R_s = \frac{S_{xy}}{\sqrt{S_{xx}S_{yy}}} = \frac{61}{\sqrt{82.5 \times 80}} = 0.75$$

EXAMPLE 14.13 The Minitab solution for Example 14.12 is shown below. The raw data are entered into columns c1 and c2. The rank command is then used to put the ranks into columns c3 and c4. Then the Pearson correlation coefficient is requested for columns c3 and c4.

```
Row     tornado   deaths
 1        35        0
 2        85        0
 3        90        2
 4        94        1
 5        50        0
 6        76        3
 7        98        4
 8       104        4
 9        88        5
10       128        7

MTB > rank c1 put into c3
MTB > rank c2 put into c4
MTB > print c3 c4

Row    C3      C4
 1      1     2.0
 2      4     2.0
 3      6     5.0
 4      7     4.0
 5      2     2.0
 6      3     6.0
 7      8     7.5
 8      9     7.5
 9      5     9.0
10     10    10.0

MTB > correlation c3 c4
Correlation of c3 and c4 = 0.739
```

The *population Spearman rank correlation coefficient* is represented by ρ_s. In order to test the null hypothesis that $\rho_s = 0$, we use the result that $\sqrt{n-1}\,r_s$ has an approximate standard normal distribution when the null hypothesis is true and $n \geq 10$. Some books suggest a larger value of n in order to use the normal approximation.

EXAMPLE 14.14 The data in Table 14.12 may be used to test the null hypothesis $H_0: \rho_s = 0$ vs. $H_a: \rho_s \neq 0$ at level of significance $\alpha = .05$. The critical values are ± 1.96. The computed value of the test statistic is found as follows: $z^* = \sqrt{10-1} \times .75 = 2.25$. We conclude that there is a positive association between the number of tornadoes per month and the number of tornado deaths per month since $z^* > 1.96$.

RUNS TEST FOR RANDOMNESS

Suppose 3.5-ounce bars of soap are being sampled and their true weights determined. If the weight is above 3.5 ounces, then the letter A is recorded. If the weight is below 3.5 ounces, then a B is recorded. Suppose that in the last 10 bars the result was AAAAABBBBB. This would indicate nonrandomness in the variation about the 3.5 ounces. A result such as ABABABABAB would also indicate nonrandomness. The first outcome is said to have R = 2 runs. The second outcome is said to have R = 10 runs. There are $n_1 = 5$ A's and $n_2 = 5$ B's. A *run* is a sequence of the same letter written one or more times. There is a different letter (or no letter) before and after the sequence. A very large or very small number of runs would imply nonrandomness.

In general, suppose there are n_1 symbols of one type and n_2 symbols of a second type and R runs in the sequence of $n_1 + n_2$ symbols. When the occurrence of the symbols is random, the mean value of R is given by formula (*14.5*):

$$\mu = \frac{2n_1n_2}{n_1 + n_2} + 1 \qquad\qquad (14.5)$$

When the occurrence of the symbols is random, the standard deviation of R is given by formula (*14.6*):

$$\sigma = \sqrt{\frac{2n_1n_2(2n_2n_2 - n_1 - n_2)}{(n_1 + n_2)^2(n_1 + n_2 - 1)}} \qquad\qquad (14.6)$$

When both n_1 and n_2 are greater than 10, R is approximately normally distributed. The size requirements for n_1 and n_2 in order that R be normally distributed vary from book to book. We shall use the requirement that both exceed 10. From the proceeding discussion, we conclude that when the occurrence of the symbols is random and both n_1 and n_2 exceed 10, then the expression in formula (*14.7*) has a standard normal distribution.

$$Z = \frac{R - \mu}{\sigma} \qquad\qquad (14.7)$$

When the normal approximation is not appropriate, several Statistics books give a table of critical values for the runs test.

EXAMPLE 14.15 Bars of soap are selected from the production line and the letter A is recorded if the weight of the bar exceeds 3.5 ounces and B is recorded if the weight is below 3.5 ounces. The following sequence was obtained.

AAABBAAAABBABABAAABBBBAABBBABABBAA

These results are used to test the null hypothesis that the occurrence of A's and B's is random vs. the alternative hypothesis that the occurrences are not random. The level of significance is $\alpha = .01$. The critical values are ± 2.58. We note that n_1 = the number of A's = 18 and n_2 = the number of B's = 16. The number of runs is R = 17. Assuming the null hypothesis is true, the mean number of runs is $\mu = \dfrac{2n_1n_2}{n_1 + n_2} + 1 = \dfrac{2 \times 18 \times 16}{34} + 1 = 17.9411$. Assuming the null hypothesis is true, the standard deviation of the number of runs is :

$$\sigma = \sqrt{\frac{2n_1n_2(2n_2n_2 - n_1 - n_2)}{(n_1 + n_2)^2(n_1 + n_2 - 1)}} = \sqrt{\frac{2 \times 18 \times 16 \times 542}{34^2 \times 33}} = 2.8607$$

The computed value of the test statistic is $Z^* = \dfrac{R - \mu}{\sigma} = \dfrac{17 - 17.9411}{2.8607} = -.33$.

There is no reason to doubt the randomness of the observations. The process seems to be varying randomly about the 3.5 ounces, the target value for the bars of soap.

EXAMPLE 14.16 The Minitab solution to Example 14.15 is shown below. The symbols A and B are coded as 1 and –1 respectively. The 0, in the statement **Runs 0 C1**, is the mean of –1 and 1. The output gives the observed number of runs, the mean number of runs, the number of A's and B's, and the p value = 0.7422. The output is the same as that obtained in Example 14.15.

```
C1
    1     1     1    -1    -1     1     1     1     1    -1    -1     1    -1     1    -1
    1     1     1    -1    -1    -1    -1     1     1    -1    -1    -1     1    -1     1
   -1    -1     1     1

MTB > Runs 0 C1.
```

Runs Test

```
The observed number of runs = 17
The expected number of runs = 17.9412
18 Observations above K   16 below
The test is significant at 0.7422
Cannot reject at alpha = 0.05
```

Solved Problems

NONPARAMETRIC METHODS

14.1 A table (not shown) gives the number of students per computer for each of the 50 states and the District of Columbia. If the data values in the table are replaced by their ranks, what is the sum of the ranks?

Ans. The ranks will range from 1 to 51. The sum of the integers from 1 to 51 inclusive is $\dfrac{51 \times 52}{2} =$ 1,326. This same sum is obtained even if some of the ranks are tied ranks.

14.2 A group of 30 individuals are asked to taste a low fat yogurt and one that is not low fat, and to indicate their preference. The participants are unaware which one is low fat and which one is not. A plus sign is recorded if the low fat yogurt is preferred and a minus sign is recorded if the one that is not low fat is preferred. Assuming there is no difference in taste for the two types of yogurt, what is the probability that all 30 prefer the low fat yogurt?

Ans. Under the assumption that there is no difference in taste, the plus and minus signs are randomly assigned. There are 2^{30} = 1,073,741,824 different arrangements of the 30 plus and minus signs.

The probability that all 30 are plus signs is 1 divided by 1,073,741,824, or $9.313225746 \times 10^{-10}$. If this outcome occurred, we would surely reject the hypothesis of no difference in the taste of the two types of yogurt.

SIGN TEST

14.3 Find the probability that 20 or more prefer the low fat yogurt in problem 14.2 if there is no difference in the taste. Use the normal approximation to the binomial distribution to determine your answer.

Ans. Assuming that there is no difference in taste, the 30 responses represent 30 independent trials with $p = q = .5$, where p is the probability of a + and q is the probability of a –. The mean number of + signs is $\mu = np = 15$ and the standard deviation is $\sigma = \sqrt{npq} = 2.739$. The z value corresponding to 20 plus signs is $z = \dfrac{19.5 - 15}{2.739} = 1.64$. The probability of 20 or more plus signs is approximated by $P(z > 1.64) = .5 - .4495 = .0505$.

14.4 A sociological study involving married couples, where both husband and wife worked full time, recorded the yearly income for each in thousands of dollars. The results are shown in Table 14.13. Use the sign test to test the research hypothesis that husbands have higher salaries. Use level of significance $\alpha = .05$.

Table 14.13

Couple	1	2	3	4	5	6	7	8	9	10	11	12	13	14	15
Husband	35	16	17	25	30	32	28	31	27	15	19	22	33	30	18
Wife	25	20	18	20	25	25	25	19	30	20	15	20	27	28	20
Difference	10	–4	–1	5	5	7	3	12	–3	–5	4	2	6	2	–2

Ans. A high number of plus signs among the differences supports the research hypothesis. There are 10 plus signs in the 15 differences. The p value is equal to the probability of 10 or more plus signs in the 15 differences. The p value is computed assuming the probability of a plus sign is .5 for any one of the differences, since the p value is computed assuming the null hypothesis is true. From the binomial distribution, we find that the p value = $.0916 + .0417 + .0139 + .0032 + .0005' + .0000 = 0.1509$. Since the p value exceeds the preset level of significance, we cannot reject the null hypothesis.

WILCOXON SIGNED-RANK TEST FOR TWO DEPENDENT SAMPLES

14.5 Use the Wilcoxon signed-rank test and the data in Table 14.13 to test the research hypothesis that husbands have higher salaries. Determine the values for W^+ and W^- and then use the normal approximation procedure to perform the test. Use level of significance $\alpha = .05$

Ans. From Table 14.14, we see that $W^+ = 15 + 10 + 10 + 13 + 5.5 + 14 + 7.5 + 3 + 12 + 3 = 93$, and $W^- = 7.5 + 1 + 5.5 + 10 + 3 = 27$. Since Difference = Husband salary – Wife salary, large values for W^+ or small values for W^- will lend support to the research hypothesis that husbands have higher salaries. If W^+ is used, the test will be upper-tailed. If W^- is used, the test will be lower-tailed. The test statistic used in the normal approximation is:

$$Z = \frac{W - n(n+1)/4}{\sqrt{n(n+1)(2n+1)/24}} = \frac{W - 60}{17.607}$$

Table 14.14

Husband	Wife	Difference, D	|D|	Rank of |D|	Signed rank
35	25	10	10	15	15
16	20	−4	4	7.5	−7.5
17	18	−1	1	1	−1
25	20	5	5	10	10
30	25	5	5	10	10
32	25	7	7	13	13
28	25	3	3	5.5	5.5
31	19	12	12	14	14
27	30	−3	3	5.5	−5.5
15	20	−5	5	10	−10
19	15	4	4	7.5	7.5
22	20	2	2	3	3
33	27	6	6	12	12
30	28	2	2	3	3
18	20	−2	2	3	−3

If we use W^-, then $Z = \dfrac{27-60}{17.607} = -1.87$, and the p value $= .5 - .4693 = .0307$. If we use W^+, then $Z = \dfrac{93-60}{17.607} = 1.87$, and the p value $= .0307$. It is clear that we may use either W^+ or W^-. In either case, we reject the null hypothesis since the p value $< .05$.

14.6 The Minitab output for the solution to problem 14.5 is shown below. Answer the following questions by referring to this output.

(a) Explain the various parts of the command **WTest 0.0 'Diff';**

(b) Why was the subcommand **Alternative 1.** used?

(c) What is the number 93.0 shown under Wilcoxon statistic?

(d) How does the p value given in the Minitab output compare with the p value found in problem 14.5?

```
MTB > print c1
```

Data Display
```
Diff
     10      -4      -1       5       5       7       3      12      -3      -5
      4       2       6       2      -2
```

```
MTB > WTest 0.0 'Diff';
SUBC>  Alternative 1.
```

Wilcoxon Signed Rank Test
```
Test of median = 0.000000 vs. median > 0.000000

            N for   Wilcoxon              Estimated
       N    Test    Statistic      P      Median
Diff   15    15      93.0        0.032    2.750
```

Ans. (a) Wtest indicates that the Wilcoxon signed-rank test is to be performed. The value 0.0 indicates that we are testing that the median difference is assumed to be 0.0. The median difference will be 0.0 if the null hypothesis is true. Diff indicates that we are performing the test for the differences in column 1.

(b) The subcommand **Alternative 1.** indicates an upper-tailed test. Since Diff = Husband salary − Wife salary, a positive median will support the research hypothesis that the husband salaries are greater than the wife salaries.

(c) The Wilcoxon statistic is the same as W^+ found in problem 14.5.

(*d*) The p values are very close. The p value based on the normal approximation is 0.0307 and the Minitab p value is 0.032.

WILCOXON RANK-SUM TEST FOR TWO INDEPENDENT SAMPLES

14.7 A study compared the household giving for two different Protestant denominations. Table 14.15 gives the yearly household giving in hundreds of dollars for individuals in both samples. The ranks are shown in parentheses beside the household giving. Use the normal approximation to the Wilcoxon rank-sum test statistic to test the research hypothesis that the household giving differs for the two denominations. Use level of significance $\alpha = .01$.

Table 14.15

Denomination 1	Denomination 2
9.3 (5)	7.0 (1)
13.5 (20)	9.0 (4)
13.7 (21)	9.7 (8)
10.2 (10)	10.4 (11)
10.0 (9)	10.9 (13)
10.7 (12)	8.0 (2)
11.3 (15)	8.5 (3)
11.4 (16)	9.5 (6.5)
11.5 (17)	9.5 (6.5)
12.3 (18)	11.1 (14)
14.2 (22)	14.5 (23)
12.5 (19)	
$R_1 = 184$	$R_2 = 92$

Ans. The sample sizes are $n_1 = 12$ and $n_2 = 11$. The rank sum associated with sample 1 is $R_1 = 184$. The test statistic is computed as follows:

$$Z = \frac{R_1 - n_1(n_1 + n_2 + 1)/2}{\sqrt{n_1 n_2 (n_1 + n_2 + 1)/12}} = \frac{184 - 12(12 + 11 + 1)/2}{\sqrt{12 \times 11 \times (12 + 11 + 1)/12}} = 2.46$$

The critical values are ± 2.58. Since the computed value of the test statistic does not exceed the right side critical value, the null hypothesis is not rejected. The p value is computed as follows. The area to the right of 2.46 is $.5 - .4931 = .0069$, and since the alternative is two sided, the p value $= 2 \times .0069 = 0.0138$. If the p value approach to testing is used, we do not reject since the p value is not less than the preset α.

14.8 The Minitab output for the solution to problem 14.7 is shown below. Answer the following questions by referring to this output.
(*a*) Explain the command Mann-Whitney **'Denom1' 'Denom2'**;
(*b*) What does the subcommand **Alternative 0.** mean?
(*c*) What is the line W = 184 giving you?
(*d*) Explain the line Test of ETA1 = ETA2 vs. ETA1 not = ETA2 is significant at 0.0151.

Row	Denom1	Denom2	Row	Denom1	Denom2
1	9.3	7.0	7	11.3	8.5
2	13.5	9.0	8	11.4	9.5
3	13.7	9.7	9	11.5	9.5
4	10.2	10.4	10	12.3	11.1
5	10.0	10.9	11	14.2	14.5
6	10.7	8.0	12	12.5	

```
MTB > Mann-Whitney 'Denom1' 'Denom2';
SUBC> Alternative 0.
```

Mann-Whitney Confidence Interval and Test

```
Denom1      N = 12      Median = 11.450
Denom2      N = 11      Median =  9.500
Point estimate for ETA1-ETA2 is 2.000
95.5 Percent CI for ETA1-ETA2 is (0.499,3.501)
W = 184.0
Test of ETA1 = ETA2 vs. ETA1 not = ETA2 is significant at 0.0151
The test is significant at 0.0150 (adjusted for ties)
```

Ans. (*a*) This command requests a Mann-Whitney test for the data in columns named Denom1 and Denom2.

(*b*) This subcommand indicates that the hypothesis is two-tailed.

(*c*) This is the same as R_1, the sum of ranks for sample 1.

(*d*) This line gives the p value for the test as 0.0151. This value is close to the p value obtained in problem 14.7.

KRUSKAL-WALLIS TEST

14.9 Thirty individuals were randomly divided into 3 groups of 10 each. Each member of one group completed a questionnaire concerning the Internal Revenue Service (IRS). The score on the questionnaire is called the Customer Satisfaction Index (CSI). The higher the score, the greater the satisfaction. A second set of CSI scores were obtained for garbage collection from the second group, and a third set of scores were obtained for long distance telephone service. The scores are given in Table 14.16. Perform a Kruskal-Wallis test to determine if the population distributions differ. Use significance level $\alpha = .01$.

Table 14.16

IRS	Garbage collection	Long distance service
40	55	60
45	60	65
55	65	70
50	50	70
45	55	70
45	55	75
50	60	75
50	70	70
40	65	80
60	60	80

Ans. The data in Table 14.16 are combined and ranked as one group. The resulting ranks are given in Table 14.17. The following rank sums are obtained for the three groups: $R_1 = 65.0$, $R_2 = 154.0$, and $R_3 = 246.0$.

Table 14.17

IRS	Garbage collection	Long distance service
1.5	11.5	16.0
4.0	16.0	20.0
11.5	20.0	24.0
7.5	7.5	24.0
4.0	11.5	24.0
4.0	11.5	27.5
7.5	16.0	27.5
7.5	24.0	24.0
1.5	20.0	29.5
16.0	16.0	29.5

The computed value of the Kruskal-Wallis test statistic is found as follows:

$$W = \frac{12}{n(n+1)}\left[\sum_{1}^{k}\frac{R_i^2}{n_i}\right] - 3(n+1) = \frac{12}{30 \times 31}\left[\frac{65^2}{10} + \frac{154^2}{10} + \frac{246^2}{10}\right] - 3 \times 31 = 21.138$$

The critical value is found by using the Chi-square distribution table with df = 3 − 1 = 2 to equal 9.210. The null hypothesis is rejected since the computed value of the test statistic exceeds the critical value.

14.10 The Minitab analysis for problem 14.9 is shown below. Answer the following questions.
(*a*) Explain the command **Kruskal-Wallis 'CSI' 'Group'**.
(*b*) If the average ranks are multiplied by 10, what do you obtain?
(*c*) What does the row H = 21.14 DF = 2 P = 0.000 give you?

```
MTB > Kruskal-Wallis 'CSI' 'Group'.
```

Kruskal-Wallis Test
```
Kruskal-Wallis Test on CSI

Group       N     Median     Ave Rank        Z
  1        10      5.750         6.5       -3.96
  2        10     16.000        15.4       -0.04
  3        10     24.000        24.6        4.00
Overall    30                   15.5

H = 21.14   DF = 2   P = 0.000
H = 21.48   DF = 2   P = 0.000 (adjusted for ties)
```

Ans. (*a*) The command asks for a Kruskal-Wallis test for the data in the column called CSI. The column called Group identifies the three samples.
(*b*) $R_1 = 65$, $R_2 = 154$, and $R_3 = 246$.
(*c*) This row gives the computed value of the test statistic, the degrees of freedom for the Chi-square distribution, and the p value = 0.000.

RANK CORRELATION

14.11 The body mass index (BMI) for an individual is found as follows: Using your weight in pounds and your height in inches, multiply your weight by 705, divide the result by your height, and divide again by your height. The desirable body mass index varies between 19 and 25. Table 14.18 gives the body mass index and the age for 20 individuals. Find the Spearman rank correlation coefficient for data shown in the table.

Table 14.18

BMI	Age	BMI	Age
22.5	27	19.0	44
24.6	32	17.5	18
28.7	45	32.5	29
30.1	49	22.4	29
18.5	19	28.8	40
20.0	22	21.3	39
24.5	31	25.0	21
25.0	27	19.0	20
27.5	25	29.7	52
30.0	44	16.7	19

Ans. Table 14.19 contains the ranks for the BMI values and the ranks for the ages given in Table 14.18. The formula for the Pearson correlation coefficient given in formula (*14.4*) is applied to the ranks of the BMI values and the ranks of the ages. Let x represent the ranks of the BMI values and y represent the ranks of the ages.

$\Sigma x = 9.0 + 11.0 + 15.0 + \cdots + 17.0 + 1.0 = 210$, $\Sigma x^2 = 81 + 121 + 225 + \cdots + 289 + 1 = 2869$,

$\Sigma y = 8.5 + 13.0 + 18.0 + \cdots + 20 + 2.5 = 210$, $\Sigma y^2 = 72.25 + 169 + 324 + \cdots + 400 + 6.25 = 2868$,

$\Sigma xy = 9 \times 8.5 + 11 \times 13 + 15 \times 18 + \cdots + 17 \times 20 + 1 \times 2.5 = 2646.5$

$S_{xx} = \Sigma x^2 - \dfrac{(\Sigma x)^2}{n} = 2869 - 2205 = 664$, $S_{yy} = \Sigma y^2 - \dfrac{(\Sigma y)^2}{n} = 2868 - 2205 = 663$

$S_{xy} = \Sigma xy - \dfrac{(\Sigma x)(\Sigma y)}{n} = 2646.5 - 2205 = 441.5$

$$r_s = \frac{S_{xy}}{\sqrt{S_{xx}S_{yy}}} = \frac{441.5}{\sqrt{664 \times 663}} = 0.665$$

Table 14.19

BMI	Age	BMI	AGE
9.0	8.5	4.5	16.5
11.0	13.0	2.0	1.0
15.0	18.0	20.0	10.5
19.0	19.0	8.0	10.5
3.0	2.5	16.0	15.0
6.0	6.0	7.0	14.0
10.0	12.0	12.5	5.0
12.5	8.5	4.5	4.0
14.0	7.0	17.0	20.0
18.0	16.5	1.0	2.5

14.12 Use the results found in problem 14.11 to test the null hypothesis $H_0: \rho_s = 0$ vs. $H_a: \rho_s \neq 0$ at level of significance $\alpha = .05$.

Ans. The critical values are ± 1.96. The computed value of the test statistic is found as follows: $z^* = \sqrt{20-1} \times .665 = 2.90$. We conclude that there is a positive association between age and body mass index.

RUNS TEST FOR RANDOMNESS

14.13 The gender of the past 30 individuals hired by a personnel office are as follows, where M represents a male and F represents a female.

FFFMMMMFMFMFFFFMMMMMMMFFMMMFFF

Test for randomness in hiring using level of significance $\alpha = .05$.

Ans. There are $n_1 = 14$ F's and $n_2 = 16$ M's. There are $R = 11$ runs. If the hiring is random, the mean number of runs is

$$\mu = \frac{2n_1 n_2}{n_1 + n_2} + 1 = \frac{2 \times 14 \times 16}{30} + 1 = 15.933$$

and the standard deviation of the number of runs is

$$\sigma = \sqrt{\frac{2n_1 n_2 (2n_1 n_2 - n_1 - n_2)}{(n_1 + n_2)^2 (n_1 + n_2 - 1)}} = \sqrt{\frac{2 \times 14 \times 16 \times 418}{900 \times 29}} = 2.679$$

The computed value of the test statistic is $Z^* = \dfrac{R - \mu}{\sigma} = \dfrac{11 - 15.933}{2.679} = -1.84$. The critical values are ± 1.96, and since the computed value of the test statistic is not less than -1.96, we are unable to reject randomness in hiring with respect to gender.

14.14 The Minitab analysis for the runs test is shown below. Compute the p value corresponding to the value $Z^* = -1.84$ in problem 14.13 and compare it with the p value given in the Minitab output.

```
C1
0   0   0   1   1   1   1   0   1   0   1   0   0   0   0   1   1   1
1   1   1   1   0   0   1   1   1   0   0   0


MTB > Runs .5 C1.

The observed number of runs = 11
The expected number of runs = 15.9333
16 Observations above K   14 below
The test is significant at 0.0658
Cannot reject at alpha = 0.05
```

Ans. The area to the left of $Z^* = -1.84$ is $.5 - .4671 = .0329$. Since the alternative hypothesis is two-tailed, the p value $= 2 \times .0329 = 0.0658$. This is the same as the p value given in the Minitab output.

Supplementary Problems

NONPARAMETRIC METHODS

14.15 Three mathematical formulas for dealing with ranks are often utilized in nonparmetric methods. These formulas are used for summing powers of ranks. They are as follows:

$$1 + 2 + 3 + \cdots + n = \frac{n(n+1)}{2}$$

$$1^2 + 2^2 + 3^2 + \cdots + n^2 = \frac{n(n+1)(2n+1)}{6}$$

$$1^3 + 2^3 + 3^3 + \cdots + n^3 = \frac{n^2(n+1)^2}{4}$$

Use the above special formulas to find the sum, sum of squares, and sum of cubes for the ranks 1 through 10.

Ans. sum = 55, sum of squares = 385, and sum of cubes = 3,025.

14.16 Nonparametric methods often deal with arrangements of plus and minus signs. The number of different possible arrangements consisting of a plus signs and b minus signs is given by $\begin{pmatrix} a+b \\ a \end{pmatrix}$ or $\begin{pmatrix} a+b \\ b \end{pmatrix}$. Use this formula to determine the number of different arrangements consisting of 2 plus signs and 3 negative signs and list the arrangements.

Ans. The number of arrangements is $\begin{pmatrix} 5 \\ 2 \end{pmatrix} = \dfrac{5!}{2! \times 3!} = 10$. The arrangements are as follows:

$++--$, $+-+--$, $+--+-$, $+---+$, $-++--$, $-+-+-$, $-+--+$, $--++-$,
$--+-+$, $---++$

SIGN TEST

14.17   The police chief of a large city claims that the median response time for all 911 calls is 15 minutes. In a
random sample of 250 such calls it was found that 149 of the calls exceeded 15 minutes in response
time. All the other response times were less than 15 minutes. Do these results refute the claim? Use
level of significance .05. Assume a two-tailed alternative hypothesis.

   Ans. Assuming the claim is correct, we would expect $\mu = np = 250 \times .5 = 125$ calls on the average. The
standard deviation is equal $\sigma = \sqrt{npq} = 7.906$. The z value corresponding to 149 calls is 2.97.
Since the computed z value exceeds 1.96, the results refute the claim.

14.18   The claim is made that the median number of movies seen per year per person is 35. Table 14.20 gives
the number of movies seen last year by a random sample of 25 individuals.

Table 14.20

19	44	37	29	35
39	12	31	35	41
45	29	23	25	25
35	70	50	52	88
66	44	52	26	26

   Test the null hypothesis that the median is 35 vs. the alternative that the median is not 35. Use $\alpha = .05$.

   Ans. Table 14.21 gives a plus sign if the value in the corresponding cell in Table 14.20 exceeds 35, a 0
if the corresponding cell value equals 35 and a minus sign if the corresponding cell value is less
than 35. There are 12 plus signs, 10 minus signs, and 3 zeros.

Table 14.21

–	+	+	–	0
+	–	–	0	+
+	–	–	–	–
0	+	+	+	+
+	+	+	–	–

   The below Minitab output indicates a p value = 0.8318, indicating that there is no statistical
evidence to reject the claim that the median is 35.

```
MTB > print c1

19    39    45    35    66    44    12    29    70    44    37
31    23    50    52    29    35    25    52    26    35    41
25    88    26

MTB > STest 35 'movies';
SUBC> Alternative 0.
```

Sign Test for Median
```
Sign test of median = 35.00 vs.  not =  35.00

             N    Below  Equal  Above     P      Median
Movies      25      10     3      12   0.8318    35.00
```

WILCOXON SIGNED-RANK TEST FOR TWO DEPENDENT SAMPLES

14.19   Use the normal approximation to the Wilcoxon sign-rank test statistic to solve problem 14.18. Note that this application of the Wilcoxon sign-rank test does not involve dependent samples. We are testing the median value of a population using a single sample.

> *Ans.* Table 14.22 shows the computation of the signed ranks. Note that 0 differences are not ranked and are omitted from the analysis. The sum of positive ranks is $W^+ = 150.5$, and the absolute value of the sum of the negative ranks is $W^- = 102.5$. The test statistic used in the normal approximation is:

$$Z^* = \frac{W - n(n+1)/4}{\sqrt{n(n+1)(2n+1)/24}} = \frac{150.5 - 22 \times 23/4}{\sqrt{22 \times 23 \times 45/24}} = 0.779$$

> Since the computed value of the test statistic does not exceed 1.96, the null hypothesis is not rejected.

Table 14.22

X = Number	D = X – 35	|D|	Rank of |D|	Signed rank
19	–16	16	16.0	–16.0
39	4	4	2.5	2.5
45	10	10	12.0	12.0
35	0	0	Not used	
66	31	31	20.0	20.0
44	9	9	8.5	8.5
12	–23	23	19.0	–19.0
29	–6	6	5.0	–5.0
70	35	35	21.0	21.0
44	9	9	8.5	8.5
37	2	2	1.0	1.0
31	–4	4	2.5	–2.5
23	–12	12	14.0	–14.0
50	15	15	15.0	15.0
52	17	17	17.5	17.5
29	–6	6	5.0	–5.0
35	0	0	Not used	
25	–10	10	12.0	–12.0
52	17	17	17.5	17.5
26	–9	9	8.5	–8.5
35	0	0	Not used	
41	6	6	5.0	5.0
25	–10	10	12.0	–12.0
88	53	53	22.0	22.0
26	–9	9	8.5	–8.5

14.20   The Minitab output for problem 14.19 is shown below. Answer the following questions.
     (*a*) Explain the command **WTest 35 'movies'**;
     (*b*) Explain the subcommand **Alternative 0**.
     (*c*) Explain the output.
     (*d*) Compute the p value corresponding to the value $Z^* = 0.779$ in problem 14.29 and compare it with the p value given in the Minitab output.

```
MTB > print c1

  19        39        45        35        66        44        12        29        70
  44        37        31        23        50        52        29        35        25
  52        26        35        41        25        88        26
```

```
MTB > WTest 35 'movies';
SUBC> Alternative 0.
```

Wilcoxon Signed Rank Test
```
Test of median = 35.00 vs. median not = 35.00

                N for   Wilcoxon              Estimated
        N       Test    Statistic     P        Median
Movies  25       22       150.5      0.445      37.50
```

Ans. (a) Wtest indicates a Wilcoxon sign-rank test. The number 35 is the median value to be tested. Movies indicates the column containing the sample data.

(b) The subcommand indicates that the research hypothesis is two-tailed.

(c) The output indicates that the sample size is 25. However only 22 of the values were used since there were 3 differences that were 0. The value for the Wilcoxon Statistic is the same as W^+ in problem 14.19. The p value is 0.445.

(d) P value $= 2 \times (.5 - .2823) = 0.4354$

WILCOXON RANK-SUM TEST FOR TWO INDEPENDENT SAMPLES

14.21 The leading cause of posttraumatic stress disorder (PTSD) among men is wartime combat and among women the leading causes are rape and sexual molestation. The duration of symptoms was measured for a group of men and a group of women. The times of duration in years for both groups are given in Table 14.23. Use the Wilcoxon rank-sum test to test the research hypothesis that the median times of duration differ for men and women. Use level of significance $\alpha = .01$. Use the normal approximation procedure to perform the test.

Table 14.23

Men	Women
2.5	3.0
3.3	4.0
3.5	4.0
4.0	6.5
5.0	7.0
5.0	8.5
6.5	10.0
6.5	12.5
10.0	13.0
15.0	17.5

Ans. Table 14.24 gives the rankings when the two samples are combined and ranked together.

Table 14.24

Men	Women
1	2
3	6
4	6
6	11
8.5	13
8.5	14
11	15.5
11	17
15.5	18
19	20
$R_1 = 87.5$	$R_2 = 122.5$

The sample sizes are $n_1 = 10$ and $n_2 = 10$. The rank sum associated with sample 1 is $R_1 = 87.5$. The test statistic is computed as follows:

$$Z = \frac{R_1 - n_1(n_1 + n_2 + 1)/2}{\sqrt{n_1 n_2 (n_1 + n_2 + 1)/12}} = \frac{87.5 - 10(10 + 10 + 1)/2}{\sqrt{10 \times 10 \times (10 + 10 + 1)/12}} = -1.32$$

The critical values are ± 2.58. Since the computed value of the test statistic does not exceed the left side critical value, the null hypothesis is not rejected. The p value is computed as follows. The area to the left of −1.32 is .5 − .4066 = .0934, and since the alternative is two sided, the p value = 2 × .0934 = 0.1868. If the p value approach to testing is used, we do not reject the null since the p value is not less than the preset α.

14.22 The Minitab solution for problem 14.21 is shown below. Answer the following questions.
(a) Explain the command line **Mann-Whitney 95.0 'Men' 'Women';**
(b) Explain the subcommand **Alternative 0.**
(c) Discuss the output.

```
MTB > print c1 c2
Row    Men    Women
  1    2.5     3.0
  2    3.3     4.0
  3    3.5     4.0
  4    4.0     6.5
  5    5.0     7.0
  6    5.0     8.5
  7    6.5    10.0
  8    6.5    12.5
  9   10.0    13.0
 10   15.0    17.5

MTB > Mann-Whitney 95.0 'Men' 'Women';
SUBC> Alternative 0.
```

Mann-Whitney Confidence Interval and Test
```
Men            N = 10    Median =  5.000
Women          N = 10    Median =  7.750
Point estimate for ETA1-ETA2 is -2.250
95.5 Percent CI for ETA1-ETA2 is (-6.503,1.002)
W = 87.5
Test of ETA1 = ETA2 vs. ETA1 not = ETA2 is significant at 0.1988
The test is significant at 0.1971 (adjusted for ties)

Cannot reject at alpha = 0.05
```

Ans. (a) The command line requests a Mann-Whitney analysis using the data in the columns labeled Men and Women. Set a 95% confidence interval on the difference in the medians for the two populations.
(b) The subcommand indicates that the test is two-tailed.
(c) The output gives a 95% confidence interval on the difference in the two population medians. The p value is given to be 0.1988. We are unable to reject the null hypothesis.

KRUSKAL-WALLIS TEST

14.23 The cost of a meal, including drink, tax and tip was determined for ten randomly selected individuals in New York City, Chicago, Boston, and San Francisco. The results are shown in Table 14.25. Test the null hypothesis that the four population distributions of costs are the same for the four cities vs. the hypothesis that the distributions are different. Use level of significance $\alpha = .05$.

Table 14.25

New York	Chicago	Boston	San Francisco
21.00	15.00	15.50	16.50
23.00	17.50	16.00	18.25
23.50	19.00	18.50	21.25
25.00	21.50	19.50	24.25
25.50	22.00	20.00	26.25
30.00	23.75	22.50	27.75
30.00	25.00	25.25	29.00
33.50	26.00	26.50	31.25
35.00	27.50	30.00	35.50
40.00	33.00	37.50	42.25

Ans. Table 14.26 gives the results when the four samples are combined and ranked together.

Table 14.26

New York	Chicago	Boston	San Francisco
11.0	1.0	2.0	4.0
16.0	5.0	3.0	6.0
17.0	8.0	7.0	12.0
20.5	13.0	9.0	19.0
23.0	14.0	10.0	25.0
31.0	18.0	15.0	28.0
31.0	20.5	22.0	29.0
35.0	24.0	26.0	33.0
36.0	27.0	31.0	37.0
39.0	34.0	38.0	40.0
$R_1 = 259.50$	$R_2 = 164.50$	$R_3 = 163.00$	$R_4 = 233.00$

$$W^* = \frac{12}{n(n+1)}\left[\sum_1^k \frac{R_i^2}{n_i}\right] - 3(n+1) = \frac{12}{40 \times 41}\left[\frac{259.5^2}{10} + \frac{164.5^2}{10} + \frac{163^2}{10} + \frac{233^2}{10}\right] - 3 \times 41 = 5.24$$

The critical value is 7.815. We cannot conclude that the distributions are different.

14.24 The Minitab output for the Kruskal-Wallis test in problem 14.23 is shown below.
(a) Explain the command line.
(b) Explain the output.

MTB > **Kruskal-Wallis 'cost' 'city'.**

Kruskal-Wallis Test
Kruskal-Wallis Test on cost

City	N	Median	Ave Rank	Z
1	10	27.75	26.0	1.70
2	10	22.88	16.5	-1.27
3	10	21.25	16.3	-1.31
4	10	27.00	23.3	0.87
Overall	40		20.5	

```
H = 5.24   DF = 3   P = 0.155
H = 5.24   DF = 3   P = 0.155 (adjusted for ties)
```

Ans. (a) The command line **Kruskal-Wallis 'cost' 'city'.** requests a Kruskal-Wallis test for the responses in the column called cost and the cities identified in the column called city.
(b) The output lists the 4 cities, the median cost for each city, and the mean rank for each city. In addition, the computed Kruskal-Wallis test statistic is H = 5.24. The p value is 0.155.

RANK CORRELATION

14.25 Table 14.27 gives the percent of calories from fat and the micrograms of lead per deciliter of blood for a sample of preschoolers. Find the Spearman rank correlation coefficient for data shown in the table.

Table 14.27

Percent Fat Calories	Lead
40	13
35	12
33	11
29	8
35	13
36	15
30	11
36	14
33	12
28	9
39	15
26	7

Ans. The Spearman correlation coefficient = 0.919.

14.26 Test for a positive correlation between the percent of calories from fat and the level of lead in the blood using the results in problem 14.25.

Ans. The computed test statistic is 3.05 and the $\alpha = .05$ critical value is 1.65. We conclude that a positive correlation exists.

RUNS TEST FOR RANDOMNESS

14.27 The first 100 decimal places of π contain 51 even digits (0, 2, 4, 6, 8) and 49 odd digits. The number of runs is 43. Are the occurrences of even and odd digits random?

Ans. Assuming randomness, $\mu = 50.98$, $\sigma = 4.9727$, and $z^* = -1.60$. Critical values $= \pm 1.96$. The occurrences are random.

14.28 The weights in grams of the last 30 containers of black pepper selected from a filling machine are shown in Table 14.28. The weights are listed in order by rows. Are the weights varying randomly about the mean?

Table 14.28

28.2	28.2	27.9	27.8	27.8	28.0	28.3	28.1	28.2	28.3
27.9	27.8	28.1	28.2	28.1	27.7	28.4	28.2	28.3	28.3
28.1	28.0	27.9	27.8	27.6	27.8	27.9	28.1	28.3	28.5

Ans. The mean of the 30 observations is 28.06. Table 14.29 shows an A if the observation is above 28.06 and a B if the observation is below 28.06.

Table 14.29

A	A	B	B	B	B	A	A	A	A
B	B	A	A	A	B	A	A	A	A
A	B	B	B	B	B	B	A	A	A

The sequence AABBBBAAAABBAAABAAAAABBBBBBAAA has 17 A's, 13 B's, and 9 runs. If the weights are varying randomly about the mean, the mean number of runs is 15.7333 and the standard deviation is 2.6414. The computed value of z is $z^* = -2.55$. The critical values for $\alpha = .05$ are ± 1.96. Randomness about the mean is rejected.

Appendix 1

Binomial Probabilities

n	x	.05	.10	.20	.30	.40	.50	.60	.70	.80	.90	.95
1	0	.9500	.9000	.8000	.7000	.6000	.5000	.4000	.3000	.2000	.1000	.0500
	1	.0500	.1000	.2000	.3000	.4000	.5000	.6000	.7000	.8000	.9000	.9500
2	0	.9025	.8100	.6400	.4900	.3600	.2500	.1600	.0900	.0400	.0100	.0025
	1	.0950	.1800	.3200	.4200	.4800	.5000	.4800	.4200	.3200	.1800	.0950
	2	.0025	.0100	.0400	.0900	.1600	.2500	.3600	.4900	.6400	.8100	.9025
3	0	.8574	.7290	.5120	.3430	.2160	.1250	.0640	.0270	.0080	.0010	.0001
	1	.1354	.2430	.3840	.4410	.4320	.3750	.2880	.1890	.0960	.0270	.0071
	2	.0071	.0270	.0960	.1890	.2880	.3750	.4320	.4410	.3840	.2430	.1354
	3	.0001	.0010	.0080	.0270	.0640	.1250	.2160	.3430	.5120	.7290	.8574
4	0	.8145	.6561	.4096	.2401	.1296	.0625	.0256	.0081	.0016	.0001	.0000
	1	.1715	.2916	.4096	.4116	.3456	.2500	.1536	.0756	.0256	.0036	.0005
	2	.0135	.0486	.1536	.2646	.3456	.3750	.3456	.2646	.1536	.0486	.0135
	3	.0005	.0036	.0256	.0756	.1536	.2500	.3456	.4116	.4096	.2916	.1715
	4	.0000	.0001	.0016	.0081	.0256	.0625	.1296	.2401	.4096	.6561	.8145
5	0	.7738	.5905	.3277	.1681	.0778	.0312	.0102	.0024	.0003	.0000	.0000
	1	.2036	.3280	.4096	.3602	.2592	.1562	.0768	.0284	.0064	.0005	.0000
	2	.0214	.0729	.2048	.3087	.3456	.3125	.2304	.1323	.0512	.0081	.0011
	3	.0011	.0081	.0512	.1323	.2304	.3125	.3456	.3087	.2048	.0729	.0214
	4	.0000	.0004	.0064	.0283	.0768	.1562	.2592	.3601	.4096	.3281	.2036
	5	.0000	.0000	.0003	.0024	.0102	.0312	.0778	.1681	.3277	.5905	.7738
6	0	.7351	.5314	.2621	.1176	.0467	.0156	.0041	.0007	.0001	.0000	.0000
	1	.2321	.3543	.3932	.3025	.1866	.0937	.0369	.0102	.0015	.0001	.0000
	2	.0305	.0984	.2458	.3241	.3110	.2344	.1382	.0595	.0154	.0012	.0001
	3	.0021	.0146	.0819	.1852	.2765	.3125	.2765	.1852	.0819	.0146	.0021
	4	.0001	.0012	.0154	.0595	.1382	.2344	.3110	.3241	.2458	.0984	.0305
	5	.0000	.0001	.0015	.0102	.0369	.0937	.1866	.3025	.3932	.3543	.2321
	6	.0000	.0000	.0001	.0007	.0041	.0156	.0467	.1176	.2621	.5314	.7351
7	0	.6983	.4783	2097	.0824	.0280	.0078	.0016	.0002	.0000	.0000	.0000
	1	.2573	.3720	.3670	.2471	.1306	.0547	.0172	.0036	.0004	.0000	.0000
	2	.0406	.1240	.2753	.3177	.2613	.1641	.0774	.0250	.0043	.0002	.0000
	3	.0036	.0230	.1147	.2269	.2903	.2734	.1935	.0972	.0287	.0026	.0002
	4	.0002	.0026	.0287	.0972	.1935	.2734	.2903	.2269	.1147	.0230	.0036
	5	.0000	.0002	.0043	.0250	.0774	.1641	.2613	.3177	.2753	.1240	.0406
	6	.0000	.0000	.0004	.0036	.0172	.0547	.1306	.2471	.3670	.3720	.2573
	7	.0000	.0000	.0000	.0002	.0016	.0078	.0280	.0824	.2097	.4783	.6983
8	0	.6634	.4305	.1678	.0576	.0168	.0039	.0007	.0001	.0000	.0000	.0000
	1	.2793	.3826	.3355	.1977	.0896	.0312	.0079	.0012	.0001	.0000	.0000
	2	.0515	.1488	.2936	.2965	.2090	.1094	.0413	.0100	.0011	.0000	.0000
	3	.0054	.0331	.1468	.2541	.2787	.2187	.1239	.0467	.0092	.0004	.0000
	4	.0004	.0046	.0459	.1361	.2322	.2734	.2322	.1361	.0459	.0046	.0004

n	x	.05	.10	.20	.30	.40	.50	.60	.70	.80	.90	.95
	5	.0000	.0004	.0092	.0467	.1239	.2187	.2787	.2541	.1468	.0331	.0054
	6	.0000	.0000	.0011	.0100	.0413	.1094	.2090	.2965	.2936	.1488	.0515
	7	.0000	.0000	.0001	.0012	.0079	.0312	.0896	.1977	.3355	.3826	.2793
	8	.0000	.0000	.0000	.0001	.0007	.0039	.0168	.0576	.1678	.4305	.6634
9	0	.6302	.3874	.1342	.0404	.0101	.0020	.0003	.0000	.0000	.0000	.0000
	1	.2985	.3874	.3020	.1556	.0605	.0176	.0035	.0004	.0000	.0000	.0000
	2	.0629	.1722	.3020	.2668	.1612	.0703	.0212	.0039	.0003	.0000	.0000
	3	.0077	.0446	.1762	.2668	.2508	.1641	.0743	.0210	.0028	.0001	.0000
	4	.0006	.0074	.0661	.1715	.2508	.2461	.1672	.0735	.0165	.0008	.0000
	5	.0000	.0008	.0165	.0735	.1672	.2461	.2508	.1715	.0661	.0074	.0006
	6	.0000	.0001.	.0028	.0210	.0743	.1641	.2508	.2668	.1762	.0446	.0077
	7	.0000	.0000	.0003	.0039	.0212	.0703	.1612	.2668	.3020	.1722	.0629
	8	.0000	.0000	.0000	.0004	.0035	.0176	.0605	.1556	.3020	.3874	.2985
	9	.0000	.0000	.0000	.0000	.0003	.0020	.0101	.0404	.1342	.3874	.6302
10	0	.5987	.3487	.1074	.0282	.0060	.0010	.0001	.0000	.0000	.0000	.0000
	1	.3151	.3874	.2684	.1211	.0403	.0098	.0016	.0001	.0000	.0000	.0000
	2	.0746	.1937	.3020	.2335	.1209	.0439	.0106	.0014	.0001	.0000	.0000
	3	.0105	.0574	.2013	.2668	.2150	.1172	.0425	.0090	.0008	.0000	.0000
	4	.0010	.0112	.0881	.2001	.2508	.2051	.1115	.0368	.0055	.0001	.0000
	5	.0001	.0015	.0264	.1029	.2007	.2461	.2007	.1029	.0264	.0015	.0001
	6	.0000	.0001	.0055	.0368	.1115	.2051	.2508	.2001	.0881	.0112	.0010
	7	.0000	.0000	.0008	.0090	.0425	.1172	.2150	.2668	.2013	.0574	.0105
	8	.0000	.0000	.0001	.0014	.0106	.0439	.1209	.2335	.3020	.1937	.0746
	9	.0000	.0000	.0000	.0001	.0016	.0098	.0403	.1211	.2684	.3874	.3151
	10	.0000	.0000	.0000	.0000	.0001	.0010	.0060	.0282	.1074	.3487	.5987
11	0	.5688	.3138	.0859	.0198	.0036	.0005	.0000	.0000	.0000	.0000	.0000
	1	.3293	.3835	.2362	.0932	.0266	.0054	.0007	.0000	.0000	.0000	.0000
	2	.0867	.2131	.2953	.1998	.0887	.0269	.0052	.0005	.0000	.0000	.0000
	3	.0137	.0710	.2215	.2568	.1774	.0806	.0234	.0037	.0002	.0000	.0000
	4	.0014	.0158	.1107	.2201	.2365	.1611	.0701	.0173	.0017	.0000	.0000
	5	.0001	.0025	.0388	.1321	.2207	.2256	.1471	.0566	.0097	.0003	.0000
	6	.0000	.0003	.0097	.0566	.1471	.2256	.2207	.1321	.0388	.0025	.0001
	7	.0000	.0000	.0017	.0173	.0701	.1611	.2365	.2201	.1107	.0158	.0014
	8	.0000	.0000	.0002	.0037	.0234	.0806	.1774	.2568	.2215	.0710	.0137
	9	.0000	.0000	.0000	.0005	.0052	.0269	.0887	.1998	.2953	.2131	.0867
	10	.0000	.0000	.0000	.0000	.0007	.0054	.0266	.0932	.2362	.3835	.3293
	11	.0000	.0000	.0000	.0000	.0000	.0005	.0036	.0198	.0859	.3138	.5688
12	0	.5404	.2824	.0687	.0138	.0022	.0002	.0000	.0000	.0000	.0000	.0000
	1	.3413	.3766	.2062.	.0712	.0174	.0029	.0003	.0000	.0000	.0000	.0000
	2	.0988	.2301	.2835	.1678	.0639	.0161	.0025	.0002	.0000	.0000	.0000
	3	.0173	.0852	.2362	.2397	.1419	.0537	.0125	.0015	.0001	.0000	.0000
	4	.0021	.0213	.1329	.2311	.2128	.1208	.0420	.0078	.0005	.0000	.0000
	5	.0002	.0038	.0532	.1585	.2270	.1934	.1009	.0291	.0033	.0000	.0000
	6	.0000	.0005	.0155	.0792	.1766	.2256	.1766	.0792	.0155	.0005	.0000
	7	.0000	.0000	.0033	.0291	.1009	.1934	.2270	.1585	.0532	.0038	.0002
	8	.0000	.0000	.0005	.0078	.0420	.1208	.2128	.2311	.1329	.0213	.0021
	9	.0000	.0000	.0001	.0015	.0125	.0537	.1419	.2397	.2362	.0852	.0173

n	x	.05	.10	.20	30	.40	.50	.60	.70	.80	.90	.95
	10	.0000	.0000	.0000	.0002	.0025	.0161	.0639	.1678	.2835	.2301	.0988
	11	.0000	.0000	.0000	.0000	.0003	.0029	.0174	.0712	.2062	.3766	.3413
	12	.0000	.0000	.0000	.0000	.0000	.0002	.0022	.0138	.0687	.2824	.5404
13	0	.5133	.2542	.0550	.0097	.0013	.0001	.0000	.0000	.0000	.0000	.0000
	1	.3512	.3672	.1787	.0540	.0113	.0016	.0001	.0000	.0000	.0000	.0000
	2	.1109	.2448	.2680	.1388	.0453	.0095	.0012	.0001	.0000	.0000	.0000
	3	.0214	.0997	.2457	.2181	.1107	.0349	.0065	.0006	.0000	.0000	.0000
	4	.0028	.0277	.1535	.2337	.1845	.0873	.0243	.0034	.0001	.0000	.0000
	5	.0003	.0055	.0691	.1803	.2214	.1571	.0656	.0142	.0011	.0000	.0000
	6	.0000	.0008	.0230	.1030	.1968	.2095	.1312	.0442	.0058	.0001	.0000
	7	.0000	.0001	.0058	.0442	.1312	.2095	.1968	.1030	.0230	.0008	.0000
	8	.0000	.0000	.0011	.0142	.0656	.1571	.2214	.1803	.0691	.0055	.0003
	9	.0000	.0000	.0001	.0034	.0243	.0873	.1845	.2337	.1535	.0277	.0028
	10	.0000	.0000	.0000	.0006	.0065	.0349	.1107	.2181	.2457	.0997	.0214
	11	.0000	.0000	.0000	.0001	.0012	.0095	.0453	.1388	.2680	.2448	.1109
	12	.0000	.0000	.0000	.0000	.0001	.0016	.0113	.0540	.1787	.3672	.3512
	13	.0000	.0000	.0000	.0000	.0000	.0001	.0013	.0097	.0550	.2542	.5133
14	0	.4877	.2288	.0440	.0068	.0008	.0001	.0000	.0000	.0000	.0000	.0000
	1	.3593	.3559	.1539	.0407	.0073	.0009	.0001	.0000	.0000	.0000	.0000
	2	.1229	.2570	.2501	.1134	.0317	.0056	.0005	.0000	.0000	.0000	.0000
	3	.0259	.1142	.2501	.1943	.0845	.0222	.0033	.0002	.0000	.0000	.0000
	4	.0037	.0349	.1720	.2290	.1549	.0611	.0136	.0014	.0000	.0000	.0000
	5	.0004	.0078	.0860	.1963	.2066	.1222	.0408	.0066	.0003	.0000	.0000
	6	.0000	.0013	.0322	.1262	.2066	.1833	.0918	.0232	.0020	.0000	.0000
	7	.0000	.0002	.0092	.0618	.1574	.2095	.1574	.0618	.0092	.0002	.0000
	8	.0000	.0000	.0020	.0232	.0918	.1833	.2066	.1262	.0322	.0013	.0000
	9	.0000	.0000	.0003	.0066	.0408	.1222	.2066	.1963	.0860	.0078	.0004
	10	.0000	.0000	.0000	.0014	.0136	.0611	.1549	.2290	.1720	.0349	.0037
	11	.0000	.0000	.0000	.0002	.0033	.0222	.0845	.1943	.2501	.1142	.0259
	12	.0000	.0000	.0000	.0000	.0005	.0056	.0317	.1134	.2501	.2570	.1229
	13	.0000	.0000	.0000	.0000	.0001	.0009	.0073	.0407	.1539	.3559	.3593
	14	.0000	.0000	.0000	.0000	.0000	.0001	.0008	.0068	.0440	.2288	.4877
15	0	.4633	.2059	.0352	.0047	.0005	.0000	.0000	.0000	.0000	.0000	.0000
	1	.3658	.3432	.1319	.0305	.0047	.0005	.0000	.0000	.0000	.0000	.0000
	2	.1348	.2669	.2309	.0916	.0219	.0032	.0003	.0000	.0000	.0000	.0000
	3	.0307	.1285	.2501	.1700	.0634	.0139	.0016	.0001	.0000	.0000	.0000
	4	.0049	.0428	.1876	.2186	.1268	.0417	.0074	.0006	.0000	.0000	.0000
	5	.0006	.0105	.1032	.2061	.1859	.0916	.0245	.0030	.0001	.0000	.0000
	6	.0000	.0019	.0430	.1472	.2066	.1527	.0612	.0116	.0007	.0000	.0000
	7	.0000	.0003	.0138	.0811	.1771	.1964	.1181	.0348	.0035	.0000	.0000
	8	.0000	.0000	.0035	.0348	.1181	.1964	.1771	.0811	.0138	.0003	.0000
	9	.0000	.0000	.0007	.0116	.0612	.1527	.2066	.1472	.0430	.0019	.0000
	10	.0000	.0000	.0001	.0030	.0245	.0916	.1859	.2061	.1032	.0105	.0006
	11	.0000	.0000	.0000	.0006	.0074	.0417	.1268	.2186	.1876	.0428	.0049
	12	.0000	.0000	.0000	.0001	.0016	.0139	.0634	.1700	.2501	.1285	.0307
	13	.0000	.0000	.0000	.0000	.0003	.0032	.0219	.0916	.2309	.2669	.1348
	14	.0000	.0000	.0000	.0000	.0000	.0005	.0047	.0305	.1319	.3432	.3658
	15	.0000	.0000	.0000	.0000	.0000	.0000	.0005	.0047	.0352	.2059	.4633

						P						
n	x	.05	.10	.20	.30	.40	.50	.60	.70	.80	.90	.95
16	0	.4401	.1853	.0281	.0033	.0003	.0000	.0000	.0000	.0000	.0000	.0000
	1	.3706	.3294	.1126	.0228	.0030	.0002	.0000	.0000	.0000	.0000	.0000
	2	.1463	.2745	.2111	.0732	.0150	.0018	.0001	.0000	.0000	.0000	.0000
	3	.0359	.1423	.2463	.1465	.0468	.0085	.0008	.0000	.0000	.0000	.0000
	4	.0061	.0514	.2001	.2040	.1014	.0278	.0040	.0002	.0000	.0000	.0000
	5	.0008	.0137	.1201	.2099	.1623	.0667	.0142	.0013	.0000	.0000	.0000
	6	.0001	.0028	.0550	.1649	.1983	.1222	.0392	.0056	.0002	.0000	.0000
	7	.0000	.0004	.0197	.1010	.1889	.1746	.0840	.0185	.0012	.0000	.0000
	8	.0000	.0001	.0055	.0487	.1417	.1964	.1417	.0487	.0055	.0001	.0000
	9	.0000	.0000	.0012	.0185	.0840	.1746	.1889	.1010	.0197	.0004	.0000
	10	.0000	.0000	.0002	.0056	.0392	.1222	.1983	.1649	.0550	.0028	.0001
	11	.0000	.0000	.0000	.0013	.0142	.0666	.1623	.2099	.1201	.0137	.0008
	12	.0000	.0000	.0000	.0002	.0040	.0278	.1014	.2040	.2001	.0514	.0061
	13	.0000	.0000	.0000	.0000	.0008	.0085	.0468	.1465	.2463	.1423	.0359
	14	.0000	.0000	.0000	.0000	.0001	.0018	.0150	.0732	.2111	.2745	.1463
	15	.0000	.0000	.0000	.0000	.0000	.0002	.0030	.0228	.1126	.3294	.3706
	16	.0000	.0000	.0000	.0000	.0000	.0000	.0003	.0033	.0281	.1853	.4401
17	0	.4181	.1668	.0225	.0023	.0002	.0000	.0000	.0000	.0000	.0000	.0000
	1	.3741	.3150	.0957	.0169	.0019	.0001	.0000	.0000	.0000	.0000	.0000
	2	.1575	.2800	.1914	.0581	.0102	.0010	.0001	.0000	.0000	.0000	.0000,
	3	.0415	.1556	.2393	.1245	.0341	.0052	.0004	.0000	.0000	.0000	.0000
	4	.0076	.0605	.2093	.1868	.0796	.0182	.0021	.0001	.0000	.0000	.0000
	5	.0010	.0175	.1361	.2081	.1379	.0472	.0081	.0006	.0000	.0000	.0000
	6	.0001	.0039	.0680	.1794	.1839	.0944	.0242	.0026	.0001	.0000	.0000
	7	.0000	.0007	.0267	.1201	.1927	.1484	.0571	.0095	.0004	.0000	.0000
	8	.0000	.0001	.0084	.0644	.1606	.1855	.1070	.0276	.0021	.0000	.0000
	9	.0000	.0000	.0021	.0276	.1070	.1855	.1606	.0644	.0084	.0001	.0000
	10	.0000	.0000	.0004	.0095	.0571	.1484	.1927.	.1201	.0267	.0007	.0000
	11	.0000	.0000	.0001	.0026	.0242	.0944	.1839	.1784	.0680	.0039	.0001
	12	.0000	.0000	.0000	.0006	.0081	.0472	.1379	.2081	.1361	.0175	.0010
	13	.0000	.0000	.0000	.0001	.0021	.0182	.0796	.1868	.2093	.0605	.0076
	14	.0000	.0000	.0000	.0000	.0004	.0052	.0341-	.1245	.2393	.1556	.0415
	15	.0000	.0000	.0000	.0000	.0001	.0010	.0102	.0581	.1914	.2800	.1575
	16	.0000	.0000	.0000	.0000	.0000	.0001	.0019	.0169	.0957	.3150	.3741
	17	.0000	.0000	.0000	.0000	.0000	.0000	.0002	.0023	.0225	.1668	.4181
18	0	.3972	.1501	.0180	.0016	.0001	.0000	.0000	.0000	.0000	.0000	.0000
	1	.3763	.3002	.0811	.0126	.0012	.0001	.0000	.0000	.0000	.0000	.0000
	2	.1683	.2835	.1723	.0458	.0069	.0006	.0000	.0000	.0000	.0000	.0000
	3	.0473	.1680	.2297	.1046	.0246	.0031	.0002	.0000	.0000	.0000	.0000
	4	.0093	.0700	.2153	.1681	.0614	.0117	.0011	.0000	.0000	.0000	.0000
	5	.0014	.0218	.1507	.2017	.1146	.0327	.0045	.0002	.0000	.0000	.0000
	6	.0002	.0052	.0816	.1873	.1655	.0708	.0145	.0012	.0000	.0000	.0000
	7	.0000	.0010	.0350	.1376	.1892	.1214	.0374	.0046	.0001	.0000	.0000
	8	.0000	.0002	.0120	.0811	.1734	.1669	.077f	.0149	.0008	.0000	.0000
	9	.0000	.0000	.0033	.0386	.1284	.1855	.1284	.0386	.0033	.0000	.0000
	10	.0000	.0000	.0008	.0149	.0771	.1669	.1734	.0811	.0120	.0002	.0000
	11	.0000	.0000	.0001	.0046	.0374	.1214	.1892	.1376	.0350	.0010	.0000
	12	.0000	.0000	.0000	.0012	.0145	.0708	.1655	.1873	.0816	.0052	.0002

		P										
n	x	.05	.10	.20	.30	.40	.50	.60	.70	.80	.90	.95
	13	.0000	.0000	.0000	.0002	.0045	.0327	.1146	.2017	.1507	.0218	.0014
	14	.0000	.0000	.0000	.0000	.0011	.0117	.0614	.1681	.2153	.0700	.0093
	15	.0000	.0000	.0000	.0000	.0002	.0031	.0246	.1046	.2297	.1680	.0473
	16	.0000	.0000	.0000	.0000	.0000	.0006	.0069	.0458	.1723	.2835	.1683
	17	.0000	.0000	.0000	.0000	.0000	.0001	.0012	.0126	.0811	.3002	.3763
	18	.0000	.0000	.0000	.0000	.0000	.0000	.0001	.0016	.0180	.1501	.3972
19	0	.3774	.1351	.0144	.0011	.0001	.0000	.0000	.0000	.0000	.0000	.0000
	1	.3774	.2852	.0685	.0093	.0008	.0000	.0000	.0000	.0000	.0000	.0000
	2	.1787	.2852	.1540	.0358	.0046	.0003	.0000	.0000	.0000	.0000	.0000
	3	.0533	.1796	.2182	.0869	.0175	.0018	.0001	.0000	.0000	.0000	.0000
	4	.0112	.0798	.2182	.1491	.0467	.0074	.0005	.0000	.0000	.0000	.0000
	5	.0018	.0266	.1636	.1916	.0933	.0222	.0024	.0001	.0000	.0000	.0000
	6	.0002	.0069	.0955	.1916	.1451	.0518	.0085	.0005	.0000	.0000	.0000
	7	.0000	.0014	.0443	.1525	.1797	.0961	.0237	.0022	.0000	.0000	.0000
	8	.0000	.0002	.0166	.0981	.1797	.1442	.0532	.0077	.0003	.0000	.0000
	9	.0000	.0000	.0051	.0514	.1464	.1762	.0976	.0220	.0013	.0000	.0000
	10	.0000	.0000	.0013	.0220	.0976	.1762	.1464	.0514	.0051	.0000	.0000
	11	.0000	.0000	.0003	.0077	.0532	.1442	.1797	.0981	.0166	.0002	.0000
	12	.0000	.0000	.0000	.0022	.0237	.0961	.1797	.1525	.0443	.0014	.0000
	13	.0000	.0000	.0000	.0005	.0085	.0518	.1451	.1916	.0955	.0069	.0002
	14	.0000	.0000	.0000	.0001	.0024	.0222	.0933	.1916	.1636	.0266	.0018
	15	.0000	.0000	.0000	.0000	.0005	.0074	.0467	.1491	.2182	.0798	.0112
	16	.0000	.0000	.0000	.0000	.0001	.0018	.0175	.0869	.2182	.1796	.0533
	17	.0000	.0000	.0000	.0000	.0000	.0003	.0046	.0358	.1540	.2852	.1787
	18	.0000	.0000	.0000	.0000	.0000	.0000	.0008	.0093	.0685	.2852	.3774
	19	.0000	.0000	.0000	.0000	.0000	.0000	.0001	.0011	.0144	.1351	.3774
20	0	.3585	.1216	.0115	.0008	.0000	.0000	.0000	.0000	.0000	.0000	.0000
	1	.3774	.2702	.0576	.0068	.0005	.0000	.0000	.0000	.0000	.0000	.0000
	2	.1887	.2852	.1369	.0278	.0031	.0002	.0000	.0000	.0000	.0000	.0000
	3	.0596	.1901	.2054	.0716	.0123	.0011	.0000	.0000	.0000	.0000	.0000
	4	.0133	.0898	.2182	.1304	.0350	.0046	.0003	.0000	.0000	.0000	.0000
	5	.0022	.0319	.1746	.1789	.0746	.0148	.0013	.0000	.0000	.0000	.0000
	6	.0003	.0089	.1091	.1916	.1244	.0370	.0049	.0002	.0000	.0000	.0000
	7	.0000	.0020	.0545	.1643	.1659	.0739	.0146	.0010	.0000	.0000	.0000
	8	.0000	.0004	.0222	.1144	.1797	.1201	.0355	.0039	.0001	.0000	.0000
	9	.0000	.0001	.0074	.0654	.1597	.1602	.0710	.0120	.0005	.0000	.0000
	10	.0000	.0000	.0020	.0308	.1171	.1762	.1171	.0308	.0020	.0000	.0000
	11	.0000	.0000	.0005	.0120	.0710	.1602	.1597	.0654	.0074	.0001	.0000
	12	.0000	.0000	.0001	.0039	.0355	.1201	.1797	.1144	.0222	.0004	.0000
	13	.0000	.0000	.0000	.0010	.0146	.0739	.1659	.1643	.0545	.0020	.0000
	14	.0000	.0000	.0000	.0002	.0049	.0370	.1244	.1916	.1091	.0089	.0003
	15	.0000	.0000	.0000	.0000	.0013	.0148	.0746	.1789	.1746	.0319	.0022
	16	.0000	.0000	.0000	.0000	.0003	.0046	.0350	.1304	.2182	.0898	.0133
	17	.0000	.0000	.0000	.0000	.0000	.0011	.0123	.0716	.2054	.1901	.0596
	18	.0000	.0000	.0000	.0000	.0000	.0002	.0031	.0278	.1369	.2852	.1887
	19	.0000	.0000	.0000	.0000	.0000	.0000	.0005	.0068	.0576	.2702	.3774
	20	.0000	.0000	.0000	.0000	.0000	.0000	.0000	.0008	.0115	.1216	.3585

Appendix 2

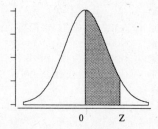

Areas Under the Standard Normal Curve from 0 to Z

Z	.00	.01	.02	.03	.04	.05	.06	.07	.08	.09
0.0	.0000	.0040	.0080	.0120	.0160	.0199	.0239	.0279	.0319	.0359
0.1	.0398	.0438	.0478	.0517	.0557	.0596	.0636	.0675	.0714	.0753
0.2	.0793	.0832	.0871	.0910	.0948	.0987	.1026	.1064	.1103	.1141
0.3	.1179	.1217	.1255	.1293	.1331	.1368	.1406	.1443	.1480	.1517
0.4	.1554	.1591	.1628	.1664	.1700	.1736	.1772	.1808	.1844	.1879
0.5	.1915	.1950	.1985	.2019	.2054	.2088	.2123	.2157	.2190	.2224
0.6	.2257	.2291	.2324	.2357	.2389	.2422	.2454	.2486	.2517	.2549
0.7	.2580	.2611	.2642	.2673	.2704	.2734	.2764	.2794	.2823	.2852
0.8	.2881	.2910	.2939	.2967	.2995	.3023	.3051	.3078	.3106	.3133
0.9	.3159	.3186	.3212	.3238	.3264	.3289	.3315	.3340	.3365	.3389
1.0	.3413	.3438	.3461	.3485	.3508	.3531	.3554	.3577	.3599	.3621
1.1	.3643	.3665	.3686	.3708	.3729	.3749	.3770	.3790	.3810	.3830
1.2	.3849	.3869	.3888	.3907	.3925	.3944	.3962	.3980	.3997	.4015
1.3	.4032	.4049	.4066	.4082	.4099	.4115	.4131	.4147	.4162	.4177
1.4	.4192	.4207	.4222	.4236	.4251	.4265	.4279	.4292	.4306	.4319
1.5	.4332	.4345	.4357	.4370	.4382	.4394	.4406	.4418	.4429	.4441
1.6	.4452	.4463	.4474	.4484	.4495	.4505	.4515	.4525	.4535	.4545
1.7	.4554	.4564	.4573	.4582	.4591	.4599	.4608	.4616	.4625	.4633
1.8	.4641	.4649	.4656	.4664	.4671	.4678	.4686	.4693	.4699	.4706
1.9	.4713	.4719	.4726	.4732	.4738	.4744	.4750	.4756	.4761	.4767
2.0	.4772	.4778	.4783	.4788	.4793	.4798	.4803	.4808	.4812	.4817
2.1	.4821	.4826	.4830	.4834	.4838	.4842	.4846	.4850	.4854	.4857
2.2	.4861	.4864	.4868	.4871	.4875	.4878	.4881	.4884	.4887	.4890
2.3	.4893	.4896	.4898	.4901	.4904	.4906	.4909	.4911	.4913	.4916
2.4	.4918	.4920	.4922	.4925	.4927	.4929	.4931	.4932	.4934	.4936
2.5	.4938	.4940	.4941	.4943	.4945	.4946	.4948	.4949	.4951	.4952
2.6	.4953	.4955	.4956	.4957	.4959	.4960	.4961	.4962	.4963	.4964
2.7	.4965	.4966	4967	.4968	.4969	.4970	.4971	.4972	.4973	.4974
2.8	.4974	.4975	.4976	.4977	.4977	.4978	.4979	.4979	.4980	.4981
2.9	.4981	.4982	.4982	.4983	.4984	.4994	.4985	.4985	.4986	.4986
3.0	.4987	.4987	.4987	.4998	.4988	.4989	.4989	.4989	.4990	.4990

Appendix 3

The entries in the table are the critical values of t for the specified number of degrees of freedom and areas in the right tail.

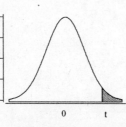

df	Areas in the Right Tail under the t Distribution Curve					
	.01	.05	.025	.01	.005	.001
1	3.078	6.314	12.706	31.821	63.657	318.309
2	1.886	2.920	4.303	6.965	9.925	22.327
3	1.638	2.353	3.182	4.541	5.841	10.215
4	1.533	2.132	2.776	3.747	4.604	7.173
5	1.476	2.015	2.571	3.365	4.032	5.893
6	1.440	1.943	2.447	3.143	3.707	5.208
7	1.415	1.895	2.365	2.998	3.499	4.785
8	1.397	1.860	2.306	2.896	3.355	4.501
9	1.385	1.833	2.262	2.821	3.250	4.297
10	1.372	1.812	2.228	2.764	3.169	4.144
11	1.363	1.796	2.201	2.718	3.106	4.025
12	1.356	1.782	2.179	2.681	3.055	3.930
13	1.350	1.771	2.160	2.650	3.012	3.852
14	1.345	1.761	2.145	2.624	2.977	3.787
15	1.341	1.753	2.131	2.602	2.947	3.733
16	1.337	1.746	2.120	2.583	2.921	3.686
17	1.333	1.740	2.110	2.567	2.898	3.646
18	1.330	1.734	2.101	2.552	2.878	3.610
19	1.328	1.729	2.093	2.539	2.861	3.579
20	1.325	1.725	2.086	2.528	2.845	3.552
21	1.323	1.721	2.080	2.518	2.831	3.527
22	1.321	1.717	2.074	2.508	2.819	3.505
23	1.319	1.714	2.069	2.500	2.807	3.485
24	1.318	1.711	2.064	2.492	2.797	3.467
25	1.316	1.708	2.060	2.485	2.787	3.450
26	1.315	1.706	2.056	2.479	2.779	3.435
27	1.314	1.703	2.052	2.473	2.771	3.421
28	1.313	1.701	2.048	2.467	2.763	3.408
29	1.311	1.699	2.045	2.462	2.756	3.396
30	1.310	1.697	2.042	2.457	2.750	3.385

Appendix 4

The entries in the table are the critical values of χ^2 for the specified degrees of freedom and areas in the right tail.

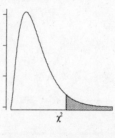

df	Area in the Right Tail under the Chi-square Distribution Curve									
	.995	.990	.975	.950	.900	.100	.050	.025	.010	.005
1	0.000	0.000	0.001	0.004	0.016	2.706	3.841	5.024	6.635	7.879
2	0.010	0.020	0.051	0.103	0.211	4.605	5.991	7.378	9.210	10.597
3	0.072	0.115	0.216	0.352	0.584	6.251	7.815	9.348	11.345	12.838
4	0.207	0.297	0.484	0.711	1.064	7.779	9.488	11.143	13.277	14.860
5	0.412	0.554	0.831	1.145	1.610	9.236	11.070	12.833	15.086	16.750
6	0.676	0.872	1.237	1.635	2.204	10.645	12.592	14.449	16.812	18.548
7	0.989	1.239	1.690	2.167	2.833	12.017	14.067	16.013	18.475	20.278
8	1.344	1.646	2.180	2.733	3.490	13.362	15.507	17.535	20.090	21.955
9	1.735	2.088	2.700	3.325	4.168	14.684	16.919	19.023	21.666	23.589
10	2.156	2.558	3.247	3.940	4.865	15.987	18.307	20.483	23.209	25.188
11	2.603	3.053	3.816	4.575	5.578	17.275	19.675	21.920	24.725	26.757
12	3.074	3.571	4.404	5.226	6.304	18.549	21.026	23.337	26.217	28.300
13	3.565	4.107	5.009	5.892	7.042	19.812	22.362	24.736	27.688	29.819
14	4.075	4.660	5.629	6.571	7.790	21.064	23.685	26.119	29.141	31.319
15	4.601	5.229	6.262	7.261	8.547	22.307	24.996	27.488	30.578	32.801
16	5.142	5.812	6.908	7.962	9.312	23.542	26.296	28.845	32.000	34.267
17	5.697	6.408	7.564	8.672	10.085	24.769	27.587	30.191	33.409	35.718
18	6.265	7.015	8.231	9.390	10.865	25.989	28.869	31.526	34.805	37.156
19	6.844	7.633	8.907	10.117	11.651	27.204	30.144	32.852	36.191	38.582
20	7.434	8.260	9.591	10.851	12.443	28.412	31.410	34.170	37.566	39.997
21	8.034	8.897	10.283	11.591	13.240	29.615	32.671	35.479	38.932	41.401
22	8.643	9.542	10.982	12.338	14.041	30.813	33.924	36.781	40.289	42.796
23	9.260	10.196	11.689	13.091	14.848	32.007	35.172	38.076	41.638	44.181
24	9.886	10.856	12.401	13.848	15.659	33.196	36.415	39.364	42.980	45.559
25	10.520	11.524	13.120	14.611	16.473	34.382	37.652	40.646	44.314	46.928
26	11.160	12.198	13.844	15.379	17.292	35.563	38.885	41.923	45.642	48.290
27	11.808	12.879	14.573	16.151	18.114	36.741	40.113	43.195	46.963	49.645
28	12.461	13.565	15.308	16.928	18.939	37.916	41.337	44.461	48.278	50.993
29	13.121	14.256	16.047	17.708	19.768	39.087	42.557	45.722	49.588	52.336
30	13.787	14.953	16.791	18.493	20.599	40.256	43.773	46.979	50.892	53.672
40	20.707	22.164	24.433	26.509	29.051	51.805	55.758	59.342	63.691	66.766
50	27.991	29.707	32.357	34.764	37.689	63.167	67.505	71.420	76.154	79.490
60	35.534	37.485	40.482	43.188	46.459	74.397	79.082	83.298	88.379	91.952
70	43.275	45.442	48.758	51.739	55.329	85.527	90.531	95.023	100.425	104.215
80	51.172	53.540	57.153	60.391	64.278	96.578	101.879	106.629	112.329	116.321

Appendix 5

The area in the right tail under the F distribution curve is equal to 0.01.

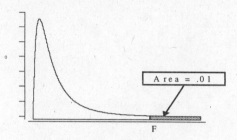

df₂	df₁													
	1	2	3	4	5	6	7	8	9	10	11	12	15	20
1	4052	5000	5403	5625	5764	5859	5928	5981	6022	6056	6083	6106	6157	6209
2	98.50	99.00	99.17	99.25	99.30	99.33	99.36	99.37	99.39	99.40	99.41	99.42	99.43	99.45
3	34.12	30.82	29.46	28.71	28.24	27.91	27.67	27.49	27.35	27.23	27.13	27.05	26.87	26.69
4	21.20	18.00	16.69	15.98	15.52	15.21	14.98	14.80	14.66	14.55	14.45	14.37	14.20	14.02
5	16.26	13.27	12.06	11.39	10.97	10.67	10.46	10.29	10.16	10.05	9.96	9.89	9.72	9.55
6	13.75	10.92	9.78	9.15	8.75	8.47	8.26	8.10	7.98	7.87	7.79	7.72	7.56	7.40
7	12.25	9.55	8.45	7.85	7.46	7.19	6.99	6.84	6.72	6.62	6.54	6.47	6.31	6.16
8	11.26	8.65	7.59	7.01	6.63	6.37	6.18	6.03	5.91	5.81	5.73	5.67	5.52	5.36
9	10.56	8.02	6.99	6.42	6.06	5.80	5.61	5.47	5.35	5.26	5.18	5.11	4.96	4.81
10	10.04	7.56	6.55	5.99	5.64	5.39	5.20	5.06	4.94	4.85	4.77	4.71	4.56	4.41
11	9.65	7.21	6.22	5.67	5.32	5.07	4.89	4.74	4.63	4.54	4.46	4.40	4.25	4.10
12	9.33	6.93	5.95	5.41	5.06	4.82	4.64	4.50	4.39	4.30	4.22	4.16	4.01	3.86
13	9.07	6.70	5.74	5.21	4.86	4.62	4.44	4.30	4.19	4.10	4.02	3.96	3.82	3.66
14	8.86	6.51	5.56	5.04	4.69	4.46	4.28	4.14	4.03	3.94	3.86	3.80	3.66	3.51
15	8.68	6.36	5.42	4.89	4.56	4.32	4.14	4.00	3.89	3.80	3.73	3.67	3.52	3.37
16	8.53	6.23	5.29	4.77	4.44	4.20	4.03	3.89	3.78	3.69	3.62	3.55	3.41	3.26
17	8.40	6.11	5.18	4.67	4.34	4.10	3.93	3.79	3.68	3.59	3.52	3.46	3.31	3.16
18	8.29	6.01	5.09	4.58	4.25	4.01	3.84	3.71	3.60	3.51	3.43	3.37	3.23	3.08
19	8.18	5.93	5.01	4.50	4.17	3.94	3.77	3.63	3.52	3.43	3.36	3.30	3.15	3.00
20	8.10	5.85	4.94	4.43	4.10	3.87	3.70	3.56	3.46	3.37	3.29	3.23	3.09	2.94
21	8.02	5.78	4.87	4.37	4.04	3.81	3.64	3.51	3.40	3.31	3.24	3.17	3.03	2.88
22	7.95	5.72	4.82	4.31	3.99	3.76	3.59	3.45	3.35	3.26	3.18	3.12	2.98	2.83
23	7.88	5.66	4.76	4.26	3.94	3.71	3.54	3.41	3.30	3.21	3.14	3.07	2.93	2.78
24	7.82	5.61	4.72	4.22	3.90	3.67	3.50	3.36	3.26	3.17	3.09	3.03	2.89	2.74
25	7.77	5.57	4.68	4.18	3.85	3.63	3.46	3.32	3.22	3.13	3.06	2.99	2.85	2.70
30	7.56	5.39	4.51	4.02	3.70	3.47	3.30	3.17	3.07	2.98	2.91	2.84	2.70	2.55
40	7.31	5.18	4.31	3.83	3.51	3.29	3.12	2.99	2.89	2.80	2.73	2.66	2.52	2.37
50	7.17	5.06	4.20	3.72	3.41	3.19	3.02	2.89	2.78	2.70	2.63	2.56	2.42	2.27
100	6.90	4.82	3.98	3.51	3.21	2.99	2.82	2.69	2.59	2.50	2.43	2.37	2.22	2.07

The area in the right tail under the F distribution curve is equal to 0.05.

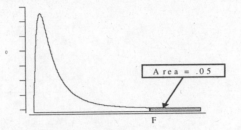

df$_2$	df$_1$													
	1	2	3	4	5	6	7	8	9	10	11	12	15	20
1	161.5	199.5	215.7	224.6	230.2	234.0	236.8	238.9	240.5	241.9	243.0	243.9	246.0	248.0
2	18.51	19.00	19.16	19.25	19.30	19.33	19.35	19.37	19.38	19.40	19.40	19.41	19.43	19.45
3	10.13	9.55	9.28	9.12	9.01	8.94	8.89	8.85	8.81	8.79	8.76	8.74	8.70	8.66
4	7.71	6.94	6.59	6.39	6.26	6.16	6.09	6.04	6.00	5.96	5.94	5.91	5.86	5.80
5	6.61	5.79	5.41	5.19	5.05	4.95	4.88	4.82	4.77	4.74	4.70	4.68	4.62	4.56
6	5.99	5.14	4.76	4.53	4.39	4.28	4.21	4.15	4.10	4.06	4.03	4.00	3.94	3.87
7	5.59	4.74	4.35	4.12	3.97	3.87	3.79	3.73	3.68	3.64	3.60	3.57	3.51	3.44
8	5.32	4.46	4.07	3.84	3.69	3.58	3.50	3.44	3.39	3.35	3.31	3.28	3.22	3.15
9	5.12	4.26	3.86	3.63	3.48	3.37	3.29	3.23	3.18	3.14	3.10	3.07	3.01	2.94
10	4.96	4.10	3.71	3.48	3.33	3.22	3.14	3.07	3.02	2.98	2.94	2.91	2.85	2.77
11	4.84	3.98	3.59	3.36	3.20	3.09	3.01	2.95	2.90	2.85	2.82	2.79	2.72	2.65
12	4.75	3.89	3.49	3.26	3.11	3.00	2.91	2.85	2.80	2.75	2.72	2.69	2.62	2.54
13	4.67	3.81	3.41	3.18	3.03	2.92	2.83	2.77	2.71	2.67	2.63	2.60	2.53	2.46
14	4.60	3.74	3.34	3.11	2.96	2.85	2.76	2.70	2.65	2.60	2.57	2.53	2.46	2.39
15	4.54	3.68	3.29	3.06	2.90	2.79	2.71	2.64	2.59	2.54	2.51	2.48	2.40	2.33
16	4.49	3.63	3.24	3.01	2.85	2.74	2.66	2.59	2.54	2.49	2.46	2.42	2.35	2.28
17	4.45	3.59	3.20	2.96	2.81	2.70	2.61	2.55	2.49	2.45	2.41	2.38	2.31	2.23
18	4.41	3.55	3.16	2.93	2.77	2.66	2.58	2.51	2.46	2.41	2.37	2.34	2.27	2.19
19	4.38	3.52	3.13	2.90	2.74	2.63	2.54	2.48	2.42	2.38	2.34	2.31	2.23	2.16
20	4.35	3.49	3.10	2.87	2.71	2.60	2.51	2.45	2.39	2.35	2.31	2.28	2.20	2.12
21	4.32	3.47	3.07	2.84	2.68	2.57	2.49	2.42	2.37	2.32	2.28	2.25	2.18	2.10
22	4.30	3.44	3.05	2.82	2.66	2.55	2.46	2.40	2.34	2.30	2.26	2.23	2.15	2.07
23	4.28	3.42	3.03	2.80	2.64	2.53	2.44	2.37	2.32	2.27	2.24	2.20	2.13	2.05
24	4.26	3.40	3.01	2.78	2.62	2.51	2.42	2.36	2.30	2.25	2.22	2.18	2.16	2.03
25	4.24	3.39	2.99	2.76	2.60	2.49	2.40	2.34	2.28	2.24	2.20	2.16	2.09	2.01
30	4.17	3.32	2.92	2.69	2.53	2.42	2.33	2.27	2.21	2.16	2.13	2.09	2.01	1.93
40	4.08	3.23	2.84	2.61	2.45	2.34	2.25	2.18	2.12	2.08	2.04	2.00	1.92	1.84
50	4.03	3.18	2.79	2.56	2.40	2.29	2.20	2.13	2.07	2.03	1.99	1.95	1.87	1.78
100	3.94	3.09	2.70	2.46	2.31	2.19	2.10	2.03	1.97	1.93	1.89	1.85	1.77	1.68

Index